Reading Essentials
Course 1

The McGraw·Hill Companies

Send all inquiries to:
McGraw-Hill Education
8787 Orion Place
Columbus, OH 43240-4027

ISBN: 978-0-07-889388-9
MHID: 0-07-889388-7

Printed in the United States of America.

1 2 3 4 5 6 7 8 9 10 REL 15 14 13 12 11 10

Table of Contents

To the Student

In today's world, a knowledge of science is important for thinking critically, solving problems, and making decisions. But understanding science can sometimes be a challenge.

Reading Essentials takes the stress out of reading, learning, and understanding science. This book covers important concepts in science, offers ideas for how to learn the information, and helps you review what you have learned.

In each lesson you will find:

Before You Read

- **What do you think?** asks you to agree or disagree with statements about topics that are discussed in the lesson.

Read to Learn

This section describes important science concepts with words and graphics. In the margins you can find a variety of study tips and ideas for organizing and learning information.

- **Study Coach and Mark the Text** offer tips for finding the main ideas in the text.
- **Foldables® Study Organizers** help you divide the information into smaller, easier-to-remember concepts.
- **Reading Check** questions ask about concepts in the lesson.
- **Think It Over** elements help you consider the material in-depth, giving you an opportunity to use your critical thinking skills.
- **Visual Check** questions relate specifically to the art and graphics used in the text. You will find questions that get you actively involved in illustrating the information you have just learned.
- **Math Skills** reinforces the connection between math and science.
- **Academic Vocabulary** defines important words that help you build a strong vocabulary.
- **Word Origin** explains the English background of a word.
- **Key Concept Check** features ask the Key Concept questions from the beginning of the lesson.
- **Interpreting Tables** includes questions or activities that help you interact with the information presented.

After You Read

This final section reviews key terms and asks questions about what you have learned.

- The **Mini Glossary** assists you with science vocabulary.
- Review questions focus on the key concepts of the lesson.
- **What do you think now?** gives you an opportunity to revisit the *What do you think?* statements to see if you changed your mind after reading the lesson.

See for yourself—***Reading Essentials*** makes science enjoyable and easy to understand.

Methods of Science

Understanding Science

Read to Learn

What is science?

Did you ever hear a bird sing and then look in nearby trees to find the singing bird? Have you ever noticed how the Moon changes from a thin crescent to a full moon each month? When you do these things, you are doing science. **Science** *is the investigation and exploration of natural events and of the new information that results from those investigations.*

For thousands of years, men and women of all countries and cultures have studied the natural world and recorded their observations. They have shared their knowledge and findings and have created a vast amount of scientific information. Scientific knowledge has been the result of a great deal of debate and confirmation within the science community.

People use science in their everyday lives and careers. For example, firefighters wear clothing that has been developed and tested to withstand extreme temperatures and not catch fire. Parents use science when they set up an aquarium for their children's pet fish. Athletes use science when they use high-performance gear or wear high-performance clothing.

Without thinking about it, you use science or the results of science in almost everything you do. Most likely, your clothing, food, hair products, electronic devices, athletic equipment, and almost everything else you use are all results of science.

Branches of Science

There are many different parts of the natural world. Because there is so much to study, a scientist often focuses his or her work in one branch of science or on one topic within that branch of science. There are three main branches of science—Earth science, life science, and physical science. ✓

Key Concepts 

- What is scientific inquiry?
- How do scientific laws and scientific theories differ?
- What is the difference between a fact and an opinion?

Study Coach

Identify the Main Ideas As you read, write one sentence to summarize the main idea in each paragraph. Write the main ideas on a sheet of paper or in your notebook to study later.

 **Reading Check**

1. Name the three main branches of science.

Earth Science The study of Earth, including rocks, soils, oceans, the atmosphere, and surface features, is Earth science. Earth scientists might ask questions such as, How do different shorelines react to tsunamis? Why do planets orbit the Sun? What is the rate of climate change?

Life Science The study of living things is life science, or biology. Biologists ask questions such as, Why do some trees lose their leaves in winter? How do birds know which direction they are going? How do mammals control their body temperature?

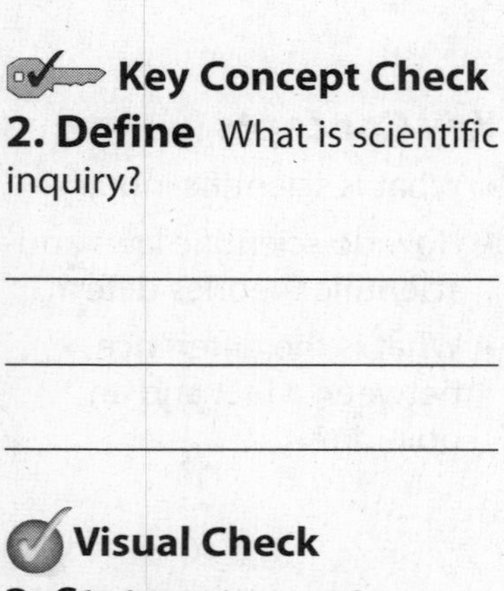

Physical Science The study of matter and energy is physical science. It includes both physics and chemistry. Physicists and chemists ask questions such as, What chemical reactions must take place to launch a spaceship into space? Is it possible to travel faster than the speed of light? What makes up matter?

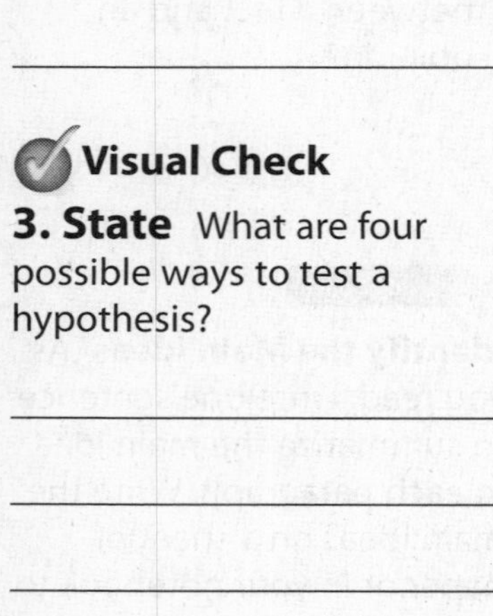

Scientific Inquiry

When scientists conduct scientific investigations, they use scientific inquiry. Scientific inquiry is a process that uses a set of skills to answer questions or to test ideas about the natural world. There are many kinds of scientific investigations and many ways to conduct them. The series of steps used in each investigation often varies. The flowchart in the figure below and on the next page shows an example of the skills used in scientific inquiry.

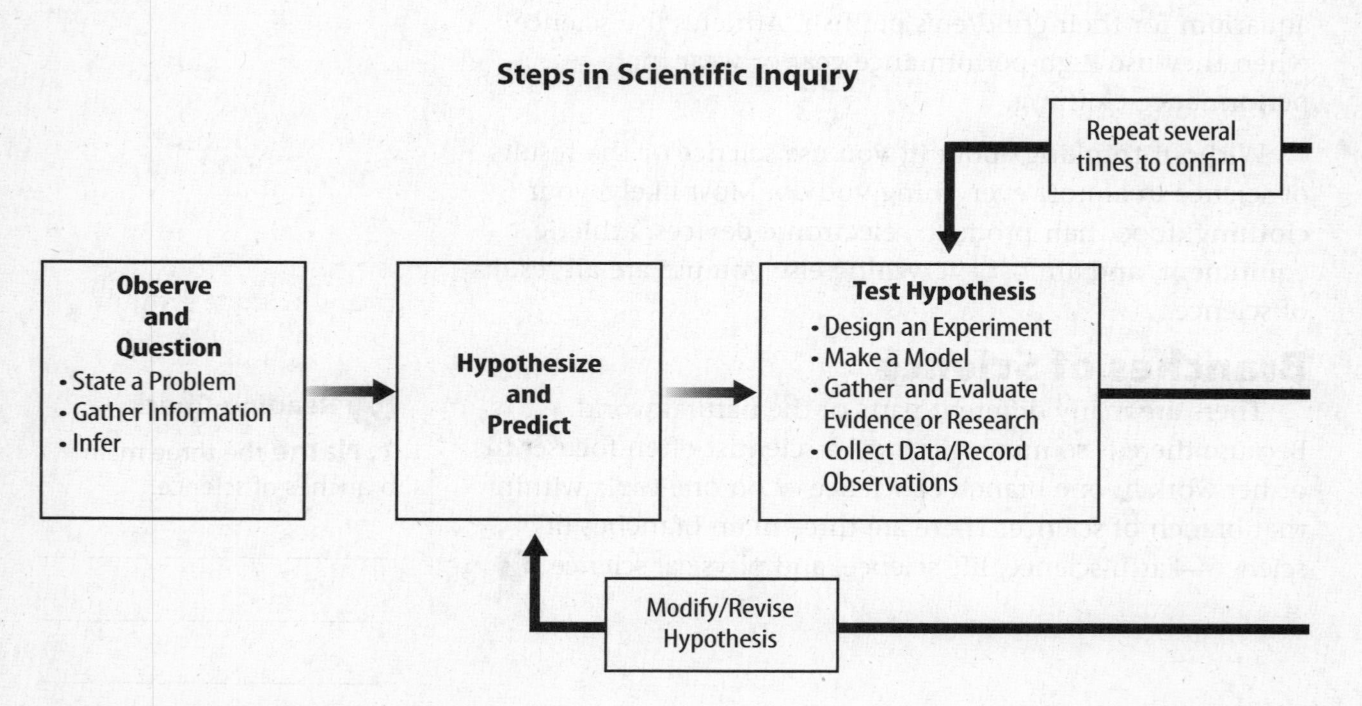

Ask Questions

One way to begin a scientific inquiry is to observe the natural world and ask questions. **Observation** *is the act of using one or more of your senses to gather information and taking note of what occurs*. Suppose you observe that the banks of a river have eroded more this year than last year. You want to know why. You note that there was an increase in rainfall this year. After these observations, you make an inference based on these observations. *An* **inference** *is a logical explanation of an observation that is drawn from prior knowledge or experience.* You infer that the increase in rainfall caused the increase in erosion. You decide to investigate further. You develop a hypothesis and a method to test it.

Reading Check

4. Select Which of the following is a way to begin a scientific inquiry? (Circle the correct answer.)

a. observe the natural world

b. ask questions

c. both a and b above

Hypothesize and Predict

A **hypothesis** *is a possible explanation for an observation that can be tested by scientific investigations*. A hypothesis states an observation and provides an explanation. You might make the following hypothesis: More of the riverbank eroded this year because the amount, the speed, and the force of the river water increased.

Scientists often use a hypothesis to make predictions. *A* **prediction** *is a statement of what will happen next in a sequence of events*. Scientists make predictions based on what information they think they will find when testing their hypothesis. A prediction for the hypothesis above might be: If rainfall increases, then the amount, the speed, and the force of river water will increase. If the amount, the speed, and the force of river water increase, then there will be more erosion.

Reading Check

5. Identify What do scientists often use to make a prediction?

____

Steps in Scientific Inquiry

Hypothesis supported

Analyze Results
- Graph Results
- Classify Information
- Make Calculations
- Other Processes

Draw Conclusions
- Infer
- Summarize

Communicate Results
- Write Science Journal Articles
- Speak at Science Conferences
- Exchange Information on Internet
- Other Ways of Exchanging Information

Hypothesis not supported

Test Hypothesis

When you test a hypothesis, you often test whether your predictions are true. If your prediction is confirmed, then it supports your hypothesis. If your prediction is not confirmed, you might need to modify your hypothesis and retest it.

There are several ways to test a hypothesis when performing a scientific investigation. You might design an experiment, make a model, gather and evaluate evidence or research, or collect data and record your observations. For example, you might make a model of a riverbank in which you change the speed and the amount of water and record observations and results. ✓

✓ Reading Check

6. Apply When should you modify your hypothesis?

____

____

____

Analyze Results

After testing your hypothesis, you analyze your results using different methods. Often, it is hard to see trends or relationships in data while collecting it. Data should be sorted, graphed, or classified in some way. After analyzing the data, additional inferences can be made.

FOLDABLES®

Make a six-tab book to organize your notes about a scientific inquiry.

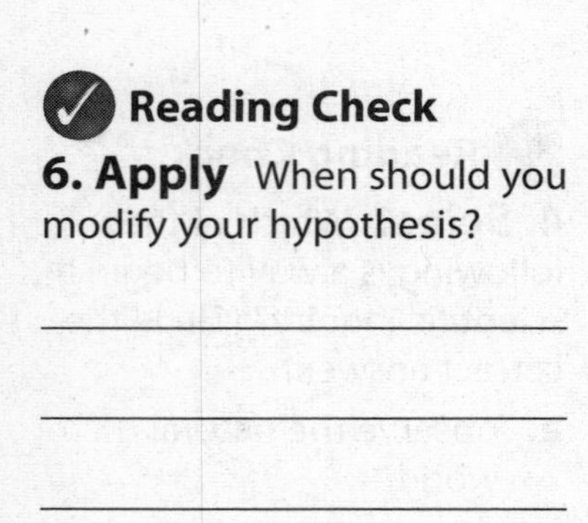

Draw Conclusions

Once you find the relationships among data and make several inferences, you can draw conclusions. A conclusion is a summary of the information gained from testing a hypothesis. Scientists study the available information and draw conclusions based on that information.

Communicate Results

An important part of the scientific inquiry process is communicating results. Ways to communicate results include writing science journal articles, speaking at science conferences, and exchanging information on the Internet. Scientists might share their information in other ways, too.

Scientists communicate results of investigations to inform other scientists about their research and the conclusions of their research. Scientists might apply each other's conclusions to their own work to help support their hypotheses.

✓ Reading Check

7. Describe What is the next step if a scientist's hypothesis is supported?

____

____

____

____

____

Further Scientific Inquiry

Scientific inquiry is not completed once one scientific investigation is completed. If predictions are correct and the hypothesis is supported, scientists will retest the predictions several times to make sure the conclusions are the same each time. If the hypothesis is not supported, any new information gained can be used to revise the hypothesis. Hypotheses can be revised and tested many times. ✓

Results of Science

The results and conclusions from an investigation can lead to many outcomes, such as the answers to a question, more information on a specific topic, or support for a hypothesis. Other outcomes are described in the following paragraphs.

Technology A technical solution can be the answer to a scientific question, such as, "How can the hearing impaired hear better?" After investigation, experimentation, and research, the conclusion might be the development of a new technology. ***Technology** is the practical use of scientific knowledge, especially for industrial or commercial use*. Technology, such as a cochlear implant, can help some deaf people hear. ✓

New Materials Space travel has unique challenges. Astronauts must carry oxygen to breathe. They also must be protected against temperature and pressure extremes, as well as small, high-speed flying objects. A spacesuit consists of 14 layers of material. The outer layer is made of a blend of three materials. One material is waterproof. Another protects against high-speed flying objects. The third material is heat and fire-resistant.

Possible Explanations Scientists often perform investigations to find explanations as to why or how something happens, such as, "How do stars form?" For example, to help answer this question, NASA's *Spitzer Space Telescope* took photos showing a cloud of gas and dust with newly formed stars. ✓

Scientific Theory and Scientific Law

Another outcome of science is the development of scientific theories and laws. Recall that a hypothesis is a possible explanation about an observation that can be tested by scientific investigations.

What happens when a hypothesis or a group of hypotheses has been tested many times and has been supported by the repeated scientific investigations? The hypothesis can become a scientific theory. ✓

✓ **Reading Check**

8. Define What is technology?

✓ **Reading Check**

9. Express What are some results of science?

✓ **Reading Check**

10. Explain How can a hypothesis become a scientific theory?

Scientific Theory

Often, the word *theory* is used in casual conversations to mean an untested idea or an opinion. However, scientists use the word *theory* differently. *A* **scientific theory** *is an explanation of observations or events that is based on knowledge gained from many observations and investigations.*

Scientists question scientific theories and test them for validity. A scientific theory generally is accepted as true until it is disproved. An example of a scientific theory is the theory of plate tectonics. The theory explains how Earth's crust moves and why earthquakes and volcanoes occur.

Key Concept Check
11. Differentiate How do scientific laws and theories differ?

____

____

____

Visual Check
12. Name two possible results from new information opposing a theory.

____

____

____

Scientific Law

A scientific law is different from a social law, which is an agreement among people concerning a behavior. *A* **scientific law** *is a rule that describes a pattern in nature.* Unlike a scientific theory that explains why an event occurs, a scientific law only states that an event will occur under certain circumstances. For example, Newton's law of gravitational force implies that if you drop an object, it will fall toward Earth. Newton's law does not explain why the object moves toward Earth when dropped, only that it will.

New Information

Scientific information constantly changes as new information is discovered or as previous hypotheses are retested. New information can lead to changes in scientific theories as explained below. When new facts are revealed, a current scientific theory might be revised to include the new facts, or it might be disproved and rejected.

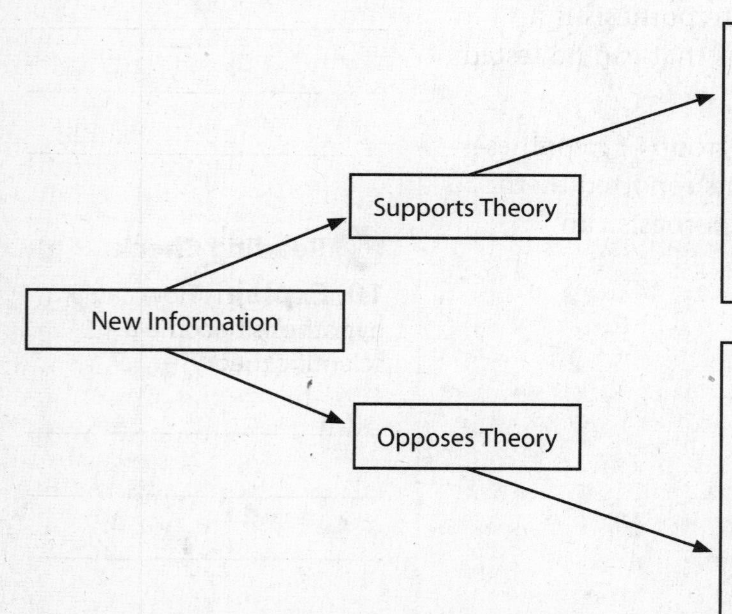

If new information supports a current scientific theory, then the theory is not changed. The information might be published in a scientific journal to show further support of the theory. The new information might also lead to advancements in technology or spark new questions that lead to new scientific investigations.

If new information opposes, or does not support a current scientific theory, the theory might be modified or rejected altogether. Often, new information will lead scientists to look at the original observations in a new way. This can lead to new investigations with new hypotheses. These investigations can lead to new theories.

Evaluating Scientific Evidence

Did you ever read an advertisement that made extraordinary claims, such as the one shown below? If so, you probably practice critical thinking. **Critical thinking** *is comparing what you already know with the information you are given in order to decide whether you agree with it.* To determine whether information is true and scientific or pseudoscience (information incorrectly represented as scientific), you should be skeptical and identify facts and opinions. This helps you evaluate the strengths and weaknesses of information and make informed decisions. Critical thinking is important in all decision making—from everyday decisions to community, national, and international decisions.

Key Concept Check

13. Contrast Refer to the figure below. How do a fact and an opinion differ?

____

____

____

____

Skepticism
To be skeptical is to doubt the truthfulness or accuracy of something. Because of skepticism, science can be self-correcting. If someone publishes results or if an investigation gives results that don't seem accurate, a skeptical scientist usually will challenge the information and test the results for accuracy.

Identifying Facts
The prices of the pillows and the savings are facts. A fact is a measurement, observation, or statement that can be strictly defined. Many scientific facts can be evaluated for their validity through investigations.

Learn Algebra *While You Sleep!*

Have you struggled to learn algebra? Struggle no more.

Math-er-ific's new algebra pillow is scientifically proven to transfer math skills from the pillow to your brain while you sleep. This revolutionary scientific design improved the algebra test scores of laboratory mice by 150 percent.

Dr. Tom Equation says, "I have never seen students or mice learn algebra so easily. This pillow is truly amazing."

For only $19.95, those boring hours spent studying are a thing of the past. So act fast! If you order today, you can get the algebra pillow and the equally amazing geometry pillow for only $29.95. That is a $10 savings!

Identifying Opinions
An opinion is a personal view, feeling, or claim about a topic. Opinions are neither true nor false.

Mixing Facts and Opinions
Sometimes people mix facts and opinions. You must read carefully to determine which information is fact and which is opinion.

Science cannot answer all questions.

Scientists recognize that some questions cannot be studied using scientific inquiry. Questions that deal with opinions, beliefs, values, and feelings cannot be answered through scientific investigation. For example, questions that cannot be answered through scientific investigation might include

- Are comedies the best kinds of movies?
- Is it ever okay to lie?
- Which food tastes best?

The answers to all of these questions are based on opinions, not facts.

Safety in Science

It is very important for anyone performing scientific investigations to use safe practices. You should always follow your teacher's instructions. If you have questions about potential hazards, use of equipment, or the meaning of safety symbols, ask your teacher. Always wear protective clothing and equipment while performing scientific investigations. If you are using live animals in your investigations, provide appropriate care and ethical treatment to them. Your teacher can help you find more information about the treatment of live animals.

ACADEMIC VOCABULARY
potential
(adjective) possible, likely, or probable

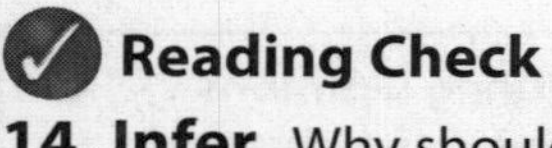

14. Infer Why should you wear protective clothing and equipment when performing scientific investigations?

After You Read

Mini Glossary

critical thinking: comparing what you already know with the information you are given in order to decide whether you agree with it

hypothesis: a possible explanation for an observation that can be tested by scientific investigations

inference: a logical explanation of an observation that is drawn from prior knowledge or experience

observation: the act of using one or more of your senses to gather information and taking note of what occurs

prediction: a statement of what will happen next in a sequence of events

science: the investigation and exploration of natural events and of the new information that results from those investigations

scientific law: a rule that describes a pattern in nature

scientific theory: an explanation of observations or events that is based on knowledge gained from many observations and investigations

technology: the practical use of scientific knowledge, especially for industrial or commercial use

1. Review the terms and their definitions in the Mini Glossary. Write a sentence that explains the purpose of a hypothesis as a step in scientific inquiry.

2. Complete the graphic organizer to show how scientists begin a scientific inquiry.

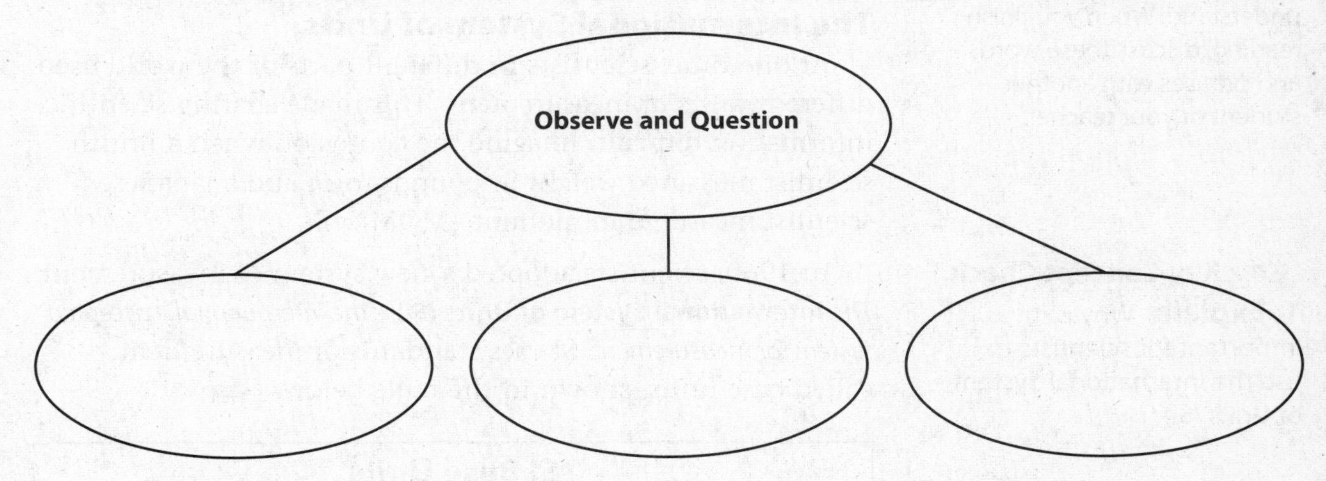

3. Why is critical thinking important when evaluating scientific information?

Log on to ConnectED.mcgraw-hill.com and access your textbook to find this lesson's resources.

Methods of Science

Measurement and Scientific Tools

Key Concepts
- Why is it important for scientists to use the International System of Units?
- What causes measurement uncertainty?
- What are mean, median, mode, and range?

Building Vocabulary As you read, underline the words and phrases that you do not understand. When you finish reading, discuss these words and phrases with another student or your teacher.

Key Concept Check
1. Explain Why is it important for scientists to use the International System of Units (SI)?

Interpreting Tables
2. Analyze Which unit is used to measure temperature?

Read to Learn

Description and Explanation

Imagine that a scientist is observing an erupting volcano. He describes in his journal that the flowing lava is bright red with a black crust, and it has a temperature of about 630°C. *A* **description** *is a spoken or written summary of observations.*

There are two types of descriptions. A qualitative description, such as *bright red,* uses senses (sight, sound, smell, touch, taste) to describe an observation. A quantitative description, such as *630°C,* uses numbers and measurements to describe an observation. Later, the scientist might explain his observations. *An* **explanation** *is an interpretation of observations.* Because the lava was bright red and about 630°C, the scientist might explain the lava is cooling, and the volcano did not recently erupt.

The International System of Units

At one time, scientists in different parts of the world used different units of measurement. This made sharing scientific information difficult. Imagine the confusion when a British scientist measured weight in pounds-force and a Japanese scientist measured in momme (MOM ee).

In 1960, scientists adopted a new system of measurement. *The* **International System of Units (SI)** *is the internationally accepted system for measurement.* SI uses standards of measurement, called base units, shown in the table below.

SI Base Units

Quantity Measured	Unit	Symbol
Length	meter	m
Mass	kilogram	kg
Time	second	s
Electric current	ampere	A
Temperature	Kelvin	K
Amount of substance	mole	mol
Intensity of light	candela	cd

SI Unit Prefixes

In addition to base units, SI uses prefixes to identify the size of the unit, as shown in the table below. Prefixes are used to indicate a fraction of ten or a multiple of ten. In other words, each unit is either ten times smaller than the next larger unit or ten times larger than the next smaller unit. For example, the prefix *deci–* means 10^{-1}, or 1/10. A decimeter is 1/10 of a meter. The prefix *kilo–* means 10^3, or 1,000. A kilometer is 1,000 m.

Prefixes

Prefix	Meaning
Mega- (M)	1,000,000 (10^6)
Kilo- (k)	1,000 (10^3)
Hecto- (h)	100 (10^2)
Deka- (da)	10 (10^1)
Deci- (d)	0.1 (10^{-1})
Centi- (c)	0.01 (10^{-2})
Milli- (m)	0.001 (10^{-3})
Micro- (μ)	0.000 001 (10^{-6})

Converting Between SI Units

Because SI is based on ten, it is easy to convert from one SI unit to another. To convert SI units, you must multiply or divide by a factor of ten. You also can use proportions, as shown in the Math Skills activity on this page.

Measurement and Uncertainty

Have you ever measured an object, such as a paper clip? The tools used to take measurements can limit the accuracy of the measurements. All measurements have some uncertainty.

If you measured a paper clip with a ruler divided into centimeters, you would know that the paper clip is between 4 cm and 5 cm, because only whole centimeters are shown. You might guess that the paper clip is 4.5 cm long.

With a ruler that has measurements divided into millimeters, you could say with more certainty that the paper clip is about 4.7 cm long. This measurement is more accurate than the first measurement.

Math Skills 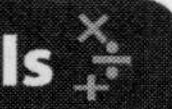

A book has a mass of 1.1 kg. Using a proportion, find the mass of the book in grams.

a. Use the table to determine the correct relationship between the units. One kg is 1,000 times greater than 1 g. So, there are 1,000 g in 1 kg.

b. Then set up a proportion.

$$\left(\frac{x}{1.1\text{ kg}}\right) = \left(\frac{1{,}000\text{ g}}{1\text{ kg}}\right)$$

$$x = \left(\frac{(1{,}000\text{ g})(1.1\text{ kg})}{1\text{ kg}}\right)$$

$$= 1{,}100\text{ g}$$

c. Check your units. The answer is 1,100 g.

3. Use Proportions Two towns are separated by 15,328 m. What is the distance in kilometers?

____

A dosage of medicine is 325 mg. What is the dosage in grams?

____

____

 **Key Concept Check**

4. Relate What causes measurement uncertainty?

____

____

____

FOLDABLES®

Make a vertical two-tab book to organize your notes about SI conversions and rounding significant digits.

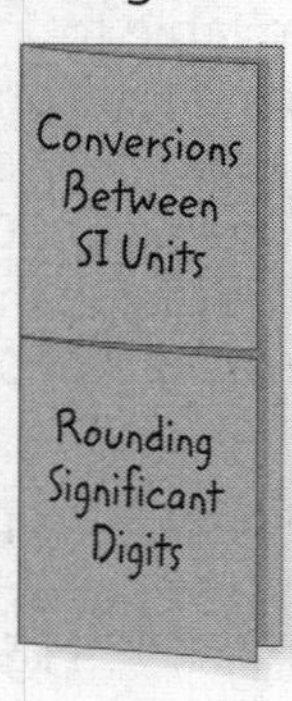

Significant Digits and Rounding

Because scientists duplicate each other's work, they must record numbers with the same degree of precision as the original data. Significant digits allow scientists to do this. **Significant digits** *are the number of digits in a measurement that you know with a certain degree of reliability.*

In order to achieve the same degree of precision as a previous measurement, it often is necessary to round a measurement to a certain number of significant digits. Suppose you need to round the number below to four significant digits.

1,34**8**.527 g

To round to four significant digits, you need to round the 8. If the digit to the right of the 8 is 0, 1, 2, 3, or 4, the digit being rounded (8) remains the same. If the digit to the right of the 8 is 5, 6, 7, 8, or 9, the digit being rounded (8) increases by one. The rounded number is 1,349 g.

What if you need to round 1,348.527 g to two significant digits? You would look at the number to the right of the 3 to determine how to round. 1,348.527 rounded to two significant digits would be 1,300 g. The 4 and 8 become zeros. The table below shows some rules for expressing and determining significant digits.

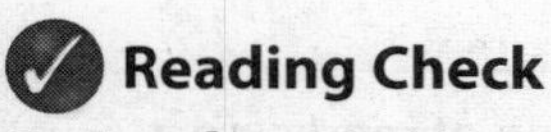
Reading Check

5. Apply Round the following number to the place value shown in bold:

2,45**1**.225

____

Significant Digits Rules

1. All nonzero numbers are significant.
2. Zeros between nonzero digits are significant.
3. One or more final zeros used after the decimal point are significant.
4. Zeros used solely for spacing the decimal point are NOT significant. The zeros only indicate the position of the decimal point.

Note: The bold numbers in the examples are the significant digits.

Number	Significant Digits	Applied Rules
1.234	4	1
1.02	3	1, 2
0.0**23**	2	1, 4
0.**200**	3	1, 3
1,002	4	1, 2
3.07	3	1, 2
0.00**1**	1	1, 4
0.0**12**	2	1, 4
50,600	3	1, 2, 4

Interpreting Tables

6. Assess Which zeros in a number are not significant?

____

____

____

____

Mean, Median, Mode, and Range

A rain gauge measures the amount of rain that falls on a location over a period of time. A rain gauge can be used to collect data in scientific investigations, such as the data shown in the table below. Scientists often need to analyze their data to obtain information. Four values often used when analyzing numbers are median, mean, mode, and range. ✓

Rainfall Data

Month Order		Numerical Order
January	7.11 cm	1.47 cm
February	11.89 cm	7.11 cm
March	9.58 cm	7.11 cm
April	8.18 cm	**8.18 cm**
May	7.11 cm	**8.84 cm**
June	1.47 cm	9.58 cm
July	18.21 cm	11.89 cm
August	8.84 cm	18.21 cm

Median The median is the middle number in a data set when the data are arranged in numerical order. The rainfall data are listed in numerical order in the right column of the table above. The items in bold are the two middle numbers. If you have an even number of data items, add the two middle numbers together and divide by two to find the median.

$$\text{median} = \frac{8.18 \text{ cm} + 8.84 \text{ cm}}{2} = 8.51 \text{ cm}$$

Mean The mean, or average, of a data set is the sum of the numbers in a data set divided by the number of entries in the set. To find the mean, add the numbers in your data set and then divide the total by the number of items in your data set.

$$\text{mean} = \frac{(\text{sum of numbers})}{(\text{number of items})} = \frac{72.39 \text{ cm}}{8 \text{ months}} = \frac{9.05 \text{ cm}}{\text{month}}$$

Mode The mode of a data set is the number or item that appears most often. The number 7.11 occurs twice. All other numbers only appear once.

$$\text{mode} = 7.11$$

Reading Check

7. State What values do scientists often use when evaluating data sets?

Interpreting Tables

8. Interpret Which month had the lowest rainfall? the highest rainfall?

Key Concept Check

9. Summarize What are median, mean, and mode?

Reading Check
10. Explain How do you find the range for a data set?

______

______

______

Reading Check
11. Define What is a science journal? How is it used?

______

______

______

______

______

Reading Check
12. Name Scientists measure the volume of a liquid in which units of measure?

______

Range The range is the difference between the greatest number and the least number in the data set.

$$\text{range} = 18.21 - 1.47 = 16.74$$

Scientific Tools

As you engage in scientific inquiry, you will need tools to help you take quantitative measurements. Always follow appropriate safety procedures when using scientific tools.

Science Journal

Use a science journal to record observations, questions, hypotheses, data, and conclusions from your scientific investigations. A science journal is any notebook that you use to take notes or record information and data while you conduct a scientific investigation.

Keep your journal organized so you can find information easily. Write the date whenever you record new information in the journal. Make sure you are recording your data honestly and accurately.

Rulers and Metersticks

Use rulers and metersticks to measure lengths and distances. The SI unit of measurement for length is the meter (m).

For small objects, such as pebbles or seeds, use a metric ruler with centimeter and millimeter markings. To measure larger objects, such as the length of your bedroom, use a meterstick.

To measure long distances, such as the distance between cities, use an instrument that measures in kilometers. Be careful when carrying rulers and metersticks, and never point them at anyone.

Glassware

Use beakers to hold and pour liquids. The lines on a beaker are not very precise measurements, so you should use a graduated cylinder to measure the volume of a liquid. Liquid volume is typically measured in liters (L) or milliliters (mL).

Triple-Beam Balance

Use a triple-beam balance to measure the mass of an object. The mass of a small object is measured in grams. The mass of a large object is usually measured in kilograms.

Triple-beam balances are instruments that require some care during use. Follow your teacher's instructions so that you do not damage the instrument. Digital balances also might be used.

Thermometer

Use a thermometer to measure the temperature of a substance. Although Kelvin is the SI unit for temperature, you will use a thermometer to measure temperature in degrees Celsius (°C).

To use a thermometer, place a room-temperature thermometer into the substance for which you want to measure temperature. Do not let the thermometer touch the bottom of the container that holds the substance or you will get an inaccurate reading.

When you finish, remember to place your thermometer in a secure place. Do not lay it on a table, because it can roll off the table. Never use a thermometer as a stirring rod.

Computers and the Internet

Use a computer to collect, organize, and store information about a research topic or scientific investigation. Computers are useful tools to scientists for several reasons.

Scientists use computers to record and analyze data and to research new information. They also can quickly share their results with others worldwide over the Internet.

Reading Check

13. Identify What can you measure using a triple-beam balance?

____

Reading Check

14. State How will you measure temperature when conducting scientific investigations?

____

Reading Check

15. Consider How do scientists use computers and the Internet?

____

____

____

____

____

Tools Used by Earth Scientists

Binoculars

Binoculars are instruments that enable people to view faraway objects more clearly. Earth scientists use them to view distant landforms, animals, or even incoming weather.

Reading Check
16. Describe how scientists might use binoculars.

Compass

A compass is an instrument that shows magnetic north. Earth scientists use compasses to navigate when they are in the field and to determine the direction of distant landforms or other natural objects.

Wind Vane and Anemometer

A wind vane is a device, often attached to the roof of a building, that rotates to show the direction of the wind. An anemometer, or wind-speed gauge, is used to measure the speed and the force of wind.

Streak Plate

A streak plate is a piece of hard, unglazed porcelain that helps you identify minerals. When you scrape a mineral along a streak plate, the mineral leaves behind powdery marks. The color of the mark is the mineral's streak.

Reading Check
17. Point Out How do scientists use a streak plate?

After You Read

Mini Glossary

description: a spoken or written summary of observations

explanation: an interpretation of observations

International System of Units (SI): the internationally accepted system for measurement

significant digits: the number of digits in a measurement that are known with a certain degree of reliability

1. Review the terms and their definitions in the Mini Glossary. Write a sentence that explains how a scientific description and explanation are related.

____

____

2. Complete the following table to describe SI base units.

Unit	Quantity Measured
kilogram	
	time
Kelvin	
	electric current
candela	

3. Explain the difference between qualitative and quantitative descriptions.

____

____

Log on to ConnectED.mcgraw-hill.com and access your textbook to find this lesson's resources.

Nature of Science

LESSON 3

Methods of Science

Case Study

Key Concepts 

- How are independent variables and dependent variables related?
- How is scientific inquiry used in a real-life scientific investigation?

Study Coach

Make an Outline As you read, summarize the information in the lesson by making an outline. Use the main headings in the lesson as the main headings in your outline. Use your outline to review the lesson.

Reading Check

1. State What is a variable?

____

____

____

Read to Learn

The Iceman's Last Journey

The Tyrolean Alps border western Austria, northern Italy, and eastern Switzerland. They are popular with tourists, hikers, mountain climbers, and skiers. In 1991, two hikers discovered the remains of a man in a melting glacier on the border between Austria and Italy. They thought the man had died in a hiking accident. So, they reported their discovery to the authorities.

Initially authorities thought the man was a music professor who disappeared in 1938. They soon learned that the music professor was buried in a nearby town. Artifacts near the frozen corpse indicated that the man died long before 1938. The artifacts were unusual. The man, nicknamed the Iceman, was dressed in leggings, a loincloth, and a goatskin jacket. He had a bearskin cap. His shoes were made of red deerskin with bearskin soles. They were stuffed with grass. A copper ax, a longbow, a quiver containing 14 arrows, a wooden backpack frame, and a dagger were found at the site.

A Controlled Experiment

The identity of the corpse was a mystery. Several people hypothesized about his identity, but controlled experiments were needed to unravel the mystery of who the Iceman was. Scientists and the public wanted to know the identity of the man, why he had died, and when he had died.

Identifying Variables and Constants

When scientists design a controlled experiment, they have to identify factors that might affect the outcome of an experiment. *A* **variable** *is any factor that can have more than one value.* In controlled experiments, there are two kinds of variables. *The* **independent variable** *is the factor that you want to test. It is changed by the investigator to observe how it affects a dependent variable. The* **dependent variable** *is the factor you observe or measure during an experiment.* When the independent variable is changed, it causes the dependent variable to change.

Controlled Experiment There are two groups in a controlled experiment—an experimental group and a control group. The experimental group is used to study how a change in the independent variable changes the dependent variable. The control group contains the same factors, but the independent variable is not changed. Without a control, it is difficult to know if your experimental observations result from the variable you are testing or from another factor.

Scientific Inquiry Scientists used inquiry to investigate the mystery of the Iceman. As you read the rest of the story, notice how scientific inquiry was used throughout the investigation. The scientific inquiry process is shown by the bold text in tables, like the ones below. These tables show what a scientist might have written in a journal.

Observations, Hypotheses, and Prediction
Scientific investigations often begin when someone asks a question about something observed in nature.
Observation: A corpse was found buried in ice in the Tyrolean Alps.
Hypothesis: The corpse found in the Tyrolean Alps is the body of a music professor because he went missing in 1938.
Observation: Artifacts found near the body suggested that the body was much older than the music professor would have been.
Revised Hypothesis: The corpse found was dead long before 1938 because the artifacts found near him date back way before the 1930s.
Prediction: If the artifacts belong to the corpse, then the corpse is not the music professor.

An Early Conclusion

Konrad Spindler was a professor of archaeology at the University of Innsbruck in Austria when the Iceman was discovered. Spindler estimated that the ax was at least 4,000 years old based on its construction. A few weeks later, however, radiocarbon dating showed that the Iceman had lived about 5,300 years ago.

Inference and Prediction
An inference is a logical explanation of observations based on past experiences.
Inference: Based on its construction, the ax is at least 4,000 years old.
Prediction: If the ax is at least 4,000 years old, then the body found near it is also at least 4,000 years old.
Test Results: Radiocarbon dating showed the man to be 5,300 years old.

Reading Check

2. Identify What two groups are needed in a controlled experiment?

Interpreting Tables

3. Relate How do scientific investigations often begin?

Interpreting Tables

4. Analyze How did the ax lead to a prediction about the age of the body?

Location of the Iceman The Iceman's body was in a mountain glacier 3,210 m above sea level. Just what could this man have been doing so high in the snow- and ice-covered mountains? Was he hunting, was he shepherding his animals, or was he looking for metal ore?

Professor Spindler's Hypothesis In 1993, Professor Spindler proposed a hypothesis based partly on the artifacts discovered with the Iceman's body. For example, he noted some of the wood used in the artifacts was from trees that grew at lower elevations. Spindler concluded that the Iceman was a seasonal visitor to the high mountains.

According to Spindler's hypothesis, the Iceman had recently driven his herds from their summer high mountain pastures to the lowland valleys. However, after arriving in the lowlands, he chose to return to the mountains. There he died of exposure to the cold, wintry weather.

Professor Spindler observed that the Iceman's body was extremely well preserved. He inferred that ice and snow covered the Iceman's body shortly after he died. Spindler concluded that the Iceman died in autumn and was quickly buried and frozen, which preserved his body and all his possessions.

Think it Over

5. Consider Why did Professor Spindler conclude that the Iceman died in autumn?

Interpreting Tables

6. State What steps are necessary before making conclusions?

Conclusion
After many observations, revised hypotheses, and tests, conclusions often can be made.
Conclusion: The Iceman is about 5,300 years old. He was a seasonal visitor to the high mountains. He died in autumn. When winter came, the Iceman's body became buried and frozen in the snow, which preserved his body.

More Observations and Revised Hypotheses

When the Iceman's body was discovered, Klaus Oeggl was an assistant professor of botany at the University of Innsbruck. His area of study was plant life during prehistoric times in the Alps. He was invited to join the research team studying the Iceman.

Reading Check

7. Name the materials Professor Oeggl initially examined.

Materials at the Discovery Site Professor Oeggl closely examined the Iceman and his belongings. He found three plant materials—grass from the Iceman's shoe, a splinter of wood from his longbow, and a tiny sloe berry fruit.

Examination of the Materials In the year following the Iceman's discovery, Professor Oeggl examined bits of charcoal wrapped in maple leaves found at the discovery site. The samples showed that the wood in the charcoal was from eight different types of trees. All but one of the trees grew only at lower elevations than where the Iceman's body was found.

A Working Hypothesis Like Professor Spindler, Professor Oeggl suspected that the Iceman had been at a lower elevation shortly before he died. This idea became his working hypothesis. However, Oeggl would need more data to prove it. He proposed that he be allowed to examine the man's digestive tract. If all went well, the study would show what the Iceman had swallowed hours before his death. ✓

Observations, Hypothesis, and Prediction
Scientific investigations often lead to new questions.
Observations: Plant matter near body to study—grass on shoe, splinter from longbow, sloe berry fruit, charcoal wrapped in maple leaves, wood in charcoal from 8 different trees—7 of 8 types of wood in charcoal grow at lower elevations
Hypothesis: The Iceman had recently been at lower elevations before he died because the plants identified near him grow only at lower elevations.
Prediction: If the identified plants are found in the digestive tract of the corpse, then the man actually was at lower elevations just before he died.
Question: What did the Iceman eat the day before he died?

Experiment to Test Hypothesis

The research teams provided Professor Oeggl with a tiny sample from the Iceman's digestive tract. He was determined to obtain as much information as he could from it. Oeggl carefully planned his scientific inquiry. He needed to work quickly to avoid the decomposition of the sample and to reduce the chances of contaminating the sample.

His plan was to divide the sample into four parts. Each sample would undergo several chemical tests. Then, the samples would be examined under an electron microscope. Professor Oeggl added several drops of saline solution to the first sample, causing it to swell. He then examined it under the microscope at low magnification. He saw a type of grain known as einkorn. This had been a common type of wheat grown in the region during prehistoric times. He also found other edible plant material. ✓

✓ **Reading Check**

8. Explain Why did Professor Oeggl want to study the Iceman's digestive tract?

Interpreting Tables

9. Evaluate How did the observations about plant matter influence the hypothesis?

✓ **Reading Check**

10. Define What is einkorn?

Interpreting Tables

11. Point Out What are the dependent variables in this example?

____

Test a Hypothesis
There is more than one way to test a hypothesis. Scientists might gather and evaluate evidence, collect data and record their observations, create a model, or design and perform an experiment. They also might perform a combination of these skills.
Test Plan: • Divide a sample of the Iceman's digestive tract into four sections. • Examine the pieces under microscopes. • Gather data from observations of the pieces and record observations.
Variables
Controlled experiments contain two types of variables.
Dependent Variables: amount of hop-hornbeam pollen grains found on slide
Independent Variable: digestive-tract sample

The sample also contained pollen grains. To see the pollen grains better, Professor Oeggl used a chemical that would separate unwanted substances from the pollen grains. He applied some alcohol to the sample and examined it under a microscope at a higher magnification. The pollen grains were more visible. Many more microscopic pollen grains could now be seen. Professor Oeggl identified these pollen grains as those from a hop-hornbeam tree. ✓

Reading Check

12. Assess Why did the professor decide to apply alcohol to the sample?

____

____

____

Analyzing Results

Professor Oeggl observed that the hop-hornbeam pollen grains had not been digested. Therefore, the Iceman must have swallowed them within hours before his death. But, hop-hornbeam trees only grow in lower valleys. Oeggl was confused. How could pollen grains from trees at low elevations be ingested within a few hours of this man dying in high, snow-covered mountains? Professor Oeggl suspected his sample might be contaminated. Oeggl knew he needed to investigate further.

Reading Check

13. Summarize At this point, what two possible conclusions could be drawn from the presence of pollen in the digestive tract sample?

____

____

____

____

____

Further Experimentation

Oeggl realized that the most likely source of contamination would be his own laboratory. He decided to test whether his lab equipment or saline solution contained the hop-hornbeam pollen grains.

Sample Slides To do this, he prepared two identical, sterile slides with saline solution. Then, on one slide, he placed a sample from the Iceman's digestive tract. The slide with the sample was the experimental group. The slide without the sample was the control group. ✓

Variables The independent variable, or the variable that Oeggl changed, is the presence of the sample on the slide. The dependent variable, or the variable that Oeggl was testing, was whether hop-hornbeam pollen grains show up on the slides. Oeggl examined the slides carefully.

Analyzing Additional Results

The experiment showed that the control group (the slide without the digestive tract sample) contained no hop-hornbeam pollen grains. Therefore, the pollen grains had not come from his lab equipment or solutions. Each sample from the Iceman's digestive tract was closely re-examined. All of the samples contained the same hop-hornbeam pollen grains. The iceman had indeed swallowed the hop-hornbeam pollen grains. ✓

Managing Error
Error is common in scientific research. Scientists are careful to document procedures and any unanticipated factors or accidents. They also are careful to document uncertainty in their measurements.
Procedure: • Sterilize laboratory equipment. • Prepare saline slides. • View saline slides under electron microscope. Results: no hop-hornbeam pollen grains • Add digestive tract sample to one slide. • View this slide under electron microscope. Result: hop-hornbeam pollen grains present
Control Group
Without a control group, it is difficult to determine the origin of some observations.
Control Group: sterilized slide
Experimental Group: sterilized slide with digestive tract sample

Mapping the Iceman's Journey

The hop-hornbeam pollen grains were helpful in determining the season the Iceman died. Because the pollen grains were whole, Professor Oeggl inferred that the Iceman swallowed the hop-hornbeam pollen grains during their blooming season. Therefore, the Iceman must have died between March and June. ✓

✓ **Reading Check**

14. State What did the professor conclude based on the control slide?

______

______

______

______

Interpreting Tables

15. Summarize How do scientists manage error in their research?

______

______

______

______

✓ **Reading Check**

16. Point Out Why did Professor Oeggl believe the Iceman died in spring?

______

______

______

______

Additional Study and Investigation Professor Oeggl was ready to map the Iceman's final trek up the mountain. Because Oeggl knew the rate at which food travels through the digestive system, Oeggl inferred that the Iceman had eaten three times in the final day-and-a-half of his life. From the digestive tract samples, Oeggl estimated where the Iceman was when he ate.

Final Hours First, the Iceman ingested pollen grains native to higher mountain regions. Then he swallowed hop-hornbeam pollen grains from the lower mountain regions several hours later. Last, the Iceman swallowed other pollen grains from trees of higher mountain areas again. Oeggl proposed the Iceman traveled from the southern region of the Italian Alps to the higher, northern region as shown in the figure below. There he died suddenly. He did this all in a period of about 33 hours.

Reading Check
17. Summarize What was Professor Oeggl able to determine about the Iceman's movements from the different types of pollen residue?

____

____

____

____

____

Visual Check
18. Draw Highlight the possible route of the Iceman. Circle the place where he was found near the border of Austria and Italy.

Iceman's Last Journey

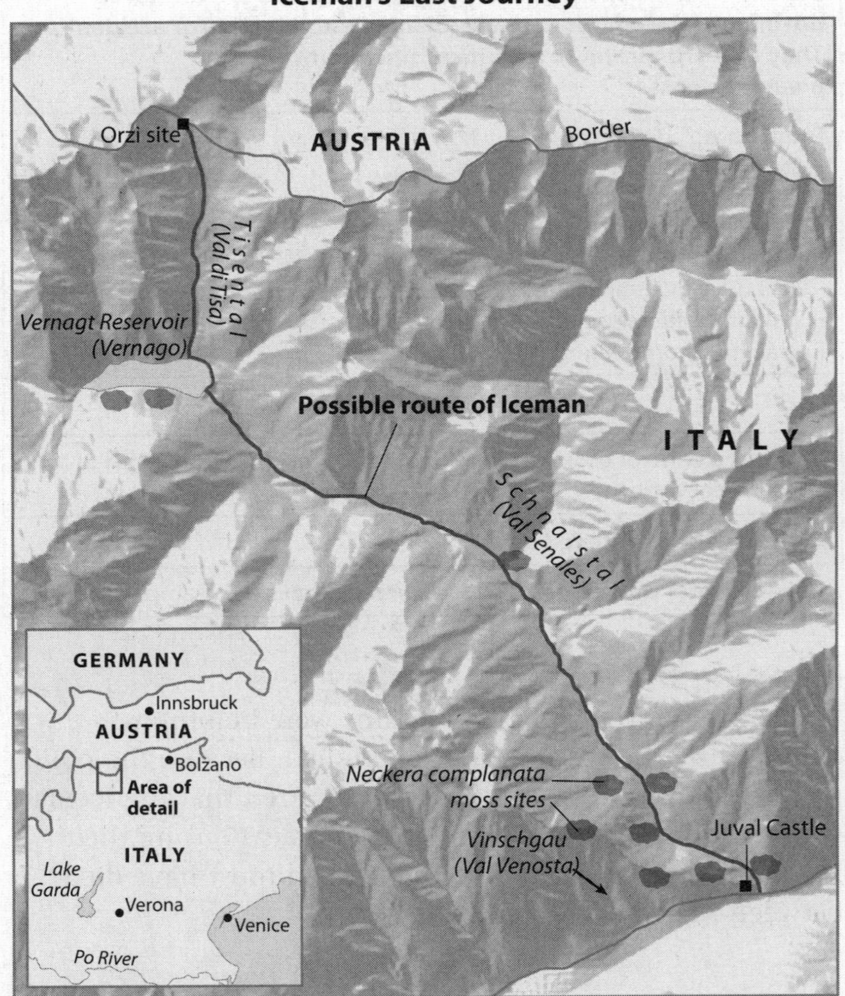

Observation, Inference, and Prediction
An inference is a logical explanation of an observation that is drawn from prior knowledge or experience. Inferences can lead to predictions, hypotheses, or conclusions.
Observation: • The Iceman's digestive tract contains pollen grains from the hop-hornbeam tree and other plants that bloom in spring. **Inference:** • Knowing the rate at which food and pollen decompose after they are swallowed, it can be inferred that the Iceman ate three times on the day that he died. **Prediction:** • The Iceman died in the spring within hours of digesting the hop-hornbeam pollen grains.

Conclusion

Researchers from around the world worked on different parts of the Iceman mystery and shared their results. Analysis of the Iceman's hair revealed his diet usually contained vegetables and meat. Examining the Iceman's one remaining fingernail, scientists determined that he had been sick three times within the last six months of his life. X-rays revealed an arrowhead under the Iceman's left shoulder. This suggested that he died from that serious injury rather than from exposure.

Finally, scientists concluded that the Iceman traveled from the high alpine region in spring to his native village in the lowland valleys. There, during a conflict, the Iceman sustained a fatal injury. He retreated back to the higher elevations, where he died. Scientists recognize their hypotheses can never be proved, only supported or not supported. However, with advances in technology, scientists are able to more thoroughly investigate mysteries of nature. ✓

Revised Conclusion
Scientific investigations may disprove early hypotheses or conclusions. However, new information can cause a hypothesis or conclusion to be revised many times.
Revised Conclusion: In spring, the Iceman traveled from the high country to the valleys. After he was involved in a violent confrontation, he climbed the mountains into a region of permanent ice where he died of his wounds.

Interpreting Tables

19. Define What is an inference?

____

____

____

✓ Reading Check

20. Describe the new evidence that led to a conclusion about why the Iceman died.

____

____

____

____

____

Interpreting Tables

21. Explain Why might a hypothesis or conclusion be revised?

____

____

After You Read

Mini Glossary

dependent variable: the factor an investigator observes or measures during an experiment

independent variable: the factor that is changed by the investigator to observe how it affects a dependent variable

variable: any factor that can have more than one value

1. Review the terms and their definitions in the Mini Glossary. Write a sentence that describes the relationship between an independent variable and a dependent variable.

____

____

2. Write the words below in the correct boxes in the flowchart to summarize the steps used in this case study.

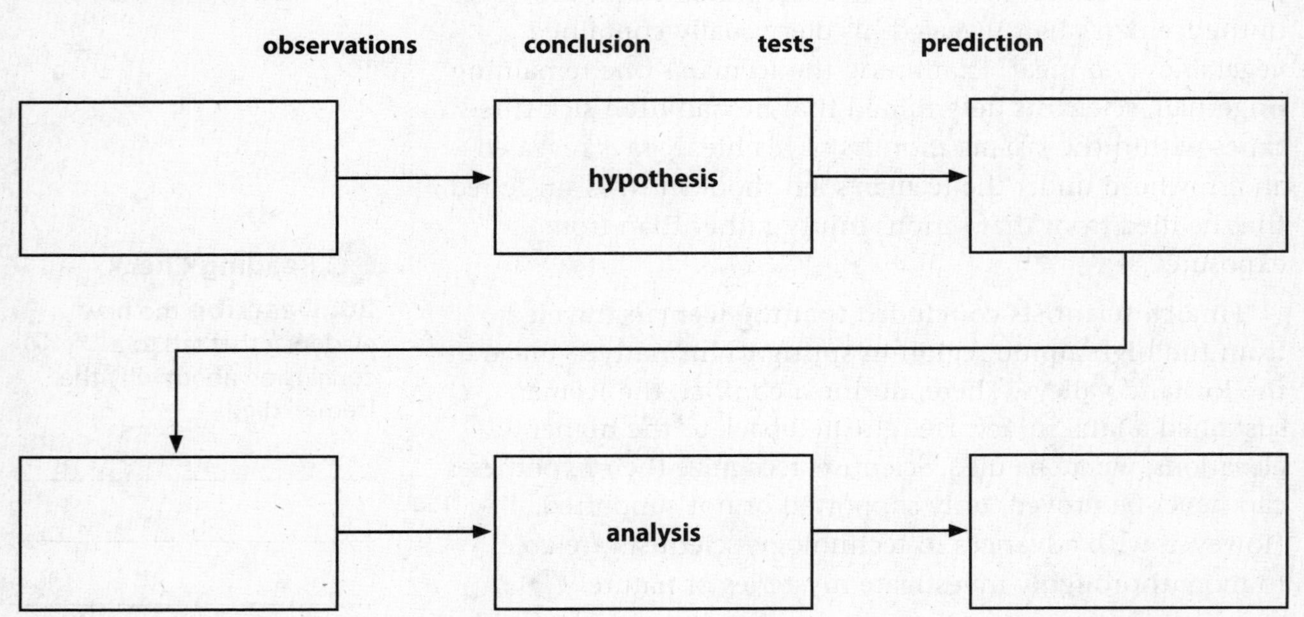

3. Explain how making an outline helped you understand this lesson.

____

____

Log on to ConnectED.mcgraw-hill.com and access your textbook to find this lesson's resources.

Mapping Earth

Maps

Before You Read

What do you think? Read the three statements below and decide whether you agree or disagree with them. Place an A in the Before column if you agree with the statement or a D if you disagree. After you've read this lesson, reread the statements to see if you have changed your mind.

Before	Statement	After
	1. Maps help determine locations on Earth.	
	2. All Earth models are spherical.	
	3. World maps are drawn accurately for every location.	

Key Concepts 

- How can a map help determine a location?
- Why are there different map projections for representing Earth's surface?

Read to Learn

Understanding Maps

You likely have used a map before to find information. You might use a map to get to a place you have never visited. You might also use a map to help you find your way. Maps help people get where they are going without getting lost.

Your school might have various kinds of maps. Perhaps you used a map on the first day of school to find all of your classrooms. Or maybe you used a map of the school to practice for a fire drill or a disaster drill. Some maps show all the exits in a building or the safest room to go to if there were a tornado.

There are many different kinds of maps, such as road maps, trail maps, and weather maps. Each type of map shows different information. Each type of map serves a different purpose. ✓

A map is a model of Earth. As you know, Earth is round. Most maps are flat. So they are a model of an area of Earth's surface. If you want to model the entire planet and its shape, you can make a globe.

Mark the Text

Identify Main Ideas
As you read, underline the main ideas under each heading. After you finish reading, review the main ideas that you have underlined.

✓ **Reading Check**

1. Name three different kinds of maps.

______

______

______

Map Views

Most maps are drawn in map view. *A* **map view** *shows an area of Earth's surface as if you are looking down from above.* A map view can also be called a plan view.

Cross sections are drawn in profile view. *A* **profile view** *is a drawing showing a vertical "slice" through the ground.* A profile view is like a side view of a house. A map view and a profile view of a house are shown below.

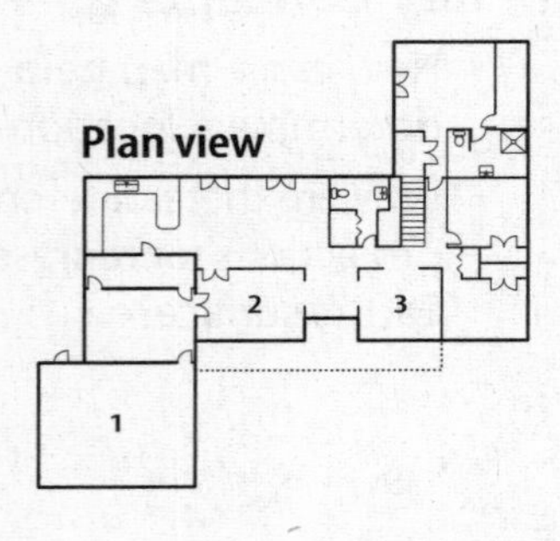

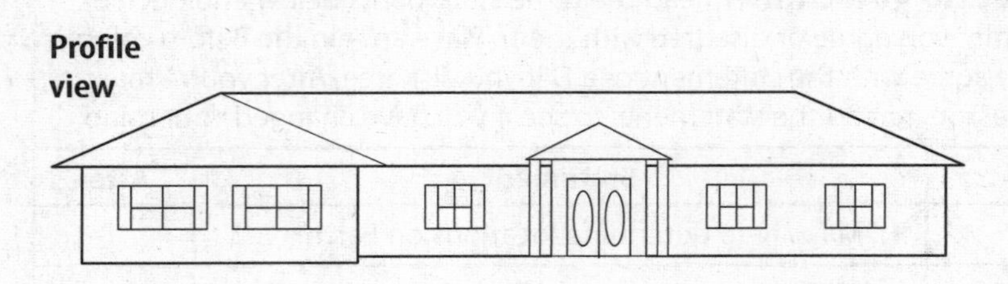

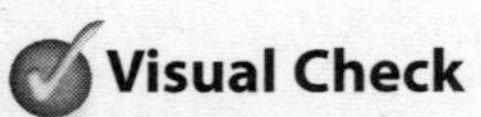

Visual Check

2. Contrast What is the difference between a map view and a profile view?

Map views and profile views are used to describe topographic maps and geologic maps. You use profile views when you study models of the inner structures of Earth.

Map Legends and Scales

Maps have two features to help you read and understand them. One feature is a series of symbols called a map legend. The other is a ratio, which establishes the map scale.

Map Legends Maps use specific symbols to represent certain features on Earth's surface, such as roads in a city or restrooms in a park. These symbols keep maps from being too cluttered. All maps include a map legend. *A* **map legend** *is a key that lists all the symbols used on the map.* It also explains what each symbol means. A map legend is shown below. In this map legend a broken line represents a trail on the map.

Reading Check

3. Explain What is the purpose of a map legend?

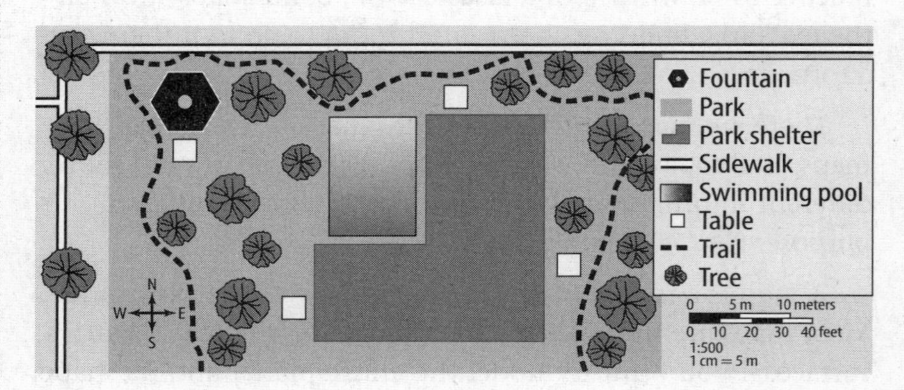

Visual Check

4. Identify Circle the tables. Highlight the trail.

Map Scales When mapmakers draw a map, they need to decide how big or small to make the map. They need to decide on the map's scale. *A* **map scale** *is the relationship between a distance on the map to the actual distance on the ground.*

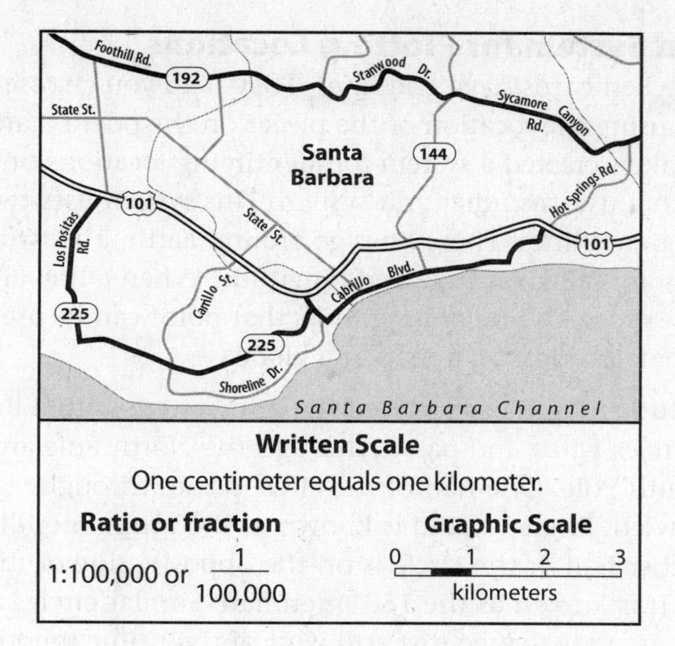

A map scale can be written with words, such as "1 centimeter is equal to 1 kilometer." A map scale also can be written with numbers as a ratio, such as 1:100. Because this is a ratio, there are no units. To explain the ratio, you would say, "Every unit on the map is equal to 100 units on the ground." If your unit were 1 cm on the map, it would be equal to 100 cm on the ground. If you drew a map of your school at a scale of 1:1, your map would be as large as your school.

The road map above gives a scale that is written with words, a ratio scale, and a scale bar in the map legend. Each one is useful in different ways. The graphic scale, or scale bar, would be useful in measuring distances on the map. You would have to measure it, however, to find that 1 cm is equal to 1 km.

Model builders usually use scale to make the model measurements accurately reflect the measurements of the real object. They do this when the model is smaller than the real objects and when it is bigger than the real object.

Reading Maps

To find your way to a specific place, you first need to know where you are on Earth. Imagine trying to describe your exact position on the snow-covered continent of Antarctica. It would be difficult to describe. Ship captains have the same problem when trying to plot their courses across the oceans. Similarly, airplane pilots have the same problem as they fly above a cloud-covered Earth.

Visual Check

5. Determine Which scale would you use to measure the distance between the rivers along Route 192?

Math Skills

A ratio is a comparison of two quantities by division. For example, a map scale is the ratio of the distance on the map to the actual distance. A map might use a scale in which 1 cm on the map represents 5 km of actual distance. This might be written as a ratio:

1 cm to 5 km or

1 cm: 5 km or

$\frac{1\text{ cm}}{5\text{ km}}$

This ratio is the map scale.

6. Use Ratios A map uses a scale of 1 cm: 1 km. If the distance between two points on the map is 3 cm, what is the actual distance between the points?

A Grid System for Plotting Locations

Checkerboards have grid lines. They help you choose your moves using the location of the pieces on the board. Early mapmakers created a system for identifying locations on Earth that uses a similar grid system. This system uses two sets of imaginary lines. These lines go around Earth. The two sets of lines are called latitude and longitude. When a line of latitude crosses a line of longitude, that point can be used to identify a location on a map or a globe.

WORD ORIGIN
longitude
from Latin *longitudo,* means "length"

Longitude Mapmakers started the grid system with a line that circles Earth and passes through the North Pole and the South Pole. One-half of the circle passes through Greenwich, England, and is known as the prime meridian. The other half of the circle is on the opposite side of the globe. It is known as the 180° meridian. Similar circles are drawn at every degree east and west of the prime meridian. **Longitude** *is the distance in degrees east or west of the prime meridian.* The prime meridian and the 180° meridian combine to divide Earth into east and west halves. The halves are known as hemispheres—the eastern hemisphere and the western hemisphere. The lines east of the prime meridian are called east longitude. The lines west of the prime meridian are called west longitude. They both meet at the 180° meridian. All the meridians pass through the North Pole and the South Pole, as shown below.

ACADEMIC VOCABULARY
prime
(adjective) first in rank

Latitude Mapmakers also drew lines east to west around Earth. These lines of latitude are somewhat perpendicular to lines of longitude. The center line, called the equator, divides Earth into the northern hemisphere and the southern hemisphere. **Latitude** *is the distance in degrees north or south of the equator.* Lines of latitude form parallel circles, as shown below. The equator is the largest circle. All other circles become smaller and smaller the closer they are to Earth's poles. The North Pole and the South Pole are each shown by a dot.

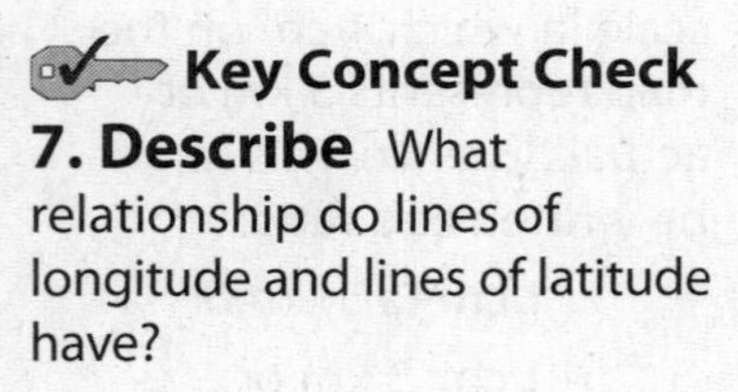

Key Concept Check
7. Describe What relationship do lines of longitude and lines of latitude have?

Visual Check
8. Identify Highlight the equator and the prime meridian.

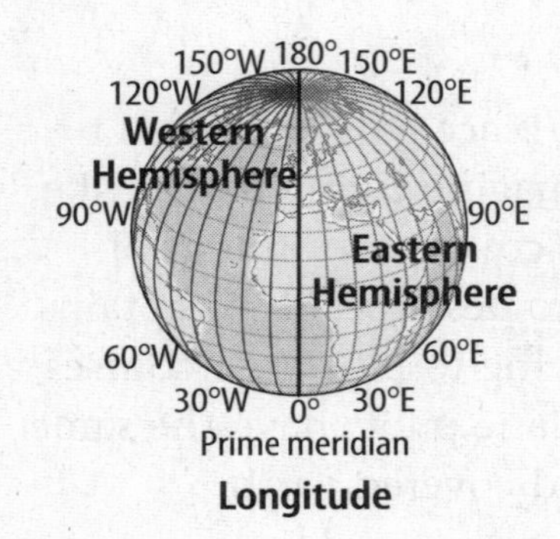

Longitude

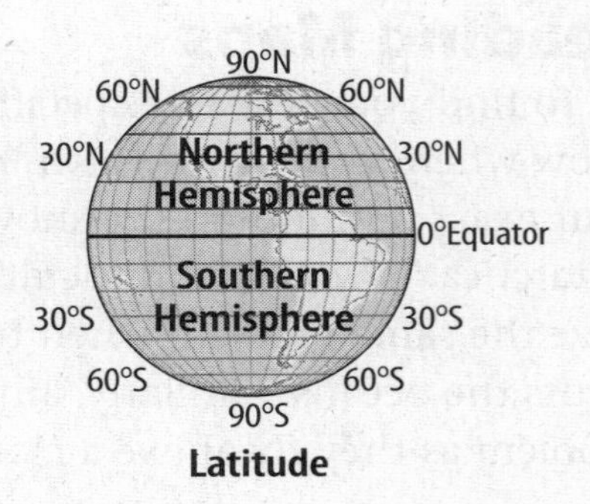

Latitude

Plotting Locations

Earth is a sphere. A sphere is a ball-shaped object. If you look straight down on a sphere, it looks like a circle. Like a circle, a sphere can be divided into 360 degrees. This is why latitude and longitude are measured in degrees.

The latitude at the equator is 0°. All other lines of latitude are measured either north or south from the equator. The North Pole is located at 90 degrees north latitude (90°N). The South Pole is located at 90 degrees south latitude (90°S). The longitude at the prime meridian is 0°. There are 180 degrees of east longitude and 180 degrees of west longitude.

Any location on Earth can be described by the intersection of the closest line of latitude and the closest line of longitude. Latitude is always stated before longitude.

Minutes and Seconds Latitude and longitude lines are far apart. To help identify locations, each degree of latitude and longitude is divided into 60 minutes (') and each minute is divided into 60 seconds (").

Time Zones It is exactly noon in your location when the Sun is directly overhead. However, it is exactly noon in only certain locations on Earth. How do you know what time it is in other places? Time zones were set up to make travel and doing business easier. *A* ***time zone*** *is the area on Earth's surface between two meridians where people use the same time.*

The reference, or starting point, for time zones is the prime meridian. The width of a time zone is 15°, but, as shown below, time zones do not follow the meridians exactly. The borders of the time zones are sometimes crooked so that whole countries can fit into them. Notice how the time changes by one hour at the boundary of each time zone.

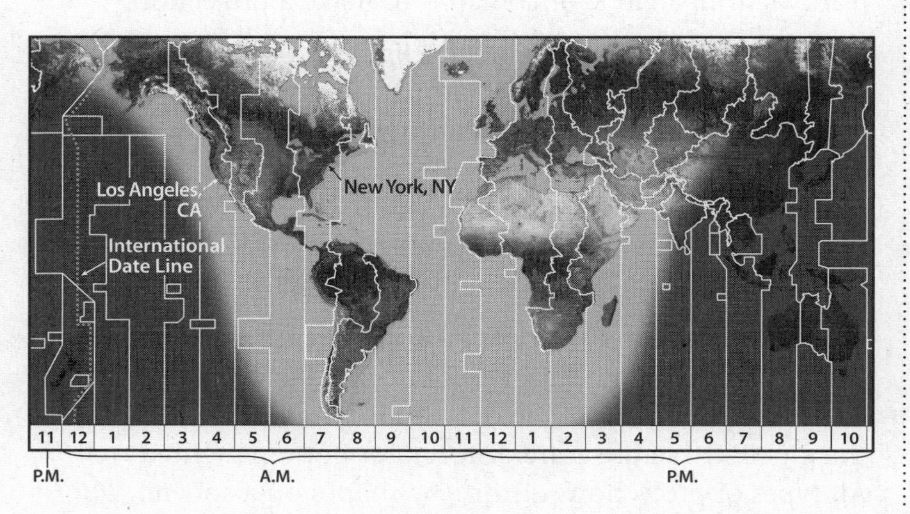

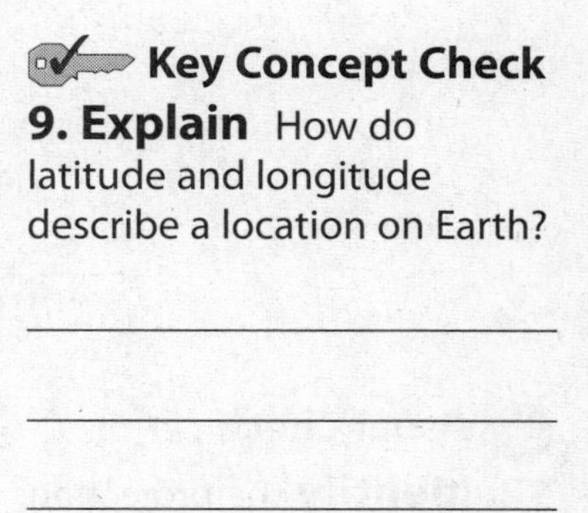

Key Concept Check

9. Explain How do latitude and longitude describe a location on Earth?

Reading Check

10. Describe Why do we need a starting point for time zones?

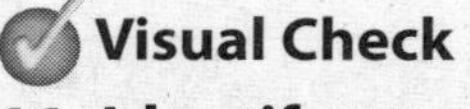

Visual Check

11. Identify If it is 2:00 P.M. in New York City, what time is it in Los Angeles, California?

International Date Line *The line of longitude 180° east or west of the prime meridian is called the* **International Date Line.** When you cross the date line from east to west, it is a day later. So, if on Sunday you were sailing west toward the International Date Line, after you crossed it, it would be Monday. If you turned around and sailed back across the International Date Line from west to east, it would be Sunday again.

The International Date Line does not follow the 180° meridian exactly. This is because some island groups would be divided by the line. It would be one day on some islands and a different day on others. So that the day is the same for all islands of one group, the date line bends around them.

FOLDABLES®

Make a folded book to record information about map projections.

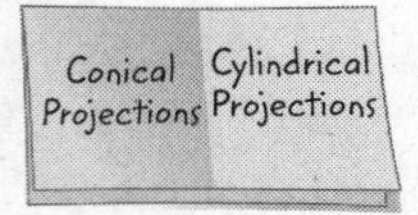

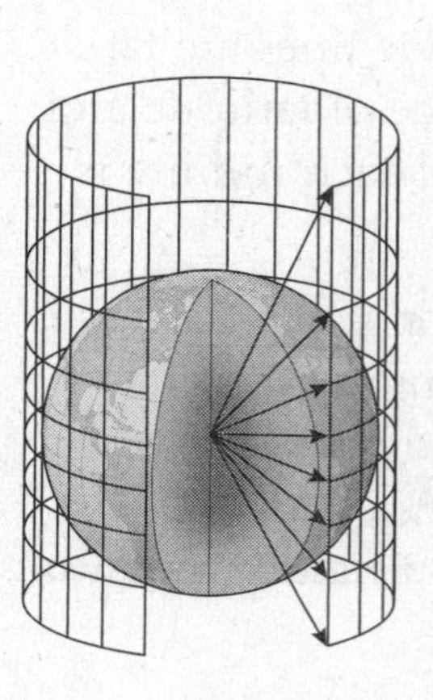

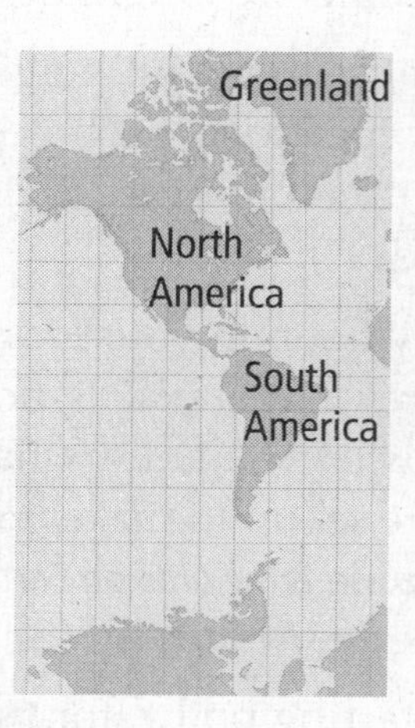

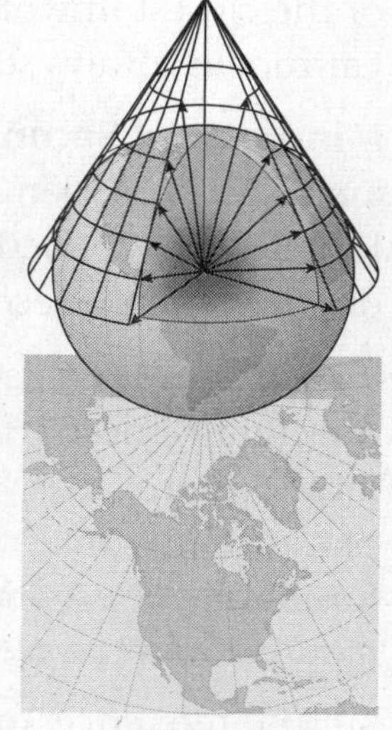

Cylindrical Projection

Conical Projection

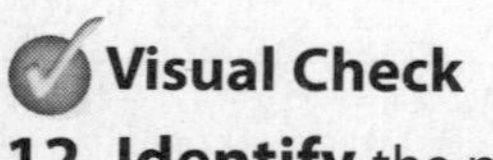

Visual Check

12. Identify the projection that shows the shape of Greenland more accurately.

____

____

Map Projections

Because a globe is a sphere like Earth, Earth's features on a globe appear as they do on Earth. Maps, however, are flat. How can a flat map be made from a sphere? One way to show features from a globe on a map is to make a projection.

Cylindrical Projections Imagine a light at the center of a globe and a sheet of paper wrapped around the globe. The light would throw shadows of the continents and the latitude and longitude lines onto the paper. Because the paper is a cylinder, as shown above, this is called a cylindrical projection. A map made in this way represents shapes near the equator very well. However, shapes near the poles are enlarged.

Conical Projections Wrapping a cone around the globe makes a conical projection. This type of map is a good representation of the shapes near the line of latitude where the cone touches the globe. Other areas are distorted or not represented well. All types of projections distort the shapes on a sphere.

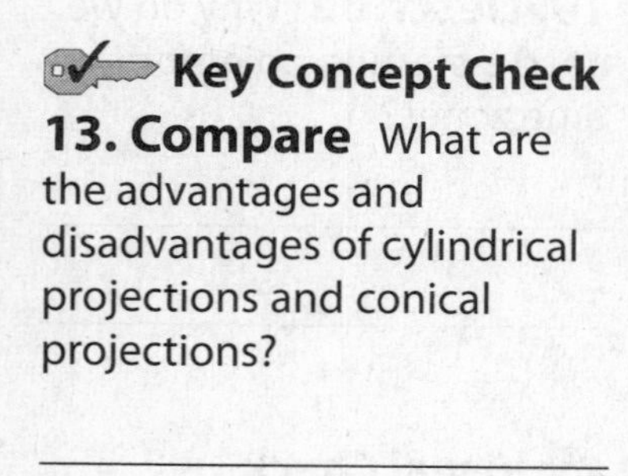

Key Concept Check

13. Compare What are the advantages and disadvantages of cylindrical projections and conical projections?

____

____

____

____

After You Read

Mini Glossary

International Date Line: the line of longitude 180° east or west of the prime meridian

latitude: the distance in degrees north or south of the equator

longitude: the distance in degrees east or west of the prime meridian

map legend: a key that lists all the symbols used on a map

map scale: the relationship between a distance on a map and the actual distance on the ground

map view: a map drawn as if you are looking down on an area from above Earth's surface

profile view: a drawing showing a vertical "slice" through the ground

time zone: the area on Earth's surface between two meridians where people use the same time

1. Review the terms and their definitions in the Mini Glossary. Write a sentence using a glossary word to describe something you find on a map that helps you interpret the map.

2. Use the map shown below to answer the following questions: How many tables are there? What is the graphic scale for this map?

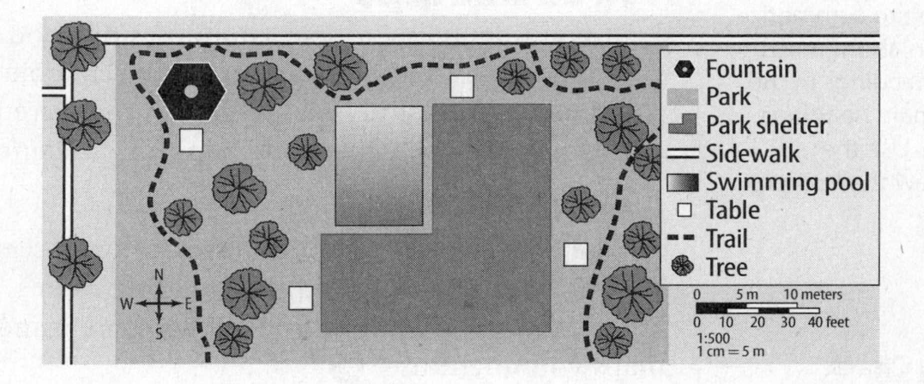

3. Name one main idea about maps you underlined in this lesson.

What do you think NOW?

Reread the statements at the beginning of the lesson. Fill in the After column with an A if you agree with the statement or a D if you disagree. Did you change your mind?

Log on to ConnectED.mcgraw-hill.com and access your textbook to find this lesson's resources.

CHAPTER 1
LESSON 2

Mapping Earth

Technology and Mapmaking

Key Concepts

- What can a topographic map tell you about the shape of Earth's surface?
- What can you learn from geologic maps about the rocks near Earth's surface?
- How can modern technology be used in mapmaking?

Before You Read

What do you think? Read the three statements below and decide whether you agree or disagree with them. Place an A in the Before column if you agree with the statement or a D if you disagree. After you've read this lesson, reread the statements to see if you have changed your mind.

Before	Statement	After
	4. Topographic maps show changes in surface elevations.	
	5. The colors on geologic maps show the colors of the surface rocks.	
	6. Satellites are too far from Earth to collect useful information about Earth's surface.	

Study Coach

Use an Outline As you read, make an outline to summarize the information about maps. Use the main headings in the lesson as the main headings in your outline. Use the outline to review the lesson.

Read to Learn

Types of Maps

If you were going to join two pieces of wood together, you might use a hammer and nails. To scramble eggs, you could use a whisk and a skillet. Just as there are tools for doing different jobs, there are maps for different purposes.

General-Use Maps

The first maps were hand drawn by explorers and sailors. They made them to record their trading routes. Today maps are used for many purposes. Following are some everyday maps you might use. ✓

- **Physical maps** use lines, shading, and color to indicate features such as mountains, lakes, and streams.
- **Relief maps** use shading and shadows to identify mountains and flat areas.
- **Political maps** show the boundaries between countries, states, counties, or townships. The boundaries can be shown as a variety of solid or dashed lines. Different colors might be used to indicate areas within the boundaries.

✓ **Reading Check**

1. Explain Who made the first maps, and what were the maps used for?

____

____

____

- **Road maps** can show interstate highways or a range of roads from four-lane expressways to gravel roads. They are all useful in finding your way to and from different locations. A city road map will show more detail than a road map of the entire United States.

Topographic Maps

If you were hiking across the United States, you might want to follow level terrain. If you were piloting an airplane across the United States, you would want to fly higher than the mountains. Showing how high or low land features are is a feature of one kind of specialty map.

The shape of the land surface is called topography. *A* **topographic map** *shows the detailed shapes of Earth's surface, along with its natural and human-made features.* A topographic map shows differences in elevation. You can form a mental picture of what the landscape looks like without seeing it.

Elevation and Relief *The height above sea level of any point on Earth's surface is its* **elevation.** For example, Mt. Rainier in Washington is 4,392 m above sea level. The city of Olympia, Washington, is about 43 m above sea level. *The difference in elevation between the highest point and the lowest point in an area is called* **relief.** You calculate relief by subtracting the height of the lowest point from the height of the highest point. For example, the relief between Mt. Rainier and Olympia is 4,349 m.

Contour lines *are lines on a topographic map that connect points of equal elevation.* Remember that lines of latitude and longitude do not really exist on Earth's surface. Similarly, contour lines do not really exist on Earth's surface. They exist only on maps.

After you learn to read contour lines, you can measure both elevation and relief on a topographical map. For example, you can use contour lines on a map to see what the elevation is for a mountain.

Interpreting Contours As you learned, contour lines represent elevation. However, the elevation might not be written on every contour line on a map. Darker contour lines, called index contours, are labeled with the elevation. To find out the elevation of the contours that are not labeled, you need to know the difference in elevation between the lines.

FOLDABLES®

Fold and label a sheet of paper to collect information about what a topographic map can show you.

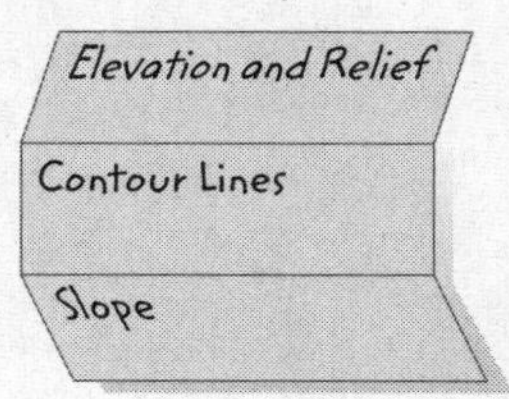

Reading Check

2. Define What is a topographic map?

____

____

____

Reading Check

3. Explain How are contour lines similar to lines of latitude and longitude?

____

____

____

Visual Check

4. Identify Highlight an area with a gentle slope and label it.

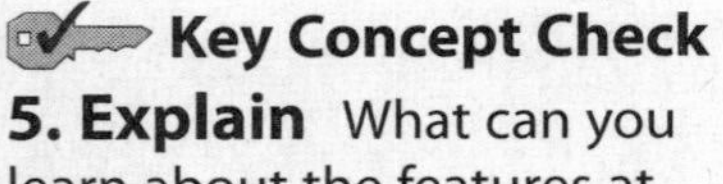

Key Concept Check

5. Explan What can you learn about the features at Earth's surface from studying contour lines?

Reading Check

6. Describe how to make a topographic profile.

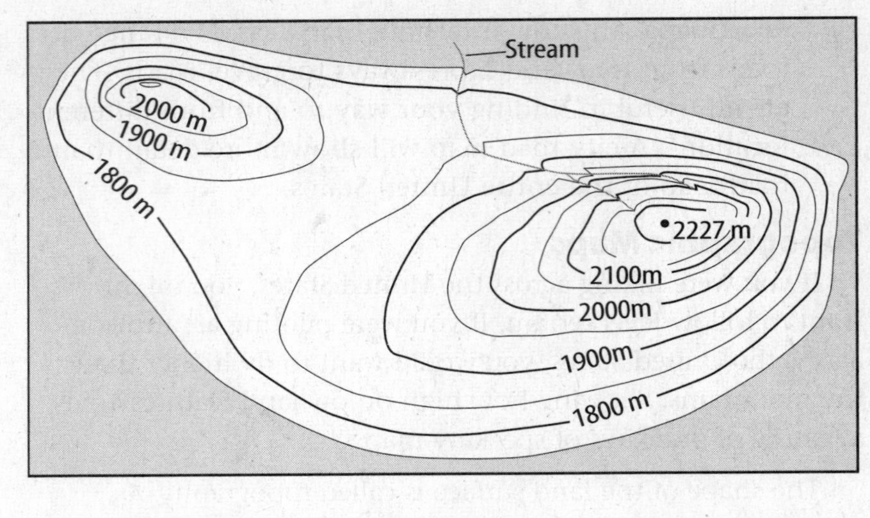

Contour Intervals *The elevation difference between contour lines that are next to each other is called the* **contour interval.** The map shown above has a contour interval of 50 m. You can find the elevation of an unlabeled contour by using the numbered index contours. First, find the closest index contour below the contour you are trying to identify. Then, count up to the unlabeled contour line by 50s from the index contour to find the elevation. Using this method, you can determine that the elevation of the first unlabeled contour line on the map above is 1,850 m. The closest index contour line below it is 1,800 m. By adding 50 m, you determine the height of the first unlabeled contour line.

Notice that a contour line at the top of the mountain forms a small loop with a dot in the middle of it. This dot represents the high point on the mountain—2,227 m. The V-shaped contours pointing downhill indicate ridges. The small V shape pointing uphill indicate a stream valley or drainage.

The spacing of the contours indicates slope. **Slope** *is a measure of the steepness of the land.* If the contours are spaced far apart, the slope is gradual or flat. If the contours are close together, the slope is steep.

Topographic Profiles The information that contour lines provide on a topographic map can be used to draw an accurate profile of the topography. To make a topographic profile graph, mapmakers first draw a profile line on the contour map. Then, they transfer the elevations of the contours crossed by the profile line to the *y*-axis of the topographic profile graph. Using topographic profiles can help you find the easiest path to take when crossing the land.

Symbols on Topographic Maps The United States Geological Survey (USGS) is the government agency responsible for mapping the United States. It has been mapping the United States since the late 1800s. Most topographic maps that you see are made by the USGS.

The table below shows some of the symbols used on the USGS topographic maps. Contour lines are brown on land and blue under water. Green represents vegetation, such as woods. Water in rivers, lakes, and oceans is shown in blue. Buildings are shown as black squares or rectangles, except in cities. In cities pink shading is used to indicate that housing is close together. Roads and railroads can be shown on topographic maps. Notice that different kinds of roads are represented by different symbols. If new information appears on a map, it is shown in purple. ✓

USGS Topographic Map Symbols

Description	Symbol
Primary highway	
Secondary highway	
Unimproved road	
Railroad	
Buildings	
Urban area	
Index contour	100
Intermediate contour	
Perennial streams	
Intermittent streams	
Wooded marsh	
Woods or brushwood	

✓ **Reading Check**

7. Explain Why is it important for a topographic map to have a legend?

✓ **Visual Check**

8. Contrast What is the difference between a primary highway and a secondary highway on a topographic map?

Geologic Maps

Another kind of specialty map is a geologic map. **Geologic maps** *show the surface geology of the mapped area*. It can show the rock types, their ages, and locations of faults.

Geologic Formations On a geologic map, different colors and symbols represent different geologic formations. A geologic formation is a volume of a particular kind of rock. The map legend lists the colors and symbols along with the age of the rock formation. The colors do not indicate the rock's true colors. The colors show the many formations on the map.

Geologic Cross Sections Sometimes geologists need to know what the rocks are like underground as well as on the surface. Because geologists cannot see the rocks under ground, they use other methods to collect information about geologic formations. Information can be gathered by drilling for samples, studying earthquake waves, or looking at cliffs. A cliff face is like a profile view of the ground. Geologists use this information to produce a profile view of the rocks below the ground. *The resulting diagram, showing a vertical slice through the rocks below the surface, is called a* **cross section.** A cross section of a geologic map is shown below.

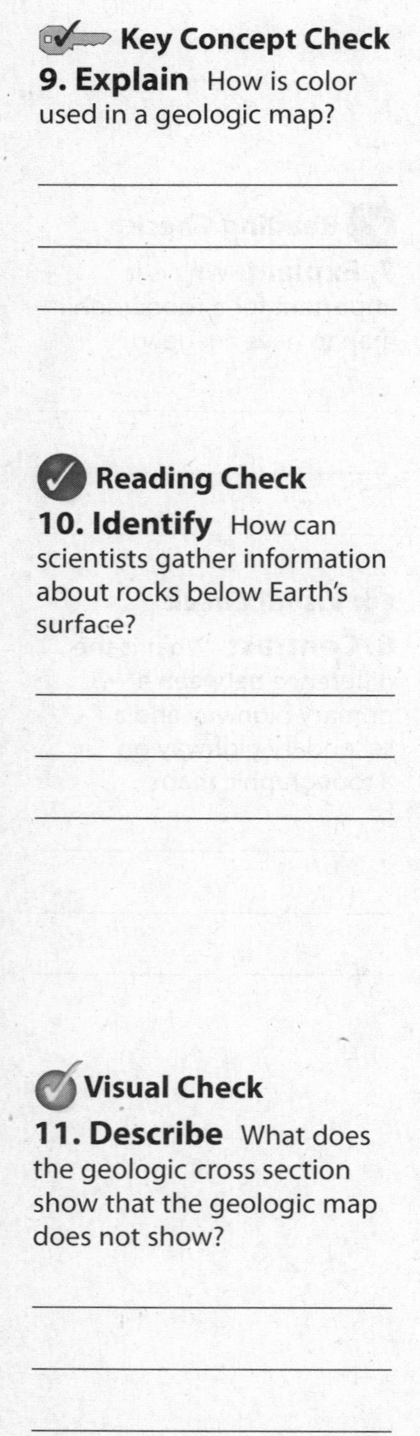

Key Concept Check

9. Explain How is color used in a geologic map?

Reading Check

10. Identify How can scientists gather information about rocks below Earth's surface?

Visual Check

11. Describe What does the geologic cross section show that the geologic map does not show?

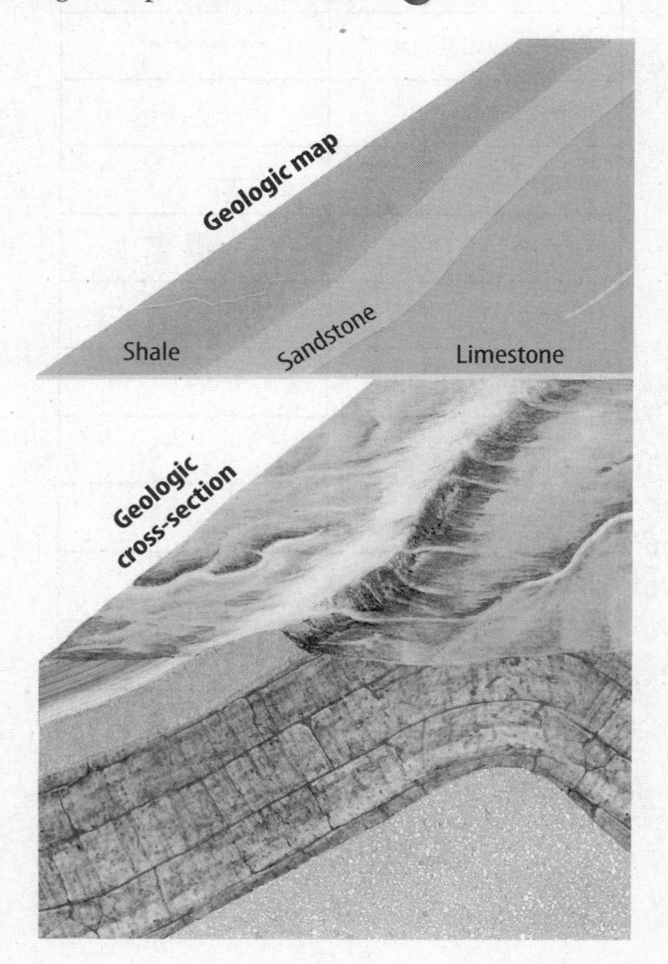

Making Maps Today

For hundreds of years, mapmakers observed Earth and gathered information from explorers. Mapmakers and explorers used tools such as compasses and telescopes to make and record measurements. Mapmakers then drew new maps by hand. Today mapmakers use computers and data from satellites to make maps.

Global Positioning System

One important resource for mapmakers today is the Global Positioning System (GPS). GPS is a group of satellites used for navigation. The 24 GPS satellites orbit Earth. They send signals back to Earth's surface. These signals are used to calculate the distance to the satellite based on the average time of the signal.

A GPS unit can receive signals from three or four different satellites, as shown below. Then the receiver calculates its location—latitude, longitude, and height above Earth's surface. GPS is used by mapmakers to locate reference points.

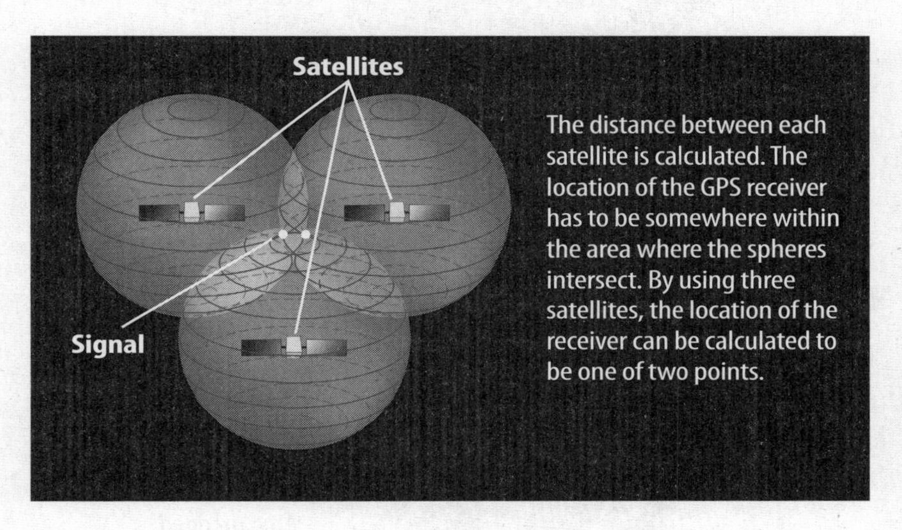

The distance between each satellite is calculated. The location of the GPS receiver has to be somewhere within the area where the spheres intersect. By using three satellites, the location of the receiver can be calculated to be one of two points.

Originally designed for military purposes, GPS is now available for everyone. Airplane pilots and ship captains use GPS to navigate. Many people use GPS technology in their cars. Other uses include tracking wildlife for scientific data collection, detecting earthquakes, hiking, biking, land surveying, and making maps. ✓

GPS technology continues to improve. Today, people and places can be located to the centimeter. Future improvements to safety and rescue operations are predicted. GPS technology eventually could be used to guide self-driven automobiles.

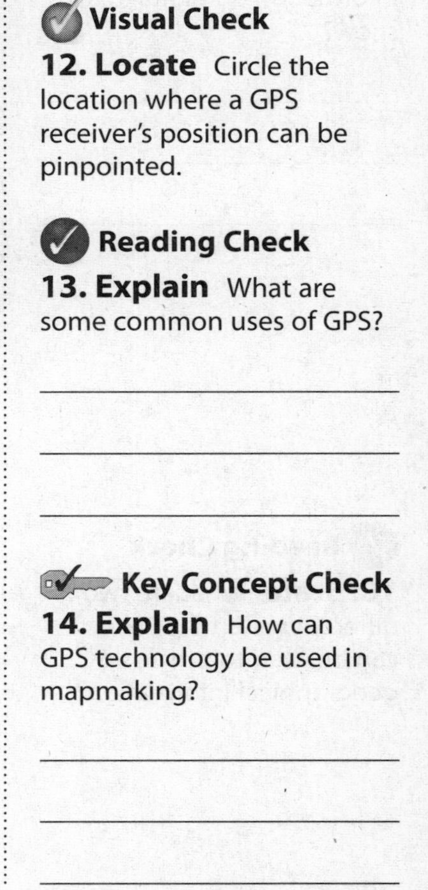

Visual Check
12. Locate Circle the location where a GPS receiver's position can be pinpointed.

Reading Check
13. Explain What are some common uses of GPS?

Key Concept Check
14. Explain How can GPS technology be used in mapmaking?

Geographic Information Systems

Geographic Information Systems (GIS) are computerized information systems used to store and analyze map data. GIS combine data collected from many different sources, including satellites, scanners, and aerial photographs. Aerial photographs are taken from above the ground. Data collection is rapid, taking only hours or minutes.

ACADEMIC VOCABULARY
aerial
(adjective) operating or occurring overhead

Mapmakers use GIS to analyze and organize those data and then make digital maps, as shown in the figure below. GIS creates different map layers of the same location. The map layers are like the layers of a cake. However, when the map layers are placed on top of one another, you can see through to the lower layers. Different layers can show land usage, elevation, roads, streams and lakes, or the type of soil on the ground.

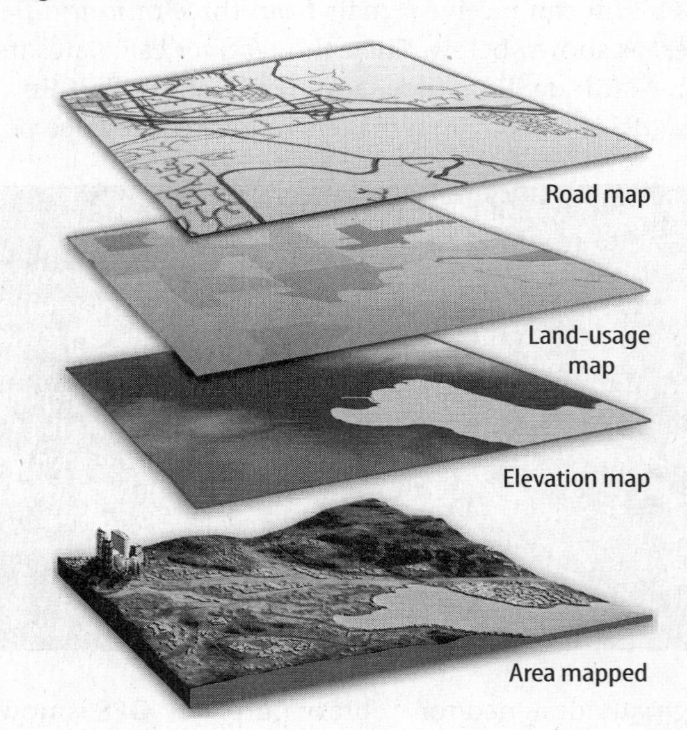

Visual Check
15. Identify What information do digital maps show?

Three Views Imagine setting up a model for a space shuttle landing under certain weather conditions using GIS.

- Database view begins the process by assembling information from existing databases on winds, airplane flights, landing procedures, and airport layouts.
- Map view draws from a set of interactive, digital maps that show features and their relationship to Earth's surface.
- Model view then pulls all the information together so that simulations can be run under changing weather conditions.

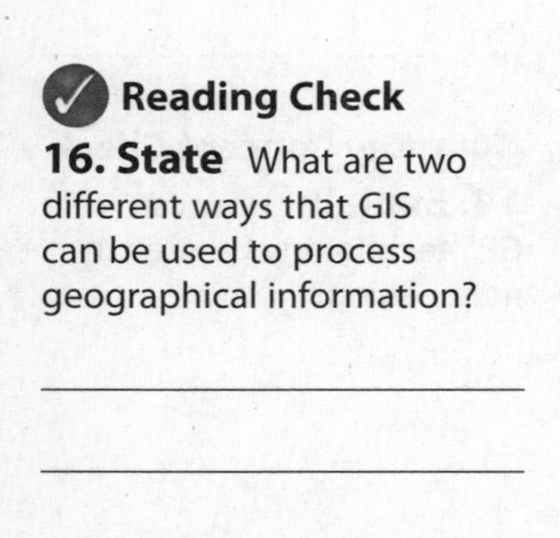

Reading Check
16. State What are two different ways that GIS can be used to process geographical information?

Remote Sensing

Remote sensing *is the process of collecting information about an area without coming into physical contact with it*. There are many uses for remote sensing. This process produces maps that show detailed information about agriculture, forestry, geology, land use, and many other subjects. Often these maps cover huge areas.

Aerial photographs taken from airplanes changed mapmaking. But now, an even more powerful type of remote sensing is being used to collect data. Since the 1970s, satellites orbiting high above Earth's surface have been used to collect data.

Monitoring Change with Remote Sensing Satellites orbit Earth repeatedly. Images of a location taken at different times can be used to study changes. These images can help people quickly make maps of areas affected by natural disasters. The maps are then used to monitor damage and help organize rescue efforts.

Landsat One series of satellites used to collect data about Earth's surface is the Landsat group. *Landsat 7,* launched in 1999, completes a scan of Earth's entire surface every 16 days. Recently it was used to map the coastal waters of the United States. Comparing the recent data to similar data collected 18 years ago, scientists recognized changes in coral reefs. Landsat has been used to contribute to the GIS database as well.

TOPEX/Jason-1 A pair of satellites—*TOPEX* and its successor, *Jason-1*—have been used to determine ocean topography, circulation, sea level, and tides. It is currently being used to identify climate change. Using radar, a signal is bounced off the ocean surface to measure bulges and valleys to within 3 m. Changes in the ocean surface due to a hurricane can be monitored in this way.

Sea Beam A device that uses sonar to map the bottom of the ocean is Sea Beam. Sea Beam is mounted onboard a ship. Computers calculate the time a sound wave takes to bounce off the ocean floor and return to the ship. This gives an accurate image of the seafloor and the depth of the ocean at that point. Sea Beam is used by fishing fleets, drilling operations, and various scientists.

Reading Check

17. Define remote sensing.

Key Concept Check

18. Explain How can remote sensing be an advantage to mapmakers?

Reading Check

19. Identify What are some methods used to collect remote-sensing data?

After You Read

Mini Glossary

contour interval: the elevation difference between contours that are next to each other

contour line: a line on a topographic map that connects points of equal elevation

cross section: a diagram showing a vertical slice through the rocks below the surface

elevation: the height above sea level of any point on Earth's surface

geologic map: shows the surface geology of the mapped area

relief: the difference in elevation between the highest and lowest points in an area

remote sensing: the process of collecting information about an area without coming into physical contact with it

slope: a measure of the steepness of the land

topographic map: shows the detailed shapes of Earth's surface, along with its natural and human-made features

1. Review the terms and their definitions in the Mini Glossary. Select two related terms and write a sentence explaining how the symbols represent Earth's features on a topographic map.

____

____

____

2. Fill in the graphic organizer to identify three things you can learn about the shape of Earth's surface from contour lines.

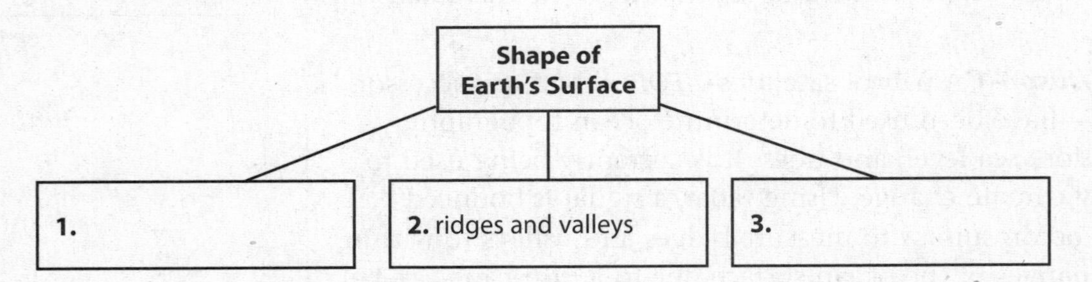

3. Compare a topographic map and a geologic map.

____

____

What do you think NOW?

Reread the statements at the beginning of the lesson. Fill in the After column with an A if you agree with the statement or a D if you disagree. Did you change your mind?

Log on to ConnectED.mcgraw-hill.com and access your textbook to find this lesson's resources.

Earth in Space

The Sun-Earth-Moon System

Before You Read

What do you think? Read the two statements below and decide whether you agree or disagree with them. Place an A in the Before column if you agree with the statement or a D if you disagree. After you've read this lesson, reread the statements to see if you have changed your mind.

Before	Statement	After
	1. Seasons are caused by the changing distance between Earth and the Sun.	
	2. The Moon has a dark side upon which the Sun never shines.	

Key Concepts 
- What causes seasons on Earth?
- How does the Moon affect Earth?
- How do solar and lunar eclipses differ?

Read to Learn

Earth and the Universe

Long ago, people studied the positions and motions of the Sun, Moon, and other objects in the sky. They used patterns in the motions to predict future positions of sky objects. But they did not understand how the objects were related. They thought Earth was the center of the universe.

Today we know that Earth is not the center of the universe. The Moon moves around, or orbits, Earth. Earth is one of eight planets that orbit the Sun. The Sun is one of billions of stars that make up the Milky Way galaxy. And the Milky Way is one of billions of galaxies in the universe.

Objects orbit the Sun because the Sun has more than 99 percent of the solar system's mass. The Sun has a huge gravitational pull. The Sun is the biggest object in the solar system. As the figure shows, the Sun's diameter is 100 times greater than Earth's diameter and 10 times greater than Jupiter's.

Size of the Sun

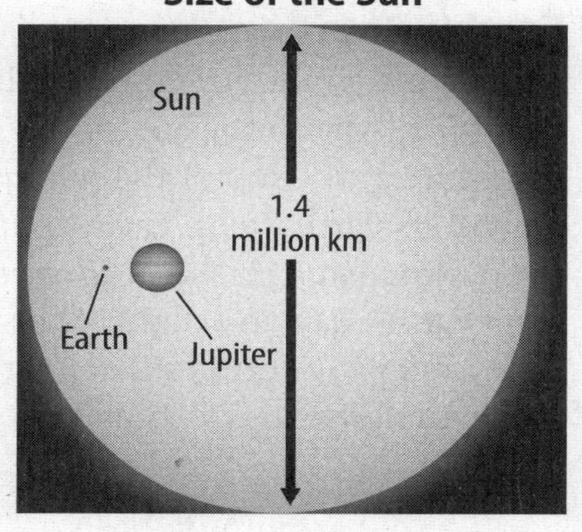

Study Coach

Building Vocabulary Work with another student to write a question about each vocabulary term in this lesson. Answer the questions and compare your answers. Reread the text to clarify the meaning of the terms.

Visual Check

1. Point Out What is the Sun's diameter?

Make an envelope book and draw an image of the Sun on the inside center. On the inside tabs, draw the position of Earth for each season.

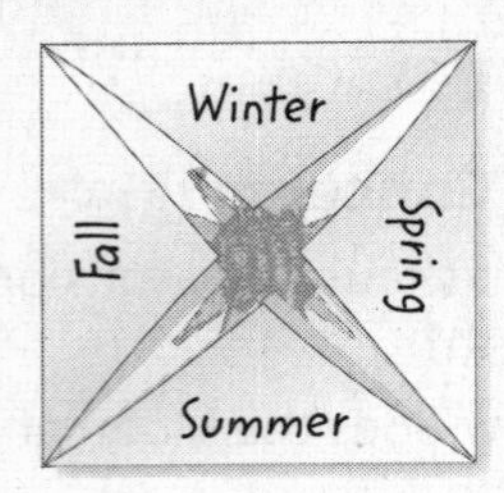

Reading Check

2. Define What is an astronomical unit?

____

____

Motions of Earth

Have you ever flown in an airplane? Airplanes travel faster than 900 km/h. Yet as you sit in one, you hardly feel like you are moving. Living on Earth is like traveling in an airplane. It seems as if Earth is still and the Sun and stars move around it. But Earth is not still. Earth moves in space.

Earth's Orbit

As you read, Earth is moving around the Sun because of the Sun's huge gravitational pull. Without the Sun's pull, Earth would move off into space in a straight line, as shown below. Earth's orbit is nearly round, or elliptical. *The orbit of an object around another object is* **revolution.** It takes Earth 365.25 days—one year—to revolve around the Sun once.

As shown below, the distance between Earth and the Sun is not always the same. An astronomical unit (AU) is the average distance between Earth and the Sun. One AU is nearly 150 million km. Scientists often use AUs to measure distances to planets and other objects within the solar system.

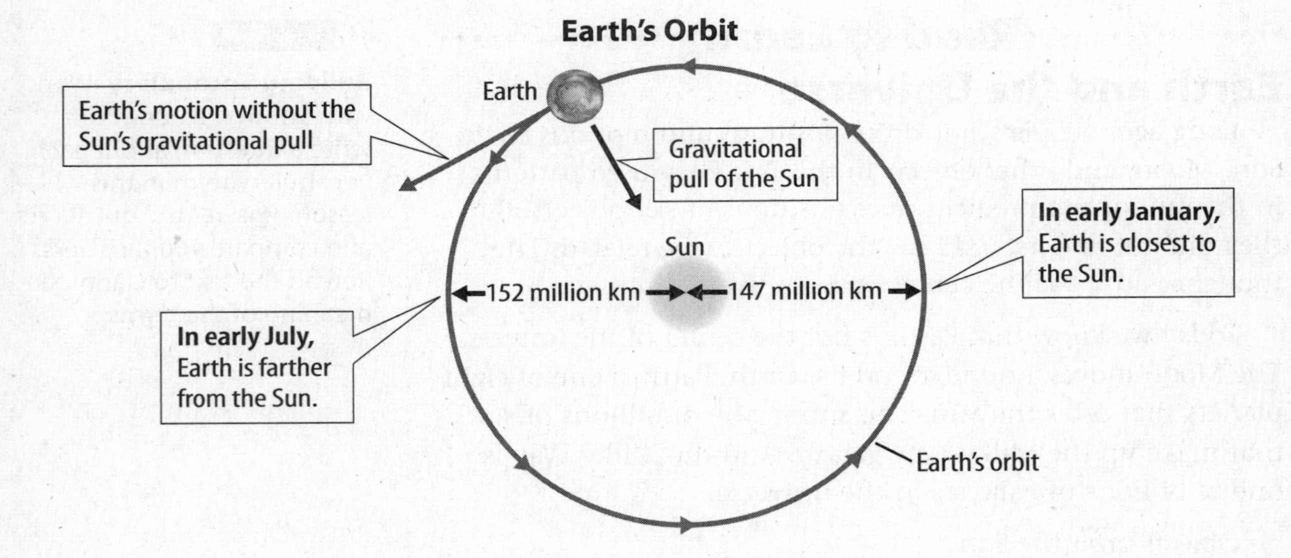

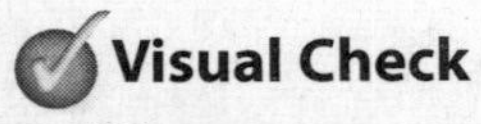

3. Discover When is Earth closest to the Sun?

____

____

Earth's Rotation

Imagine a rod pushed through the center of Earth from the North Pole to the South Pole. The images of Earth in the figure at the top of the next page show this. The rod represents Earth's axis. Earth spins, or rotates, on its axis like a top. **Rotation** *is the spin of an object around its axis.* Rotation is what causes day and night. The side of Earth facing the Sun is in daylight, and the side of Earth facing away from the Sun is in darkness. Earth makes one full rotation every 24 h.

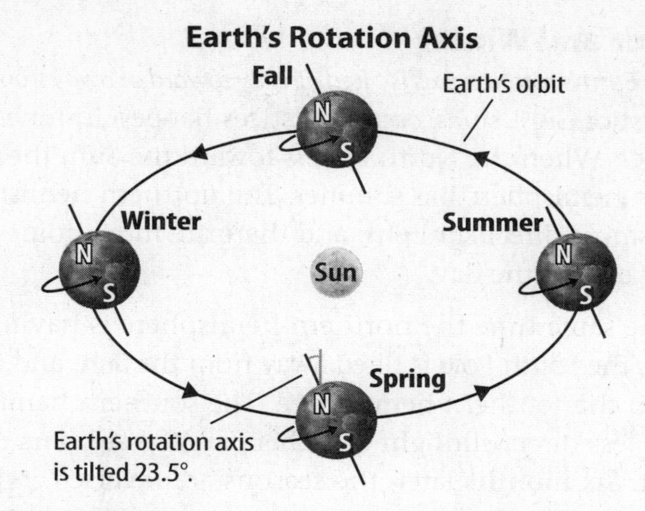

Earth's Tilt and Seasons

Earth's changing distance from the Sun does not cause the seasons. Look again at the figure on the previous page. Earth is closest to the Sun in January, when it is winter in the northern hemisphere. Seasons occur because Earth's tilt does not change as Earth orbits the Sun, as shown in the figure above.

If you drew a line perpendicular to Earth's orbital path, the angle of tilt between Earth's axis and that line would be 23.5°. As Earth moves, this angle of tilt stays the same. The North Pole and the South Pole always point in the same directions. But the position of Earth's tilt as it relates to the Sun does change, as shown in the figure below.

Spring and Fall

An **equinox** (EE kwuh nahks) *occurs when Earth's rotation axis is tilted neither toward nor away from the Sun*. *Equinox* means "equal night." Hours of daylight equal hours of darkness during an equinox. An equinox occurs two days of the year, one in March and one in September. These days are used to signify the beginning of spring or fall, as shown below.

Visual Check

4. Examine What is the degree of tilt of Earth's rotation axis?

____

Reading Check

5. Recognize The tilt of Earth's axis ___. (Circle the correct answer.)

a. does not change as Earth moves around the Sun
b. has no effect on Earth's seasons
c. changes to create the seasons on Earth

Visual Check

6. Identify How are the beginnings of spring and fall similar?

____

____

Earth's Tilt and the Seasons

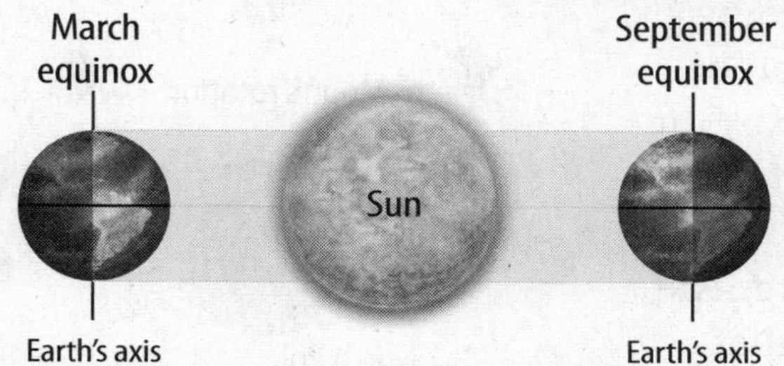

At two points in Earth's orbit—the March and September equinoxes—Earth's axis does not point either toward or away from the Sun. Light is distributed equally in the northern and southern hemispheres.

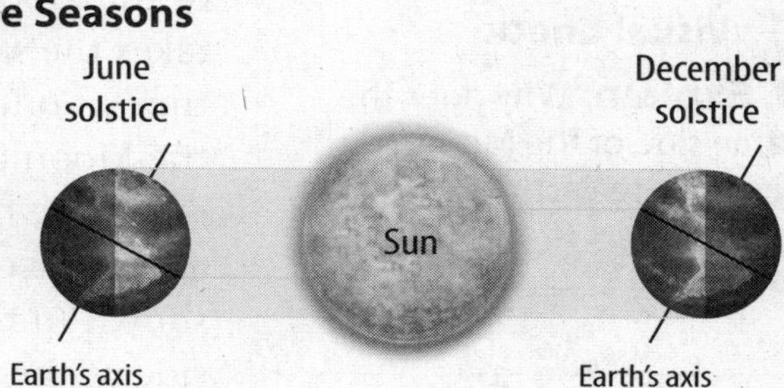

At two points in Earth's orbit—the June and December solstices—Earth's axis points the most toward or away from the Sun. Light is not distributed equally in the northern and southern hemispheres.

Summer and Winter

When Earth's rotation axis is tilted directly toward or away from the Sun a **solstice** (SAHL stuhs) *occurs.* Solstices happen in June and December. When the North Pole is toward the Sun, the northern hemisphere has summer. The northern hemisphere receives more direct sunlight, and there are more hours of sunlight during the day.

At the same time the northern hemisphere is having summer, the South Pole is tilted away from the Sun, and it is winter in the southern hemisphere. The southern hemisphere receives less direct sunlight and there are fewer hours of sunlight. Six months later, the seasons are reversed.

Key Concept Check
7. Explain What causes seasons?

Earth's Moon

You can probably guess what force holds the Moon in orbit around Earth. It's the same force that holds Earth in orbit around the Sun—gravity! The Moon is about one-fourth the size of Earth. The Moon is a dry, airless object made mostly of rock. Early in the Moon's history, many asteroids and comets crashed into it, leaving huge craters on its surface. The Moon also has mountains and smooth, dark lava plains from ancient volcanoes.

Reading Check
8. State What created the Moon's craters?

Formation of the Moon

Scientists propose that the Moon formed when a Mars-sized object collided with Earth soon after Earth formed. This collision threw debris into orbit around Earth. Gravity pulled the debris together, which formed the Moon.

Motions of the Moon

Like Earth, the Moon moves in different ways. It rotates on its axis, and it revolves around Earth. It orbits Earth once every 27.3 days. That is also how long it takes the Moon to rotate once.

Because one revolution of the Moon around Earth takes the same amount of time as one rotation of the Moon on its axis, the same side of the Moon always faces Earth, as shown in the figure. The side of the Moon that does not face Earth is called the far side. You cannot see it from Earth.

Visual Check
9. Explain Why does the same side of the Moon always face Earth?

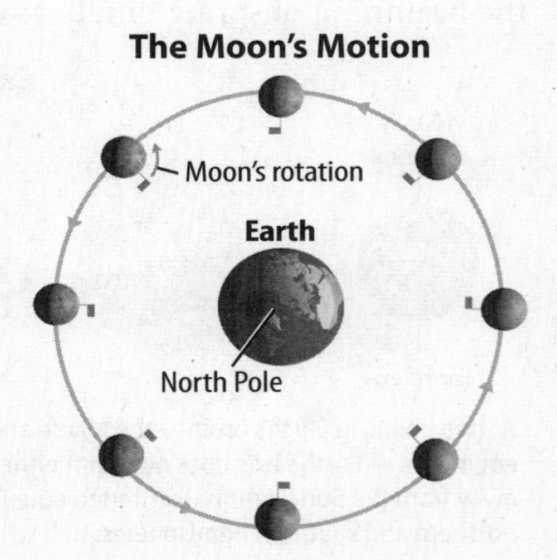

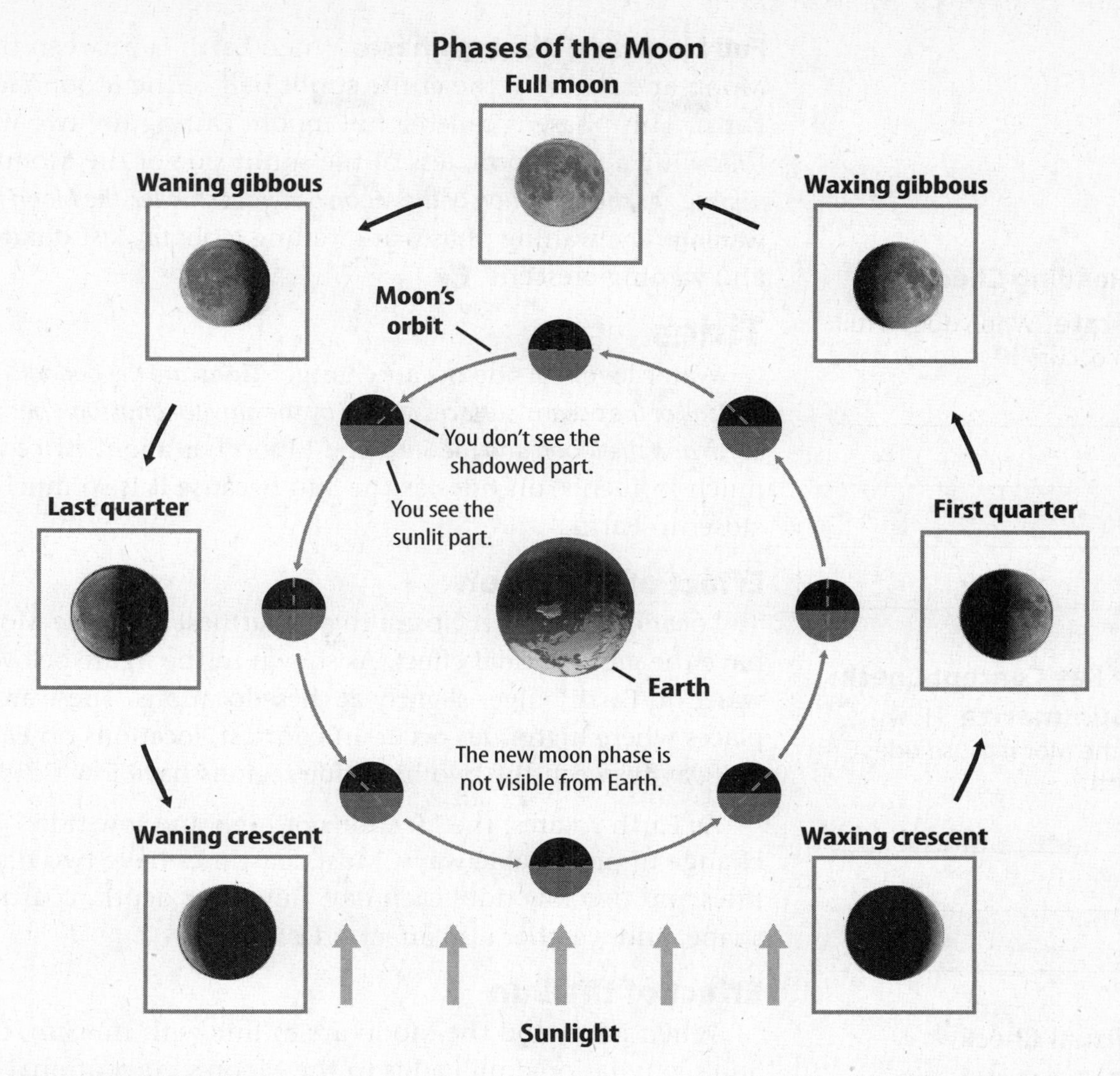

Phases of the Moon

The Moon does not create its own light. The Moon is visible only because it reflects sunlight. As the Moon orbits Earth, the half of the Moon facing the Sun is in sunlight and the half facing away is in shadow, as shown in the figure above. But as the Moon orbits Earth, the part of the Moon that can be seen from Earth seems to change shape. These shapes are the phases of the Moon. The Moon completes a cycle of phases every 29.5 days. Refer to the figure above as you read about each of the Moon's phases.

New Moon and Waxing Phases When the Moon is between Earth and the Sun, the sunlit half of the Moon faces away from Earth. The half facing Earth is dark because it is in shadow, as shown above. This phase is called a new moon. During the two weeks following a new moon, more of the Moon becomes visible. *As the lit portion of the Moon becomes larger, the Moon is* **waxing.** The waxing phases are waxing crescent, first quarter, and waxing gibbous.

Visual Check

10. Identify When does the Moon appear to get larger? When does it appear to get smaller?

Full Moon and Waning Phases When Earth is between the Moon and the Sun, the entire sunlit half of the Moon faces Earth. This phase is called a full moon. During the two weeks following a full moon, less of the sunlit side of the Moon is visible. *As the lit portion of the Moon becomes smaller, the Moon is* **waning.** The waning phases are waning gibbous, last quarter, and waning crescent.

Reading Check
11. State When does a full moon occur?

Tides

Water levels of the ocean change. **Tides** *are the periodical rise and fall of the oceans' surfaces caused by the gravitational force between Earth and the Moon and the Sun*. The Moon has about twice as much influence on tides as the Sun because it is so much closer to Earth.

Effect of the Moon

Locations on Earth closest to and farthest from the Moon have the greatest tidal effect. As shown in the figure below, water on Earth bulges slightly at these locations. These are the places where high tides occur. In contrast, locations on Earth halfway between the two high-tide regions have low tides.

As Earth rotates, the locations of high and low tide change in predictable ways. Most coastlines have two high tides and two low tides each day. But water depth, coastline shape, and weather also affect tides.

Key Concept Check
12. Summarize How does the Moon cause tides on Earth?

Effect of the Sun

When Earth and the Moon are in line with the Sun, the Sun's gravitational pull adds to the Moon's gravitational pull. As a result, high tides are higher than usual. Tides at this time are called spring tides. Spring tides occur during full moon and new moon phases as shown in the figure below.

Visual Check
13. Point Out In the figure below, highlight the parts of the diagram that show how Earth's oceans are affected by the gravitational pull of the Moon.

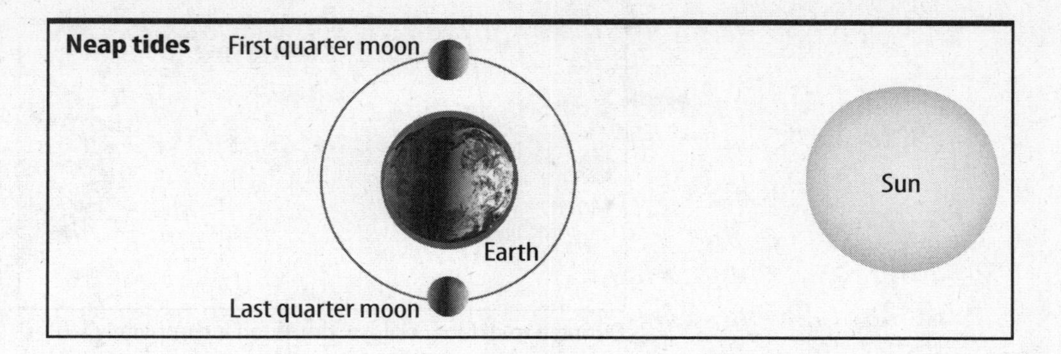

During the first and third quarter moons, the gravitational pull of the Moon is perpendicular to the gravitational pull of the Sun. High tides are lower than usual. Tides at these times are called neap tides and are shown in the figure above.

Eclipses

Throughout human history, people have interpreted eclipses as signs of war or disaster. However, there is nothing mysterious about eclipses. They are natural events.

An **eclipse** *is the movement of one solar system object into the shadow of another object*. You can view solar and lunar eclipses from Earth. As shown in the figure below, the type of eclipse depends on the positions of the Moon, Earth, and the Sun.

Solar Eclipses

A solar eclipse can only occur during a new moon. During a solar eclipse, a small part of Earth is in the Moon's shadow. The Moon appears to completely or partially cover the Sun.

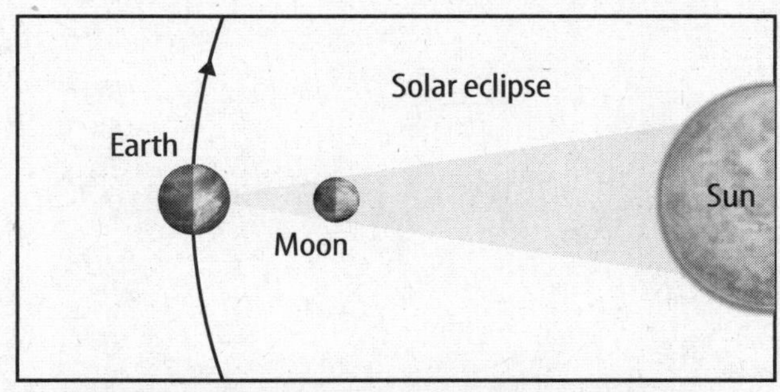

During a total solar eclipse, only a small part of Earth is covered by the Moon's shadow.

Reading Check
14. Distinguish What is the difference between spring tides and neap tides?

Reading Check
15. Define What is an eclipse?

Visual Check
16. Locate Where would you have to be on Earth to see this total solar eclipse?

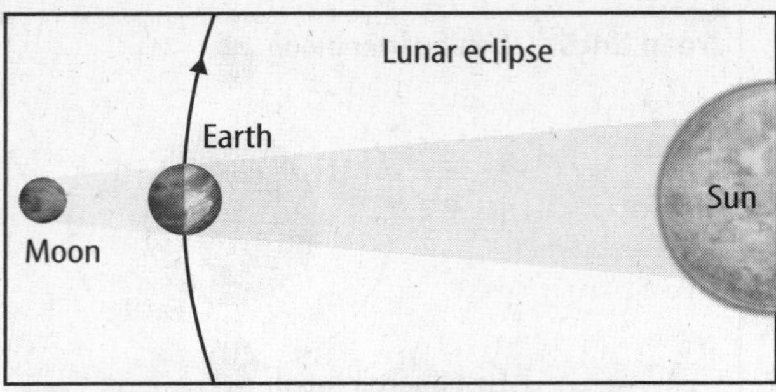

During a total lunar eclipse, the Moon is completely covered by Earth's shadow.

Lunar Eclipses

A lunar eclipse can occur only during a full moon. During a lunar eclipse, Earth's shadow completely or partially covers the Moon.

The Moon is visible during a total lunar eclipse. Light changes direction as it passes through Earth's atmosphere. The light that reaches the Moon appears red.

Key Concept Check
17. Contrast How do solar and lunar eclipses differ?

____
____
____
____
____

After You Read

Mini Glossary

eclipse: the movement of one solar system object into the shadow of another object

equinox (EE kwuh nahks): occurs when Earth's rotational axis is tilted neither toward nor away from the Sun

revolution: the orbit of an object around another object

rotation: the spin of an object around its axis

solstice (SAHL stuhs): occurs when Earth's rotation axis is tilted directly toward or away from the Sun

tides: the periodical rise and fall of the oceans' surfaces caused by the gravitational force between Earth and the Moon and the Sun

waning: Moon phase when the lit portion of the Moon becomes smaller

waxing: Moon phase when the lit portion of the Moon becomes larger

1. Review the terms and their definitions in the Mini Glossary. Write a sentence in your own words using one of the Mini Glossary terms.

___

2. Identify each model as a solar or a lunar eclipse. Write the name of each solar system object on the blank lines.

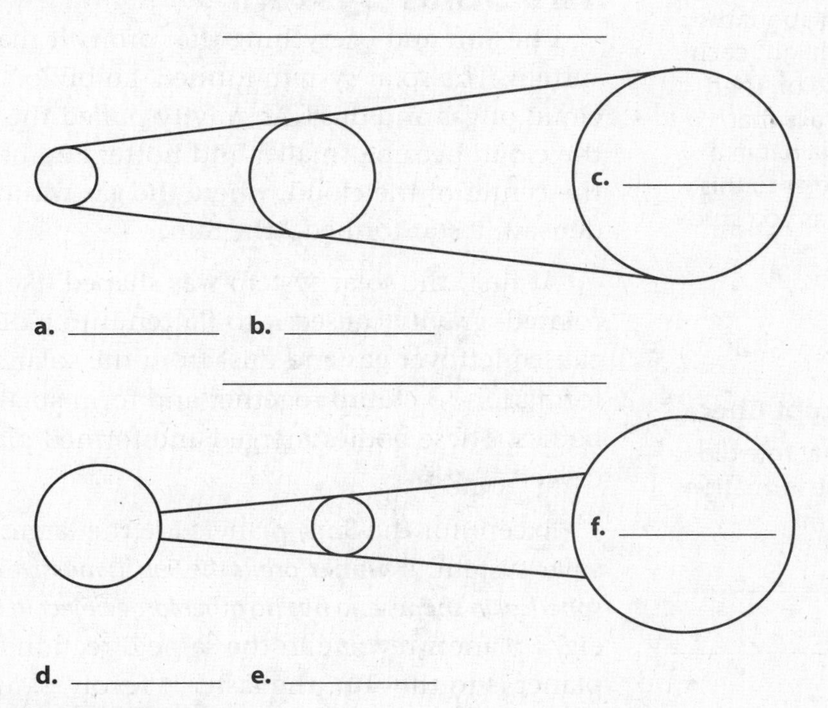

What do you think NOW?

Reread the statements at the beginning of the lesson. Fill in the After column with an A if you agree with the statement or a D if you disagree. Did you change your mind?

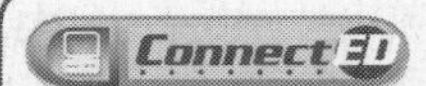

Log on to ConnectED.mcgraw-hill.com and access your textbook to find this lesson's resources.

CHAPTER 2
LESSON 2

Earth in Space

The Solar System

Key Concepts

- How does gravity influence the shape and the motion of objects in the solar system?
- What objects are in the solar system?
- How does Earth compare with other objects in the solar system?

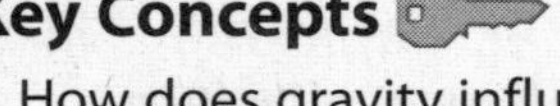

Identify the Main Ideas
To help you learn about the solar system, highlight each heading in one color. Then highlight the details that support and explain it in a different color. Refer to this highlighted text as you study the lesson.

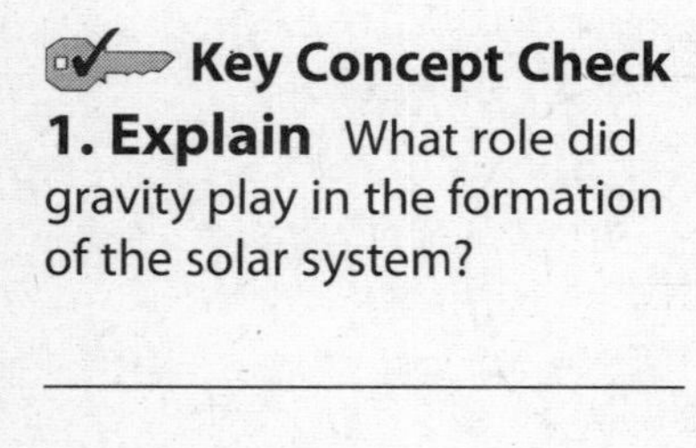

Key Concept Check
1. Explain What role did gravity play in the formation of the solar system?

Before You Read

What do you think? Read the two statements below and decide whether you agree or disagree with them. Place an A in the Before column if you agree with the statement or a D if you disagree. After you've read this lesson, reread the statements to see if you have changed your mind.

Before	Statement	After
	3. The solar system contains nine planets.	
	4. Earth is the only planet that has a moon.	

Read to Learn

The Solar System

The Sun and everything that orbits it make up the solar system. The solar system formed 4.6 billion years ago from a cloud of gas and dust. As gravity pulled the cloud together, the cloud became smaller and hotter and began to spin. In the center of the cloud, where the gas was hottest and densest, a star formed—the Sun.

At first, the solar system was shaped like a ball. As it rotated, gravity caused it to flatten into a disk. Gravity also caused leftover gas and dust from the solar system's formation to clump together and form small, rocky or icy bodies. These bodies merged and formed planets and other objects.

Except for the Sun, planets are the largest objects in the solar system. *A* **planet** *orbits the Sun, is massive enough to be nearly spherical in shape, and has no other large object in its orbital path.* All eight planets revolve in the same direction. The closer a planet is to the Sun, the faster it revolves. Mercury orbits the Sun once every 88 Earth days. The planet farthest from the Sun, Neptune, orbits the Sun once every 165 Earth years.

Recall that Earth orbits the Sun at a distance of 1 AU. Neptune is 30 times farther from the Sun. But the Sun's gravitational pull extends far beyond Neptune. Billions of small, icy objects orbit the Sun at a distance of 50,000 AU.

Objects in the Solar System

The solar system contains many different objects. These objects include planets as well as other objects that are too small to be classified as planets.

Planets and Dwarf Planets Recall that planets are massive objects that do not have other objects of similar size in their orbital paths around the Sun. Some spherical objects that orbit the Sun are similar to planets but are not massive enough to be planets. Some of these are dwarf planets. **Dwarf planets** *orbit the Sun and are nearly spherical in shape, but they share their orbital paths with other objects of similar size*. Pluto was once considered a planet but is now classified as a dwarf planet.

Other Solar System Bodies Not all spherical bodies in the solar system are planets. Many moons are massive enough to be spherical. *A* **moon** *is a natural satellite that orbits an object other than a star*. Some asteroids also are spherical. **Asteroids** *are small, rocky objects that orbit the Sun*. Most known asteroids are in the asteroid belt located between the orbits of Mars and Jupiter. **Comets** *are small, rocky, icy objects that orbit the Sun*. As comets move nearer to the Sun, the ice melts and the water forms a "tail" behind the comet. The orbital paths of comets extend to the outer solar system, beyond Neptune. **Meteoroids** *are small, rocky particles that move through space. When a meteoroid enters Earth's atmosphere, it produces a streak of light and is called a* **meteor.** A meteoroid becomes a meteorite only if it impacts Earth. Objects in the solar system are shown in the art of the solar system on the next two pages.

Inner Planets

The center of the solar system was very hot when it formed. Gases and materials with low boiling points escaped from the area closest to the Sun. The four inner planets, also called the rocky planets, formed from the rocks and the heavy elements, including metals, left behind. The cores of the inner planets are mostly iron. The inner planets are the smallest planets. They have few or no moons, no rings, and they rotate more slowly than the outer planets. Refer to the art of the solar system as you read about the inner planets.

Mercury At 0.39 AU from the Sun, Mercury is the planet closest to the Sun. It is also the smallest planet. It is only about one-third the diameter of Earth. Mercury rotates slowly. As its surface heats and cools during its long day, temperatures can vary by as much as 500°C. Mercury has almost no atmosphere. Its gray surface has many impact craters and resembles Earth's moon.

FOLDABLES®

Make a horizontal tri-fold Venn book and use it to compare and contrast characteristics of the inner and outer planets.

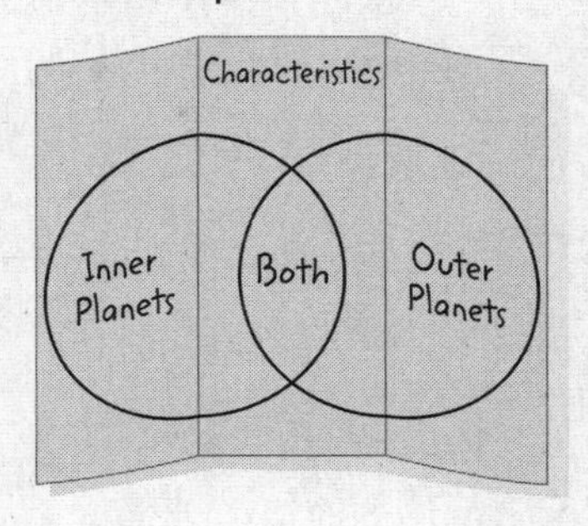

Reading Check

2. Differentiate between a planet and a dwarf planet.

Key Concept Check

3. Name What objects are in the solar system?

The Solar System

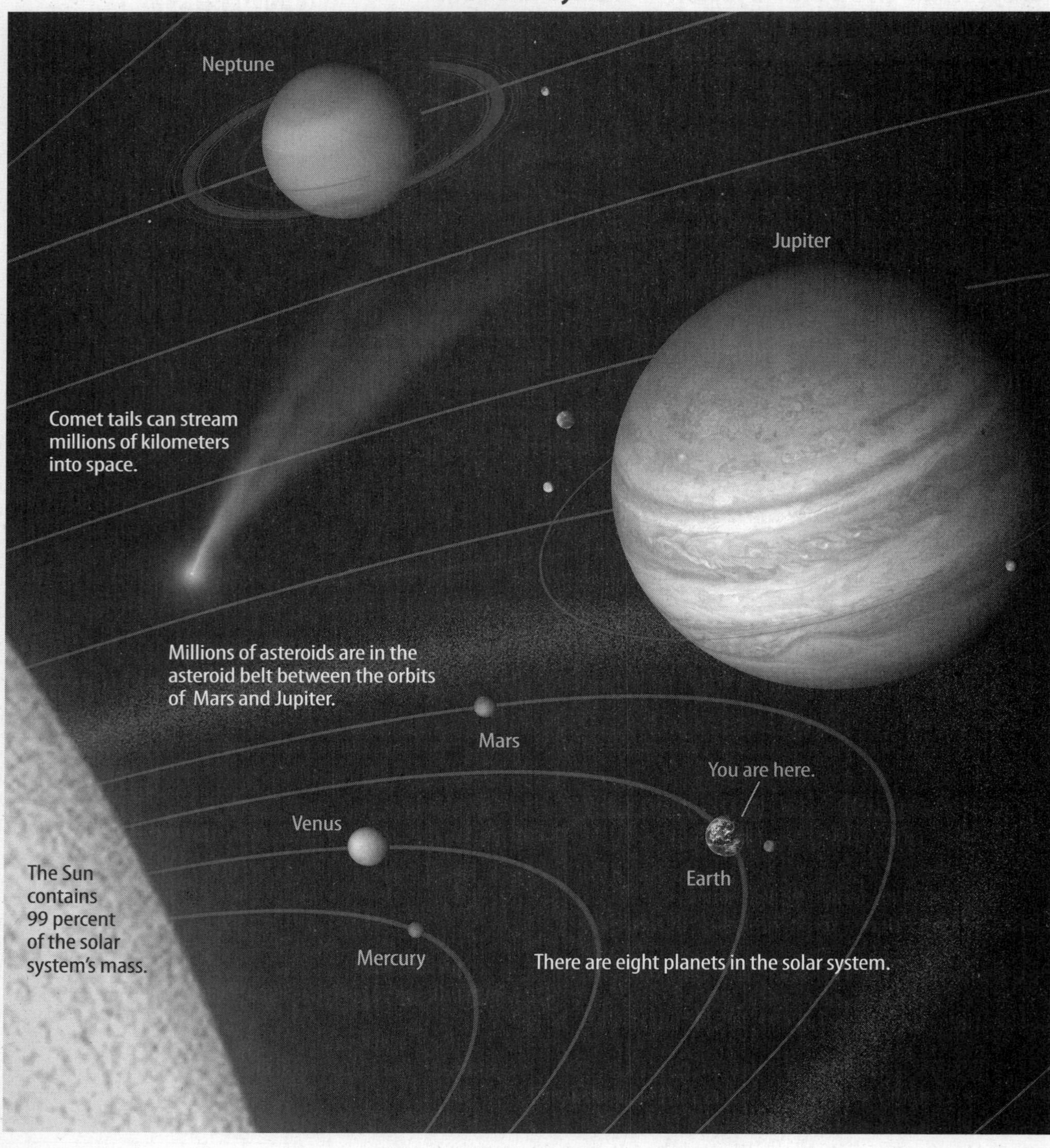

Visual Check

4. Identify What percentage of the mass in the solar system exists outside the Sun?

____

Venus Venus is 0.72 AU from the Sun. As you can see in the figure above, it is almost the same size as Earth. It also has nearly the same makeup as Earth. Venus has the slowest rotation of any planet. One day on Venus is equal to 244 Earth days. Its heavy layer of clouds and thick carbon-dioxide atmosphere trap energy from the Sun. This makes Venus the hottest planet. Scientists hypothesize that some volcanoes on its surface might be active.

The Solar System

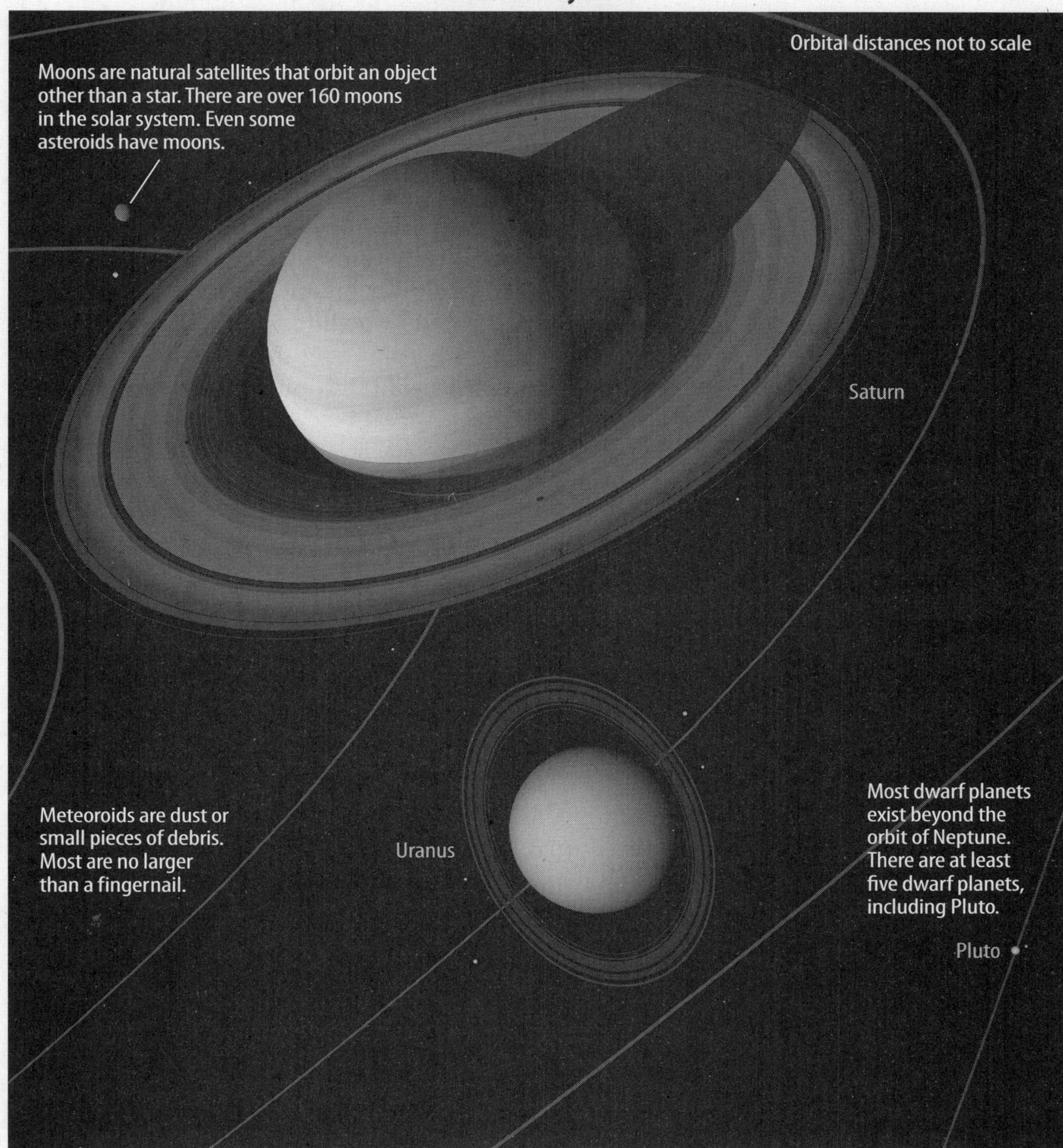

Earth Earth is 1 AU from the Sun. Earth is the largest and densest of the inner planets. It is the only planet where life is known to exist. Earth is also the only planet with large amounts of liquid water on its surface. Earth's water and water vapor appear blue and white when viewed from space. Earth's atmosphere is 78 percent nitrogen and 21 percent oxygen.

Key Concept Check

5. Contrast How does Earth differ from other inner planets?

____

____

Mars Half the size of Earth, Mars orbits at 1.5 AU from the Sun. Mars is too cold for liquid water to exist on the surface, although ice has been detected at the poles. Ice might exist below the surface of Mars. Liquid water probably flowed on Mars in the past. Rocks on Mars's surface contain iron oxides, which give Mars a reddish color. High mountains on Mars are extinct volcanoes.

Reading Check
6. Explain Why does Mars have a reddish color?

____

____

Outer Planets

The four outermost planets formed farther from the Sun than the inner planets did. As a result, they have more gases and other materials with low boiling points. They are often called the gas giants. They are larger than the inner planets, they rotate more quickly, and they each have rings. Except for Saturn's rings, the rings are barely visible. Each outer planet also has many moons. Scientists suspect that each outer planet has a small, rocky core. These planets do not have solid surfaces. They have thick atmospheres of hydrogen and helium.

Key Concept Check
7. Distinguish How do the inner and outer planets differ?

____

____

____

____

____

____

Jupiter Though it is made mostly of hydrogen and helium, Jupiter contains more mass than the rest of the planets combined. Jupiter revolves around the Sun at a distance of 5 AU. It has the fastest rotation of any planet—a day lasts just 10 Earth hours. Jupiter's clouds swirl with various colors because they contain small amounts of sulfur and phosphorus. Jupiter has strong weather systems.

Saturn At 9.5 AU from the Sun, Saturn is nearly twice as far from the Sun as Jupiter, but its makeup is similar. Saturn is the second-largest planet. It has thousands of thin rings made of billions of pieces of ice ranging in size from pebbles to boulders. Saturn's clouds form bands and spots, but they are hard to see. Saturn's hazy upper atmosphere hides its colorful lower layers.

Uranus This planet orbits the Sun at a distance of nearly 20 AU. Uranus is tilted so much that its axis sometimes points directly toward the Sun. It is a bluish-green color because of the small amount of methane in its atmosphere. Scientists think that a layer of icy liquid water, ammonia, and other compounds lies deep below Uranus's thick atmosphere.

Reading Check
8. State What makes Uranus and Neptune appear blue?

____

____

Neptune At 30 AU, Neptune is so far away that it cannot be seen from Earth without a telescope. Neptune's makeup is similar to that of Uranus, although it has more methane in its atmosphere and is deeper blue. Neptune has the fastest winds of any planet, recorded at over 1,100 km/h. The spots on its surface are hurricane-like storms, which do not last long.

After You Read

Mini Glossary

asteroid: a small, rocky object that orbits the Sun

comet: a small, rocky, icy object that orbits the Sun

dwarf planet: an object that orbits the Sun and is nearly spherical in shape but shares its orbital path with other objects of similar size

meteor: a meteoroid that enters Earth's atmosphere, producing a streak of light

meteoroid: a small, rocky particle that moves through space

moon: a natural satellite that orbits an object other than a star

planet: an object that orbits the Sun, is massive enough to be nearly spherical in shape, and has no other large object in its orbital path

1. Review the terms and their definitions in the Mini Glossary. Write a sentence explaining why the outer planets do not have solid surfaces.

____

____

2. Write the name of each inner planet in the space provided based on its size and distance from the Sun. Also, record the distance of each planet from the Sun in astronomical units (AU).

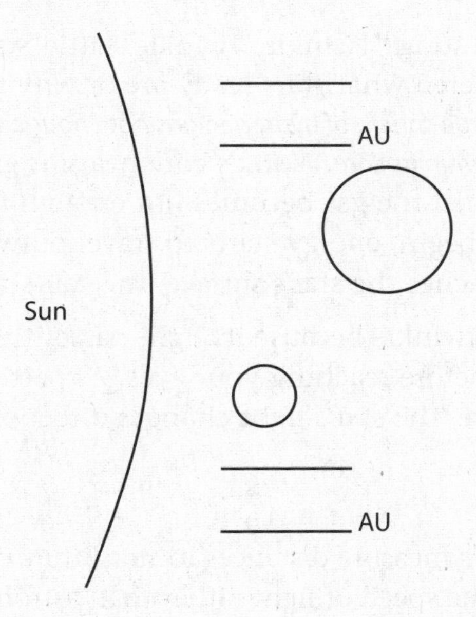

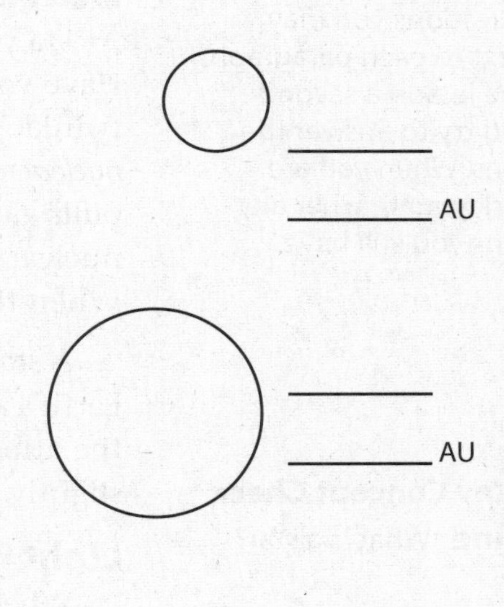

What do you think NOW?

Reread the statements at the beginning of the lesson. Fill in the After column with an A if you agree with the statement or a D if you disagree. Did you change your mind?

Log on to ConnectED.mcgraw-hill.com and access your textbook to find this lesson's resources.

CHAPTER 2
LESSON 3

Earth in Space

Stars, Galaxies, and the Universe

Key Concepts

- What are stars?
- How does the Sun compare to other stars?
- Where is Earth located in the universe?
- How is the universe structured?

Before You Read

What do you think? Read the two statements below and decide whether you agree or disagree with them. Place an A in the Before column if you agree with the statement or a D if you disagree. After you've read this lesson, reread the statements to see if you have changed your mind.

Before	Statement	After
	5. The Sun is more massive than 90 percent of other stars.	
	6. The solar system is at the center of the Milky Way.	

Mark the Text

Ask Questions As you read, write questions you may have next to each paragraph. Read the lesson a second time and try to answer the questions. When you are done, ask your teacher any questions you still have.

Read to Learn

Stars

Do you know the song "Twinkle, Twinkle, Little Star"? Have you ever wondered what stars really are or why they twinkle? *A* **star** *is a large sphere of hydrogen gas hot enough for nuclear reactions to occur in its core.* A star's core heats as gravity pulls gas inward. When the gas becomes hot enough for nuclear reactions to begin, energy starts to travel outward. When the energy reaches the star's surface, the star shines.

A star appears to twinkle because its light passes through Earth's atmosphere before reaching your eyes. As particles in the atmosphere move, the star's light changes directions slightly.

Key Concept Check
1. Define What is a star?

___

___

___

___

Light from Stars

When astronomers measure distances to stars, they often use a unit based on the speed of light rather than astronomical units. *A* **light-year** *is the distance light travels in one year.* Light travels 300,000 km/s. One light-year equals 9.46 trillion km.

Because it takes time for light to travel, stars are not seen as they are now, but as they were in the past. Proxima Centauri, the star nearest the Sun, is 4.2 light-years away. The light from Proxima Centauri that we see today left the star 4.2 years ago.

Types of Stars

When you first look at them, all stars appear white. But if you look closely at the brightest stars in the night sky, you will see that some are red, some are orange, and some appear to be blue. The color of a star indicates its temperature.

Blue stars are the hottest stars, followed by blue-white, white, yellow, and orange stars. Red stars are the coolest stars. The Sun is a yellow star.

The Star Aldebaran

When you look at stars, they appear to be the same size. But stars vary in size. The Sun is larger and more massive than 90 percent of other stars. But the Sun is tiny compared to the giant star Aldebaran shown in the figure. Aldebaran is 44 times wider than the Sun.

Visual Check

2. Assess Aldebaran is an orange star. Is Aldebaran a relatively cool or a relatively hot star? Explain.

Key Concept Check

3. Explain How does the Sun compare in size to other stars?

The Sun is a solitary, or single, star. Many other stars are members of binary star systems or multiple-star systems. In a binary star system, two stars orbit each other's center of mass. In a multiple-star system, two or more stars orbit the entire system's center of mass. Stars differ in other ways, too. For example, stars called variable stars change in brightness over time.

Earth's Star—the Sun

The Sun is the closest star to Earth. It has been shining for nearly 5 billion years. Scientists estimate that it has a lifetime of approximately 10 billion years, so it will continue to shine for 5 billion more years. When the Sun stops shining, it will become a small, dense star that emits little light. It will be a white dwarf star.

ACADEMIC VOCABULARY

estimate
(verb) to determine roughly the value, size, or extent of something

Galaxies

Stars are not randomly scattered throughout the universe. Most stars are bound by gravity into galaxies. *A* **galaxy** *is a huge collection of stars, gas, and dust.* Astronomers classify galaxies by their shapes. The three main types of galaxies are elliptical, irregular, and spiral.

Elliptical Galaxies Elliptical galaxies are shaped like basketballs or footballs. They contain older, redder stars and have little gas or dust. Because stars form from gas and dust, elliptical galaxies contain few young stars.

FOLDABLES

Make a vertical three-tab Venn book and use it to compare the Sun to other stars.

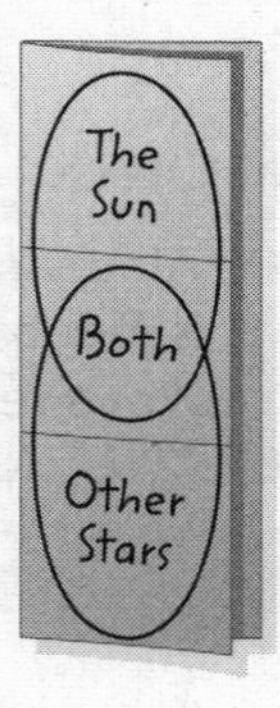

Irregular Galaxies These oddly shaped galaxies contain large amounts of gas and dust. They show the highest rate of star formation of any galaxy type. Irregular galaxies have many young stars. These galaxies do not have bright centers.

Spiral Galaxies These galaxies are shaped like disks. They contain dust, gas, and young stars in their bluish arms. Older, redder stars are in their central bulges. Spiral galaxies are surrounded by spherical halos containing older stars.

The universe contains hundreds of billions of galaxies. Each galaxy can contain hundreds of billions of stars. The solar system where you live is part of the Milky Way, a spiral galaxy. The Milky Way is larger than most galaxies in the universe. It contains more than 100 billion stars.

Because Earth is inside the Milky Way, scientists cannot see the Milky Way from the outside as they can see other galaxies. Even though they cannot see all of the Milky Way, scientists have determined that the Milky Way has at least two major spiral arms. The Sun is near one of the arms a little more than halfway from the Milky Way's center.

Key Concept Check
4. Identify In which galaxy is Earth located?

The Universe

Most galaxies are pulled by gravity into clusters of galaxies. In the clusters, the galaxies interact and sometimes merge with one another. The Milky Way is part of a cluster called the Local Group.

The Local Group contains about 30 galaxies. The Local Group, in turn, is part of a supercluster of galaxies called the Local Supercluster.

Superclusters are some of the largest structures in the universe. Some superclusters contain thousands of galaxies. But even superclusters are parts of larger structures. Superclusters form enormous, sheetlike walls in space.

Key Concept Check
5. Summarize How is the universe structured?

Astronomers study the rotations and the interactions of galaxies in clusters. In this way, they can determine how much mass the galaxies contain. Astronomers have discovered that only 5–10 percent of the mass in galaxies emits light. They hypothesize that the rest of the mass in galaxies—and in the universe—is invisible dark matter or dark energy.

Recycled Matter

Did you know that most matter that makes up your body was originally made in stars? Hydrogen is combined into more-complex elements during nuclear reactions in stars.

When a massive star explodes, it releases those elements into space. This material can then form new stars and planets. In this way, matter in the universe is recycled.

Big Bang Theory

Most scientists agree that the universe formed 13–14 billion years ago and that it had a hot and dense beginning. *The* **Big Bang theory** *states that the universe began from one point and has been expanding and cooling ever since.*

Will the universe expand forever, or will gravity eventually cause it to contract? These questions remain unanswered. Scientists have not yet been able to determine the fate of the universe.

Math Skills

Light-years (ly) describe distances to nearby stars. Astronomers often use parsecs (pc) to describe greater distances in space.

$$1 \text{ pc} = 3.26 \text{ ly}$$

$$1 \text{ ly} = 9.46 \text{ trillion km}$$

The star Proxima Centauri is 4.2 ly from Earth. What is that distance in parsecs?

a. Select a conversion factor with the unit you want in the numerator and the given unit in the denominator.

$$\frac{1 \text{ pc}}{3.26 \text{ ly}}$$

b. Multiply the starting quantity and units by the conversion factor.

$$\frac{4.2 \text{ ly} \times 1 \text{ pc}}{3.26 \text{ ly}}$$

c. Complete the calculation.

$$\frac{4.2 \text{ pc}}{3.26} = 1.3 \text{ pc}$$

6. Use Dimensional Analysis The nearest galaxy to the Milky Way is the Andromeda galaxy. It is approximately 2.5 million ly from Earth. What is that distance in parsecs?

____

After You Read

Mini Glossary

Big Bang theory: theory that the universe began from one point and has been expanding and cooling ever since

galaxy: a huge collection of stars, gas, and dust

light-year: the distance light travels in one year

star: a large sphere of hydrogen gas hot enough for nuclear reactions to occur in its core

1. Review the terms and their definitions in the Mini Glossary. Write a sentence explaining the Big Bang theory in your own words.

2. In the graphic organizer, list the color of stars from coolest to hottest. Then write *the Sun* in the correct box to show its temperature.

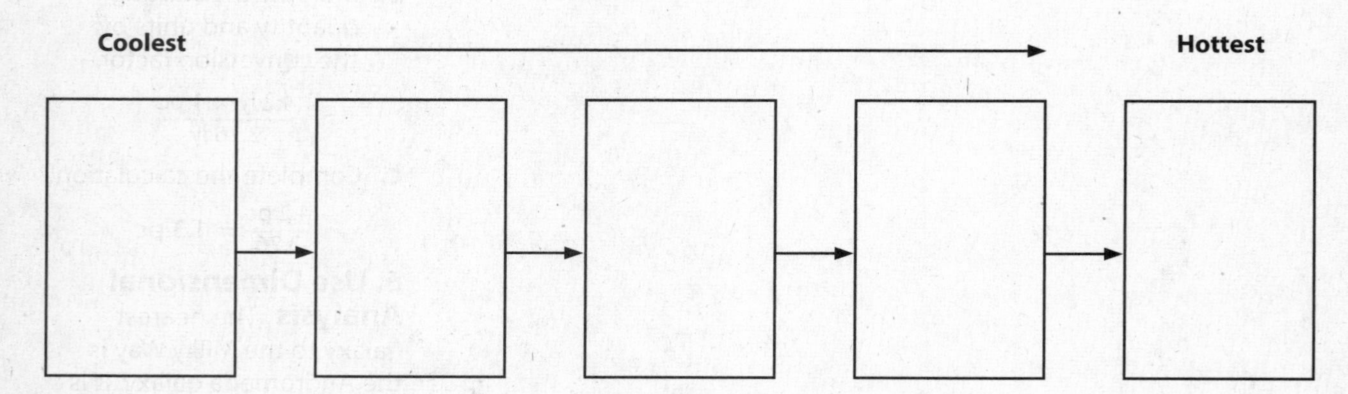

3. If the star Proxima Centauri exploded 4.2 years ago, why would we be able to observe the explosion today?

What do you think NOW?

Reread the statements at the beginning of the lesson. Fill in the After column with an A if you agree with the statement or a D if you disagree. Did you change your mind?

Log on to ConnectED.mcgraw-hill.com and access your textbook to find this lesson's resources.

Our Planet—Earth

Earth Systems

Before You Read

What do you think? Read the three statements below and decide whether you agree or disagree with them. Place an A in the Before column if you agree with the statement or a D if you disagree. After you've read this lesson, reread the statements to see if you have changed your mind.

Before	Statement	After
	1. Earth is a simple system made of rocks.	
	2. Most of Earth is covered by one large ocean.	
	3. Earth's interior is made of distinct layers.	

Key Concepts 

- What are the composition and the structure of the atmosphere?
- How is water distributed in the hydrosphere?
- What are Earth's systems?
- What are the composition and the structure of the geosphere?

Read to Learn

What is Earth?

The puffy, white clouds over your head and the hard ground under your feet are parts of Earth. The water in the oceans and the fish that live there are also parts of Earth. The planet Earth is more than a solid ball in space. It includes air molecules that float near the boundaries of outer space. It also includes molten rock that churns deep below Earth's surface.

Earth is a complex place. People often divide complex things into smaller parts to study them. Scientists divide Earth into four systems so they can understand the planet better. The systems contain different materials and they work in different ways, but they all work together, or interact. What happens in one system affects the others.

Earth's Air

The outermost Earth system is an invisible layer of gases that surrounds the planet. Even though you cannot see air, you can feel it when it is moving. Moving air is wind.

Earth's Water

Below the layer of air is the system that contains Earth's water. Like air, water can move from place to place. Some of the water is salty, and some is fresh.

Study Coach

Make Flash Cards For each head in this lesson, write a question on one side of a flash card and the answer on the other side. Quiz yourself until you know all the answers.

FOLDABLES

Make a small, horizontal four-door book and use it to organize your notes on Earth systems.

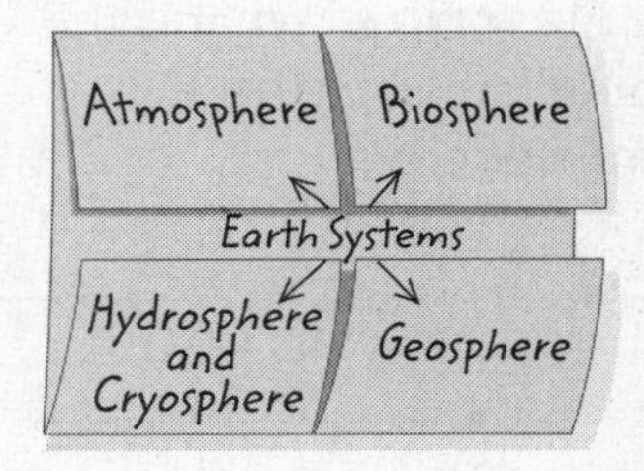

The Solid Earth

The next system is the solid part of Earth. It contains a thin layer of soil that covers a rocky center. This is by far the largest Earth system. It is solid, so materials in this system move more slowly than air or water. But they do move. Over time, landforms rise and then wear away. Large canyons like the Grand Canyon took millions of years to form.

Reading Check

1. Identify What is the largest Earth system?

Life on Earth

The Earth system that contains all living things is the **biosphere.** Living things are found in air, water, and soil. So the biosphere has no distinct boundaries. It is found within the other Earth systems. Living things like giraffes and trees are part of the biosphere. You will learn more about the biosphere when you study life science, or biology. The rest of this chapter describes the Earth systems that are made of nonliving things.

Reading Check

2. Explain Why doesn't the biosphere have distinct boundaries?

The Atmosphere

The force of Earth's gravity pulls molecules of gases into a layer surrounding the planet. *This mixture of gases forms a layer around Earth called the* **atmosphere.** The atmosphere is denser near Earth's surface and becomes thinner farther from Earth. It keeps Earth warm by trapping thermal energy from the Sun that bounces back from Earth's surface. If the atmosphere did not regulate temperature, life as it is on Earth could not exist.

What makes up the atmosphere?

The atmosphere contains a mixture of nitrogen, oxygen, and smaller amounts of other gases. The graph at the top of the next page shows the percentages of these gases. The most common gas is nitrogen, which makes up 78 percent of the atmosphere. Most of the remaining gas is oxygen.

The other gases are called trace gases because they make up only 1 percent, or a trace, of the atmosphere. But trace gases are still important. Carbon dioxide, methane, and water vapor help regulate Earth's temperature. The graph at the top of the next page shows the percentages of gases in dry air. The atmosphere also contains water vapor. The amount usually ranges from 0 to 4 percent.

Along with gases and water vapor, the atmosphere contains small amounts of solids. Particles of dust float along with the gases and water vapor. Sometimes you can see these tiny specks as sunlight reflects off them when it shines through a window.

Key Concept Check

3. Describe What is the composition of the atmosphere?

Atmospheric Gases

21% Oxygen

78% Nitrogen

1% Other Gases: Argon (Ar), Carbon dioxide (CO_2), Ozone (O_3)

Visual Check

4. Name the two gases that make up most of the atmosphere.

Layers of the Atmosphere

The composition of the atmosphere does not change much over time. But the temperature of the atmosphere does change. Thermal energy from the Sun heats the atmosphere. Different parts of the atmosphere absorb or reflect this thermal energy in different ways. The figure below shows changes in temperature as altitude increases. Scientists use temperature changes to distinguish layers in the atmosphere.

The Troposphere In the bottom layer of the atmosphere, called the troposphere, temperature decreases as you move upward from Earth's surface. Gases flow and swirl in the troposphere, causing weather. Although the troposphere does not extend very far upward, it contains most of the atmosphere's mass.

Think it Over

5. Predict How do you think the temperature might change as you hike down a mountain?

Atmospheric Layers

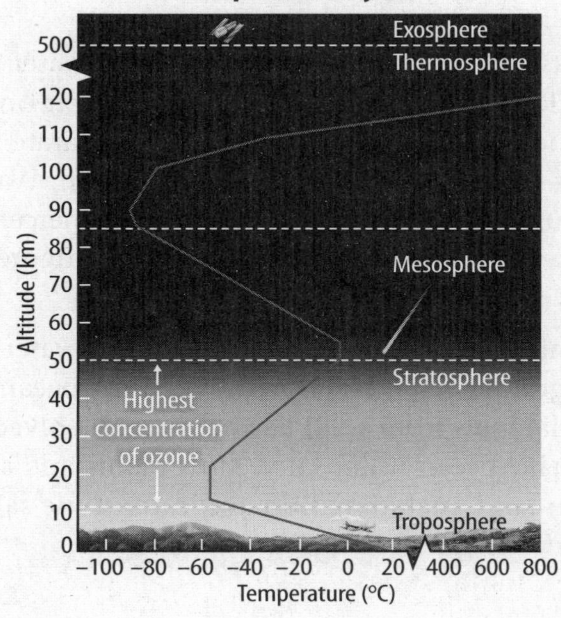

Visual Check

6. Summarize how temperature changes as altitude increases in the troposphere.

Think it Over

7. Analyze How does the ozone layer benefit life on Earth?

Key Concept Check

8. State What are the layers of the atmosphere?

Reading Check

9. Recognize How much water is in the hydrosphere?

The Stratosphere Above the troposphere is the stratosphere. Unlike gases in the troposphere, gases in the stratosphere do not swirl around. They are more stable and form flat layers.

The stratosphere has a layer of ozone, which is a form of oxygen. This ozone layer protects Earth's surface from harmful radiation from the Sun. It acts like a layer of sunscreen, protecting the biosphere. Because ozone absorbs solar radiation, temperatures increase in the stratosphere.

Upper Layers Above the stratosphere is the mesosphere. Temperature decreases in this layer, then increases again in the next layer, the thermosphere. The last layer of Earth's atmosphere is the exosphere. The lowest density of gas molecules is in this layer. Beyond the exosphere is outer space.

The Hydrosphere

Water is one of the most common and important substances on Earth. *The system containing Earth's liquid water is called the* **hydrosphere.** Most water is stored on Earth's surface, but some is located below the surface or within the atmosphere and biosphere.

The hydrosphere contains more than 1.3 billion km^3 of water. The amount of water in the hydrosphere does not change. But like the gases in the atmosphere, water in the hydrosphere flows. It moves from one location to another over time. Water also changes state. It is a liquid, a solid, and a gas on Earth.

Ocean

Scientists call the natural locations where water is stored reservoirs (REH zuh vworz). The largest reservoir on Earth is the world ocean. Though the oceans have separate names, they are all connected, making one large ocean. Water flows freely throughout the world ocean. About 97 percent of Earth's water is in the ocean, as shown in the figure on the next page.

Many minerals dissolve easily in water. As water in rivers and underground reservoirs flows toward the ocean, it dissolves materials from solid Earth. These dissolved minerals make ocean water salty. Most plants and animals that live on land, including humans, cannot use salt water. They need freshwater to survive.

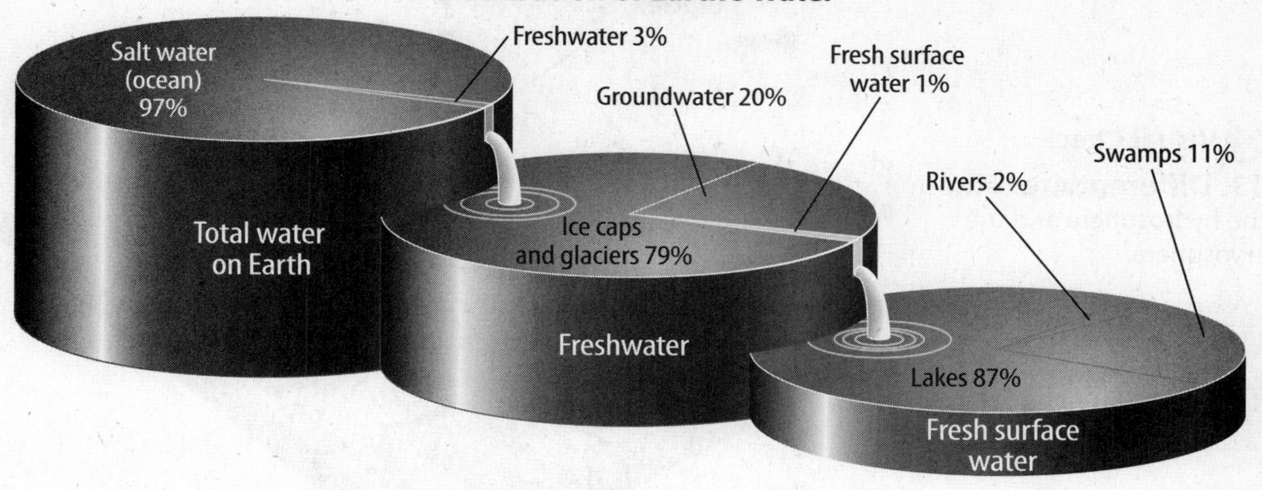

Lakes and Rivers

Less than 1 percent of freshwater is easy to get to on Earth's surface. All people and other organisms that need freshwater to live depend on this small percentage of Earth's total water. ✓

Rain and snow supply water to the surface reservoirs—lakes and rivers. Water in these surface reservoirs moves through the water cycle much faster than water frozen in glaciers and ice caps.

Groundwater

Freshwater in lakes, rivers, and icy glaciers is visible on Earth's surface. These sources hold about 80 percent of Earth's freshwater. The remaining 20 percent is beneath the ground. Large amounts of freshwater are hidden below Earth's surface.

Some rain and snow seep into the ground and collect in small cracks and open spaces called pores. **Groundwater** *is water that is stored in cracks and pores beneath Earth's surface.* Groundwater collects in layers. Many people get their water by drilling wells down into these layers of groundwater.

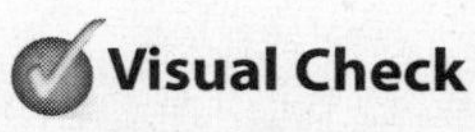

Visual Check

10. Locate Where is most water on Earth located?

Reading Check

11. Point Out All organisms on Earth are dependent upon what percentage of the total supply of freshwater?

Key Concept Check

12. Explain How is water distributed in the hydrosphere?

Visual Check

13. Differentiate between the hydrosphere and the cryosphere.

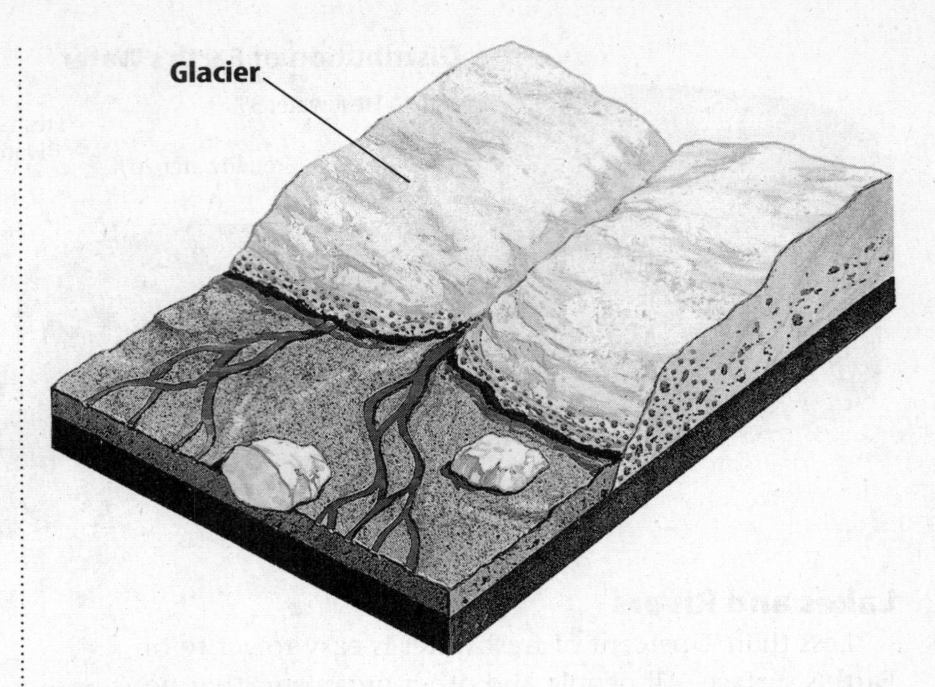

The Cryosphere

The frozen portion of water on Earth's surface is called the **cryosphere.** About 79 percent of the planet's freshwater is ice. The ice is located in glaciers, as shown above, on high mountains or in ice caps at the North Pole and the South Pole.

Water can be stored as ice for thousands of years before melting and becoming liquid water in other reservoirs.

The Geosphere

The last nonliving Earth system is the geosphere. *The* **geosphere** *is the solid part of Earth.* It includes a thin layer of soil and broken rock material along with the underlying layers of rock. The rocks and soil on land and beneath the oceans are part of the geosphere.

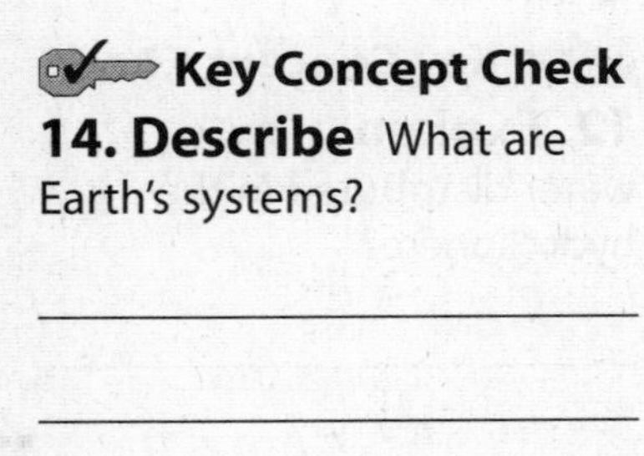

Key Concept Check

14. Describe What are Earth's systems?

Materials in the Geosphere

The geosphere is made of soil, rock, and metal. All of these materials are comprised of smaller particles.

Minerals Diamond is a mineral that is mined and then later cut and polished. **Minerals** *are naturally occurring, inorganic solids that have crystal structures and definite chemical compositions.*

Characteristics of Minerals Minerals must have all five characteristics listed in the definition above. For example, materials that are made by people are not minerals because they did not form naturally. Materials that were once alive are organic and cannot be minerals. A mineral must be solid, so liquids and gases are not minerals. The atoms in minerals must be arranged in an orderly, repeating pattern. Finally, each mineral has a unique composition made of specific elements.

Physical Properties of Minerals Minerals are identified by their physical properties, which include color, streak, hardness, luster, and crystal shape. Streak is the color of a mineral's powder. Even though different samples of some minerals have different colors, the color of the streak is the same. Hardness describes how easily a mineral can be scratched. Luster describes how a mineral reflects light. Usually, you must test several properties to identify a mineral.

Rocks Minerals are the building blocks of rocks. *A* **rock** *is a naturally occurring solid composed of minerals and sometimes other materials such as organic matter.* Scientists classify rocks according to how they formed. The three major rock types are igneous, sedimentary, and metamorphic. ✓

Characteristics of Different Rock Types Igneous rocks form when molten material, called magma, cools and then hardens. Often the magma is found deep inside Earth. But sometimes it erupts from volcanoes and flows onto Earth's surface as lava. So igneous rocks can form inside Earth or on Earth's surface.

Sedimentary rocks form when forces such as water, wind, and ice break down rocks into small pieces called sediment. These same forces carry and deposit the sediment in layers. The bottom layers of sediment are compressed and then cemented together by natural substances to form rocks. ✓

Metamorphic rocks form when extreme temperatures and pressure within Earth change existing rocks into new rocks. The rocks do not melt. Instead, their compositions or their structures change.

Structure

Earth's internal structure is layered like the layers of a hard-cooked egg. Similar to an egg, each layer of the geosphere has a different composition.

Crust The brittle outer layer of the geosphere is much thinner than the inner layers, like the shell on a hard-cooked egg. This thin layer of rock is called the crust. The crust is found under the soil on continents and under the ocean. Oceanic crust is thinner and denser than continental crust. This is due to their different compositions. Continental crust is made of igneous, sedimentary, and metamorphic rocks. Oceanic crust is made of only igneous rock. ✓

✓ **Reading Check**

15. Classify Scientists classify rocks according to _____. (Circle the correct answer.)

a. their size
b. how they formed
c. their color

✓ **Reading Check**

16. Define What is sediment?

____

____

✓ **Reading Check**

17. Differentiate Explain the difference between continental crust and oceanic crust.

____

____

____

____

____

____

Key Concept Check
18. Summarize What are the composition and the structure of the geosphere?

Visual Check
19. Label the figure to show what part of the geosphere is a dense ball of solid iron.

Mantle The middle and largest layer of the geosphere is the mantle. Like the crust, the mantle is made of rock; however, mantle rocks are hotter and denser than those in the crust. In parts of the mantle, temperatures are so high that rocks flow, a bit like partially melted plastic.

Core The center of Earth is the core. If you used a hard-cooked egg as a model of Earth, then the yolk would be the core. Unlike the crust and the mantle, the core is not made of rock. Instead, it is made mostly of the metal iron and small amounts of nickel. The core is divided into two parts. The outer core is liquid. The inner core is a dense ball of solid iron. The three basic layers of the geosphere are shown in the figure below.

Structure of the Geosphere

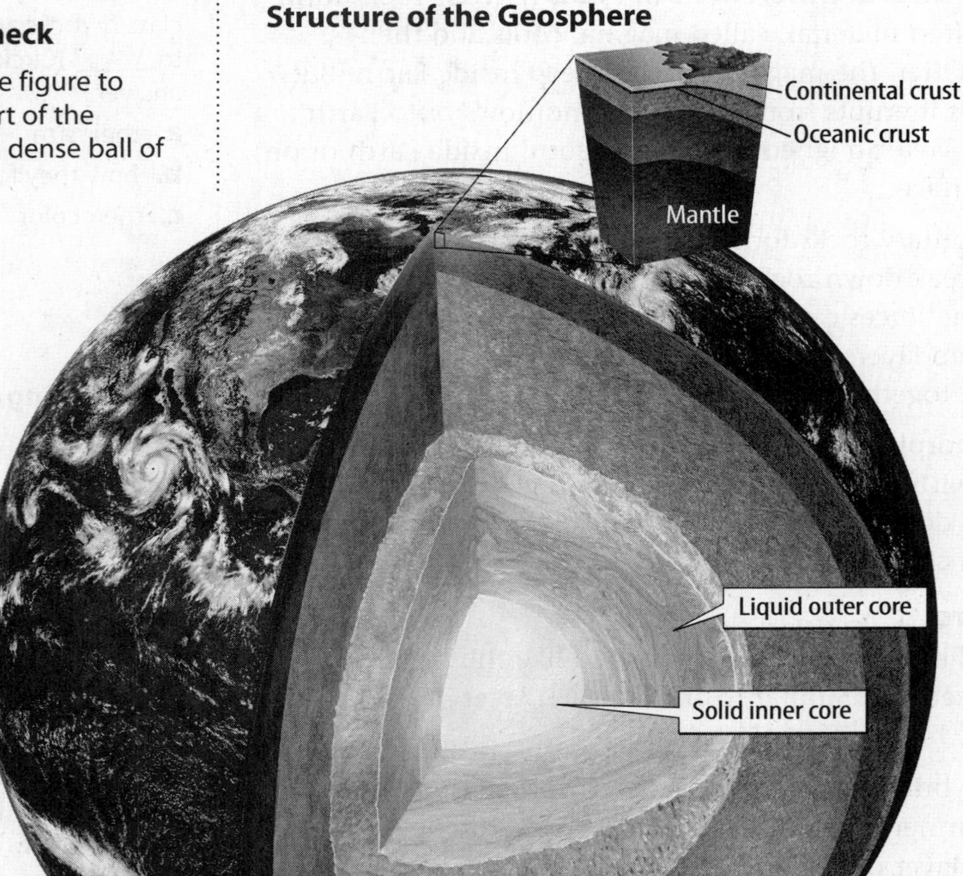

After You Read

Mini Glossary

atmosphere: the mixture of gases that forms a layer around Earth

biosphere: the Earth system that contains all living organisms

cryosphere: the frozen portion of water on Earth's surface

geosphere: the solid part of Earth

groundwater: water that is stored in cracks and pores beneath Earth's surface

hydrosphere: the system containing Earth's liquid water

mineral: a naturally occurring, inorganic solid that has a crystal structure and a definite chemical composition

rock: a naturally occurring solid composed of minerals and sometimes other materials such as organic matter

1. Review the terms and their definitions in the Mini Glossary. Write a sentence contrasting the geosphere and the hydrosphere.

___

___

2. Complete the graphic organizer to show the different layers of the atmosphere. The top and bottom layers have been completed for you.

Exosphere
Troposphere

3. Review the flash cards you created as you read the lesson. Select one and read the question. Then, without checking the lesson, write the answer below.

___

What do you think NOW?

Reread the statements at the beginning of the lesson. Fill in the After column with an A if you agree with the statement or a D if you disagree. Did you change your mind?

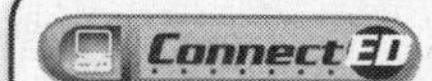

Log on to ConnectED.mcgraw-hill.com and access your textbook to find this lesson's resources.

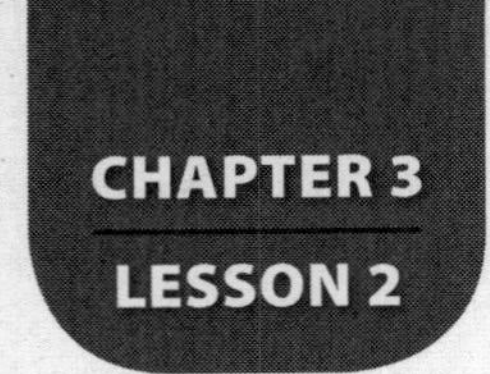

Our Planet—Earth

Interactions of Earth Systems

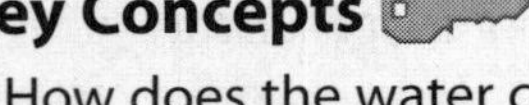

Key Concepts

- How does the water cycle show interactions of Earth systems?
- How does weather show interactions of Earth systems?
- How does the rock cycle show interactions of Earth systems?

Before You Read

What do you think? Read the three statements below and decide whether you agree or disagree with them. Place an A in the Before column if you agree with the statement or a D if you disagree. After you've read this lesson, reread the statements to see if you have changed your mind.

Before	Statement	After
	4. The water cycle begins in the ocean.	
	5. Earth's air contains solids, liquids, and gases.	
	6. Rocks are made of minerals.	

Mark the Text

Building Vocabulary As you read, underline the words and phrases that you do not understand. When you finish reading, discuss these words and phrases with another student or your teacher.

Read to Learn

The Water Cycle

You read that the amount of water on Earth does not change. The water that you drink has been on Earth for a long time. Millions of years ago, a dinosaur might have swallowed the same water that you are drinking today. Or maybe that water raged down a river, flooding an ancient city. How does water move from place to place as time passes?

*The **water cycle** is the continuous movement of water on, above, and below Earth's surface.* The Sun provides the energy that drives the water cycle and moves water from place to place. As this occurs, water can change state to a gas or a solid and then back again to a liquid. The change of state requires either an input or an output of thermal energy.

How does the movement of thermal energy affect evaporation and condensation? When water changes state from a gas to a liquid, thermal energy is released from the water. Thermal energy is absorbed by liquid water when it changes into water vapor.

The water cycle is continuous. That means it has no beginning or end. You will start your investigation of the water cycle in the hydrosphere's largest reservoir, an ocean. 

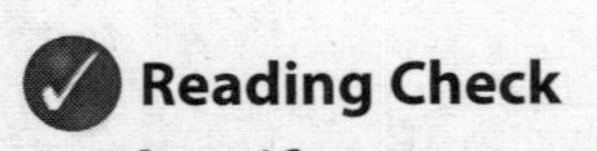

Reading Check

1. Identify What is the source of energy for the water cycle?

Evaporation

When the Sun shines on an ocean, water near the surface absorbs thermal energy and becomes warmer. As a molecule of water absorbs thermal energy, it begins to vibrate faster. When it has enough energy, it breaks away from the other water molecules in the ocean. It rises into the atmosphere as a molecule of gas called water vapor. **Evaporation** *is the process by which a liquid, such as water, changes into a gas*. Water vapor, like other gases in the atmosphere, is invisible.

Transpiration and Respiration

Oceans hold most of Earth's water, so they are major sources of water vapor. But water also evaporates from rivers, lakes, puddles, and soil. These sources, along with oceans, account for 90 percent of the water that enters the atmosphere. Most of the remaining 10 percent is produced by transpiration. **Transpiration** *is the process by which plants release water vapor through their leaves*.

Some water vapor also comes from organisms through cellular respiration. Cellular respiration takes place in many cells. Water and carbon dioxide are produced during cellular respiration. When animals breathe, they release carbon dioxide and water vapor from their lungs into the atmosphere. The wavy arrows in the figure below show water vapor entering the atmosphere.

Water Cycle

FOLDABLES

Make a pyramid book and use it to organize your notes on Earth-system interactions.

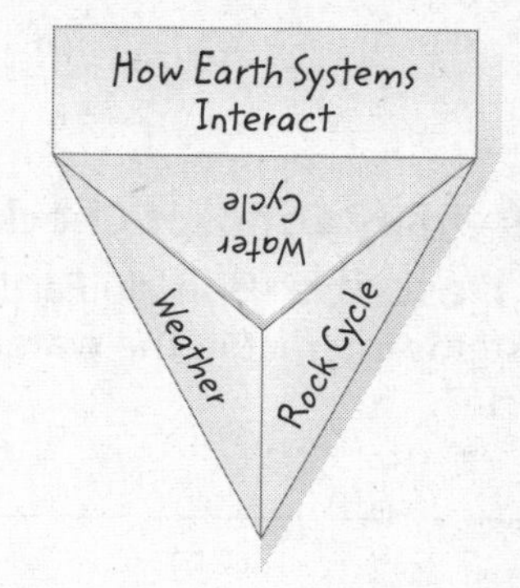

Reading Check

2. Differentiate How are transpiration and respiration similar? How are they different?

Visual Check

3. Name Through which processes does water vapor enter the atmosphere?

Condensation

Recall that temperatures in the troposphere decrease with altitude. So as water vapor rises through the troposphere, it becomes cooler. Eventually, it loses so much thermal energy that it returns to the liquid state. *The process by which a gas changes to a liquid is* **condensation.** Tiny droplets of liquid water join to form larger drops. When millions of water droplets come together, a cloud forms.

Precipitation

Eventually, drops of water in the clouds become so large and heavy that they fall to Earth's surface. *Moisture that falls from clouds to Earth's surface is* **precipitation.** Rain and snow are forms of precipitation. More than 75 percent of precipitation falls into the ocean, and the rest falls onto land. Some of this water evaporates and goes right back into the atmosphere. Some flows into lakes or rivers, and the rest seeps into soil and rocks.

In the water cycle, water continually moves between the hydrosphere, the cryosphere, the atmosphere, the biosphere, and the geosphere. As water flows across land, it interacts with soil and rocks in the geosphere. You will learn more about these interactions when you read about the rock cycle.

Key Concept Check

4. Describe How do Earth systems interact in the water cycle?

____

____

____

____

____

Changes in the Atmosphere

The atmosphere is continually changing. These changes happen mainly within the troposphere, which contains most of the gases in the atmosphere. Some changes occur within hours or days. Others can take decades or even centuries.

Weather

Weather *is the state of the atmosphere at a certain time and place.* In most places, the weather changes a bit every day. How do scientists describe weather and its changes?

Describing Weather Scientists use several factors to describe weather. These factors are shown in the figure on the next page. Air temperature is a measure of the average amount of energy produced by the motion of air molecules. Air pressure is the force exerted by air molecules in all directions. Wind is the movement of air caused by differences in air pressure. Scientists measure wind speed and wind direction. Humidity is the amount of water vapor in a given volume of air. High humidity makes it more likely that clouds will form and precipitation will fall.

Math Skills

The amount of water vapor in air is called vapor density. Relative humidity (RH) compares the actual vapor density in air to the amount of water vapor the air could contain at that temperature. For example, at 15°C, air can contain a maximum of 12.8 g/m^3 of water vapor. If the air contains 10.0 g/m^3 of water vapor, what is the RH?

a. Use the formula:

$$RH = \left(\frac{\text{actual vapor density}}{\text{maximum vapor density}}\right) \times 100\%$$

b. Work out the equation.

$$RH = \left(\frac{10.0 \text{ g}}{12.8 \text{ g}}\right) \times 100\%$$

$RH = 0.781 \times 100\% = 78.1\%$

5. Use a Formula At 0°C, air can contain 4.85 g/m^3 of water vapor. Assume the actual water vapor content is 0.970 g/m^3. What is the RH?

____

Describing the Weather

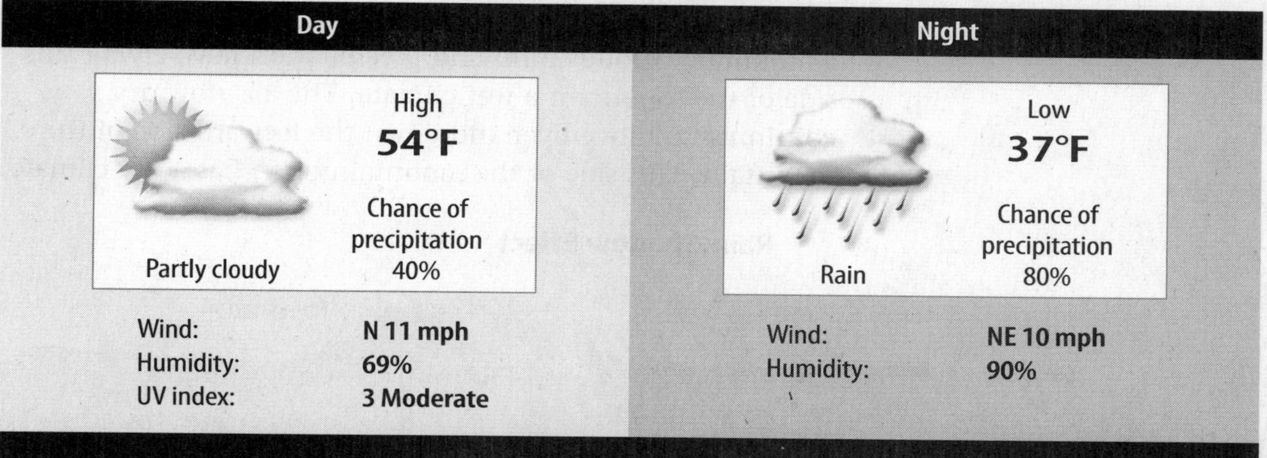

Interactions Weather is influenced by conditions in the geosphere and the hydrosphere. For example, air masses take on the characteristics of the area over which they form. If an air mass forms over a cool ocean, that air mass will bring cool, moist air.

In addition to these interactions, the hydrosphere provides much of the water for cloud formation and precipitation. Tropical waters provide the thermal energy that produces hurricanes.

Climate

What is the weather like where you live? The weather in the area where you live might change each day, but weather patterns can remain nearly the same from season to season. For example, in the summertime the weather might be different every day. One day might be a bit cool and rainy, and the next day might be hot and dry. But overall, summer is warm. These weather patterns are called climate.

Climate *is the average weather pattern for a region over a long period of time.* Earth has many climates. One reason climates are different in different regions of Earth is because of interactions between the atmosphere and other Earth systems.

Mountains Recall that air temperature decreases with altitude. So the climate near the top of a mountain often is cooler than the climate near the mountain's base. Mountains also can affect the amount of precipitation an area receives—a phenomenon known as the rain-shadow effect.

Visual Check

6. Draw a circle around the lowest air temperature shown in the figure. Draw an X through the highest humidity.

Key Concept Check

7. Explain How does weather show interactions of Earth systems?

Reading Check

8. Contrast How does weather differ from climate?

The Rain-Shadow Effect As shown in the figure below, warm, wet air rises and cools as it moves up the windward side of a mountain. Clouds form and precipitation falls, giving this side of the mountain a wet climate. The air, now dry, continues to move over and down the leeward side of the mountain. This side of the mountain often has a dry climate.

Rain-Shadow Effect

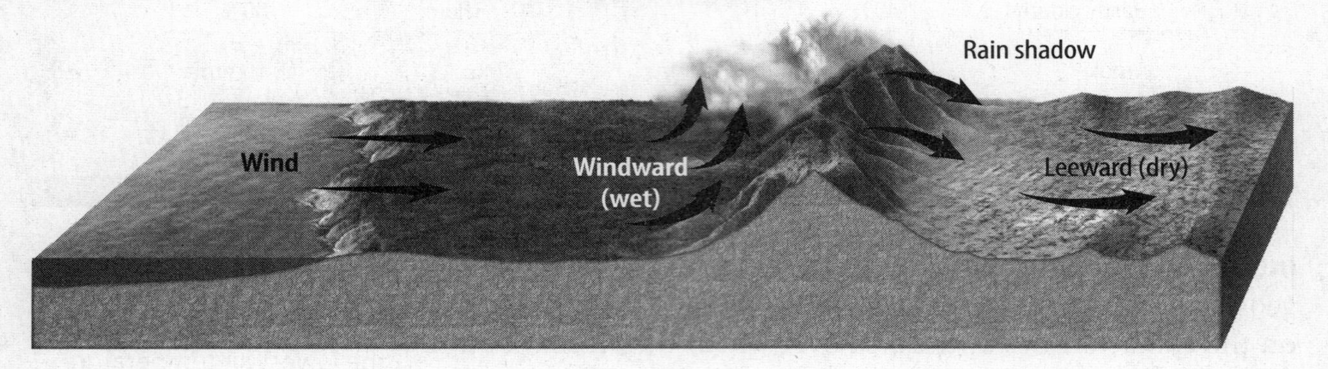

Visual Check

9. Summarize How can mountains affect the amount of precipitation an area receives?

____

____

____

____

Ocean Currents As wind blows over an ocean, it creates surface currents. Surface currents are like rivers in an ocean—the water flows in a predictable pattern. These currents move the thermal energy in water from place to place. For example, the Gulf Stream carries warm waters from tropical regions to northern Europe, making the climate of northern Europe warmer than it would be without these warm waters.

The Rock Cycle

In the water cycle, water moves throughout the hydrosphere, the cryosphere, the atmosphere, the biosphere, and the geosphere. Another natural cycle is the rock cycle. *The* ***rock cycle*** *is the series of processes that transport and continually change rocks into different forms.* The rock cycle takes place in the geosphere, but it is affected by interactions with the other Earth systems.

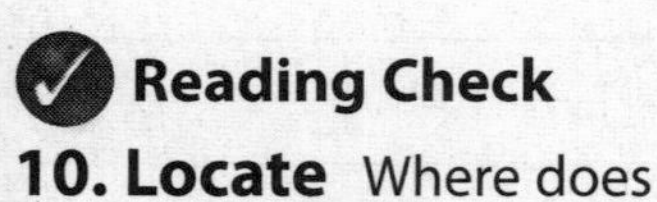

Reading Check

10. Locate Where does the rock cycle take place?

____

As rocks move through the rock cycle, they might become igneous rocks, sedimentary rocks, or metamorphic rocks. At times they might not be rocks at all. Instead, they might take the form of sediments or hot, flowing magma.

Like the water cycle, the rock cycle has no beginning or end. Some processes in this cycle take place on Earth's surface, and others take place deep within the geosphere.

Cooling and Crystallization

Magma is located inside the geosphere. When magma flows onto Earth's surface, it is called lava. Mineral crystals form as magma cools below the surface or as lava cools on the surface. This crystallization changes the molten material into igneous rock.

Uplift

Even rocks formed deep within Earth can eventually be exposed at the surface. **Uplift** *is the process that moves large bodies of Earth materials to higher elevations.*

Uplift is often associated with mountain building. After millions of years of uplift, rocks that formed deep below Earth's surface could have moved upward to the surface. ✓

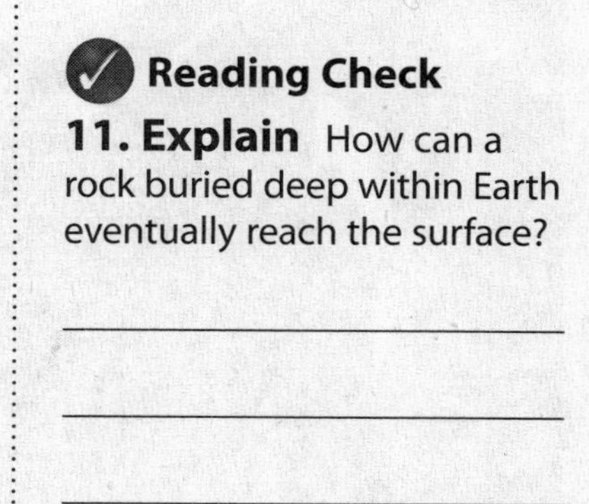

Reading Check

11. Explain How can a rock buried deep within Earth eventually reach the surface?

Weathering and Erosion

Rocks on Earth's surface are exposed to the atmosphere, the hydrosphere, the cryosphere, and the biosphere. Glaciers, wind, and rain, along with the activities of some organisms, break down rocks into sediment. This process is called weathering.

ACADEMIC VOCABULARY

process
(noun) a natural phenomenon marked by gradual changes that lead toward a particular result

Weathering can occur in the mountains, where uplift has exposed rocks. Weathering of rocks into sediments is often accompanied by erosion. Erosion occurs when the sediments are carried by agents of erosion—water, wind, or glaciers—to new locations.

Deposition

Eventually, agents of erosion lose their energy and slow down or stop. When this happens, eroded sediments are deposited, or laid down, in new places. This process is called deposition.

Deposition forms layers of sediment. Over time, more and more layers are deposited.

Compaction and Cementation

As more layers of sediment are deposited, their weight pushes down on the layers below. The deeper layers are compacted. This means they are packed down and pressed together. This process is called compaction.

Minerals dissolved in surrounding water crystallize between grains of sediment and cement the sediments together. This process is called cementation. Compaction and cementation produce sedimentary rocks. ✓

Reading Check

12. Name What kind of rocks are produced by compaction and cementation?

The Rock Cycle

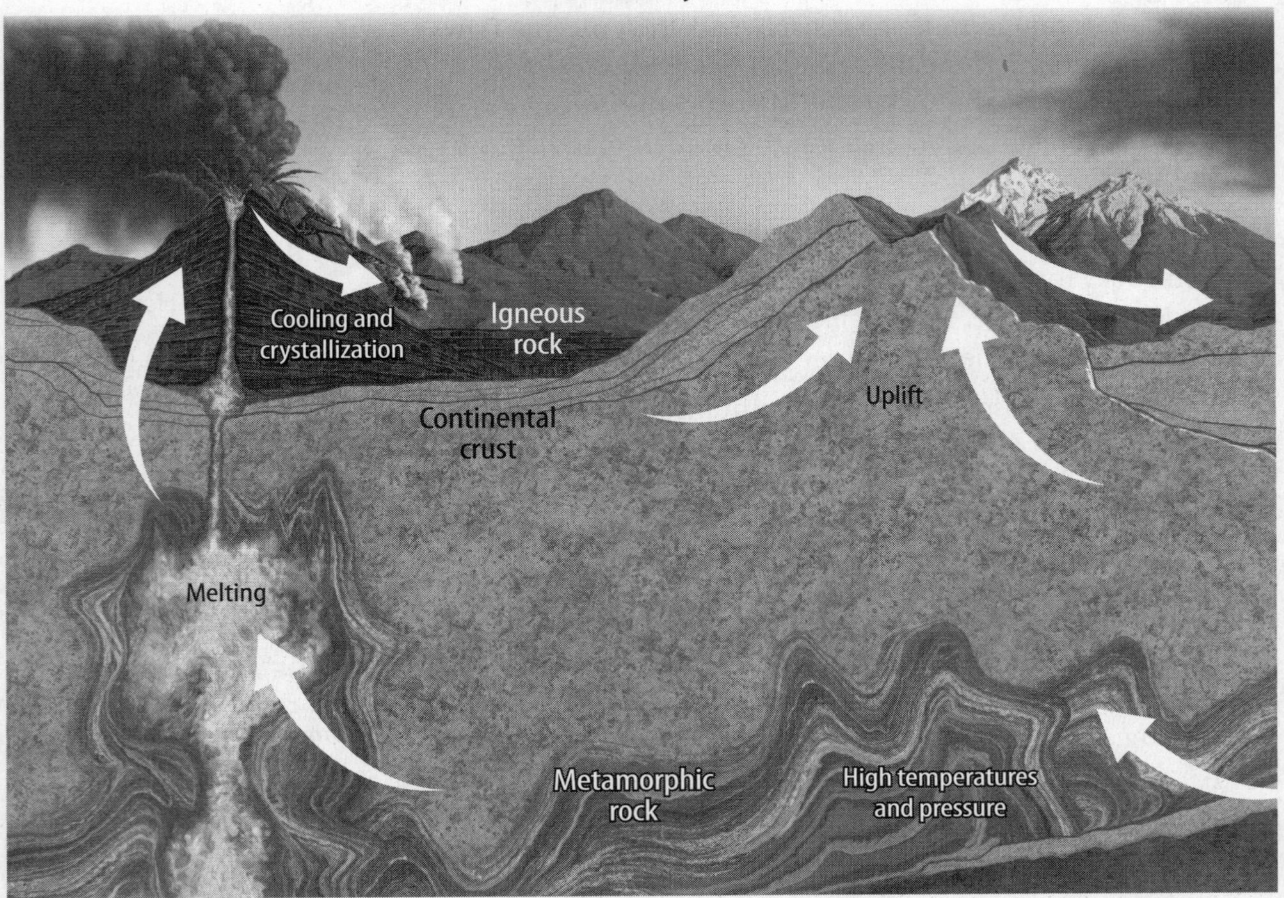

Visual Check

13. Point Out How do weathering and erosion change rocks?

____

____

____

Key Concept Check

14. Describe How do Earth systems interact in the rock cycle?

____

____

____

____

High Temperatures and Pressure

Metamorphic rocks form when rocks are subjected to high temperatures and pressure. This usually occurs far beneath Earth's surface.

Igneous, sedimentary, and even metamorphic rocks can become new metamorphic rocks. Then, uplift can bring the rocks to the surface. There, the rocks are broken down and continue moving through the rock cycle. This process along with other processes in the rock cycle is shown in the figure above and on the next page. Remember that as rocks move slowly through the rock cycle, they change from one form to another.

Endless Interactions Most interactions among the geosphere, the hydrosphere, the cryosphere, and the atmosphere occur on Earth's surface. For example, energy from the Sun—the atmosphere—reaches Earth, where it heats land and water—the hydrosphere and cryosphere. The energy is reflected by Earth's surface—the geosphere—which in turn heats the atmosphere. This affects climate. These are just a few simple examples of different interactions among Earth's systems.

The Rock Cycle

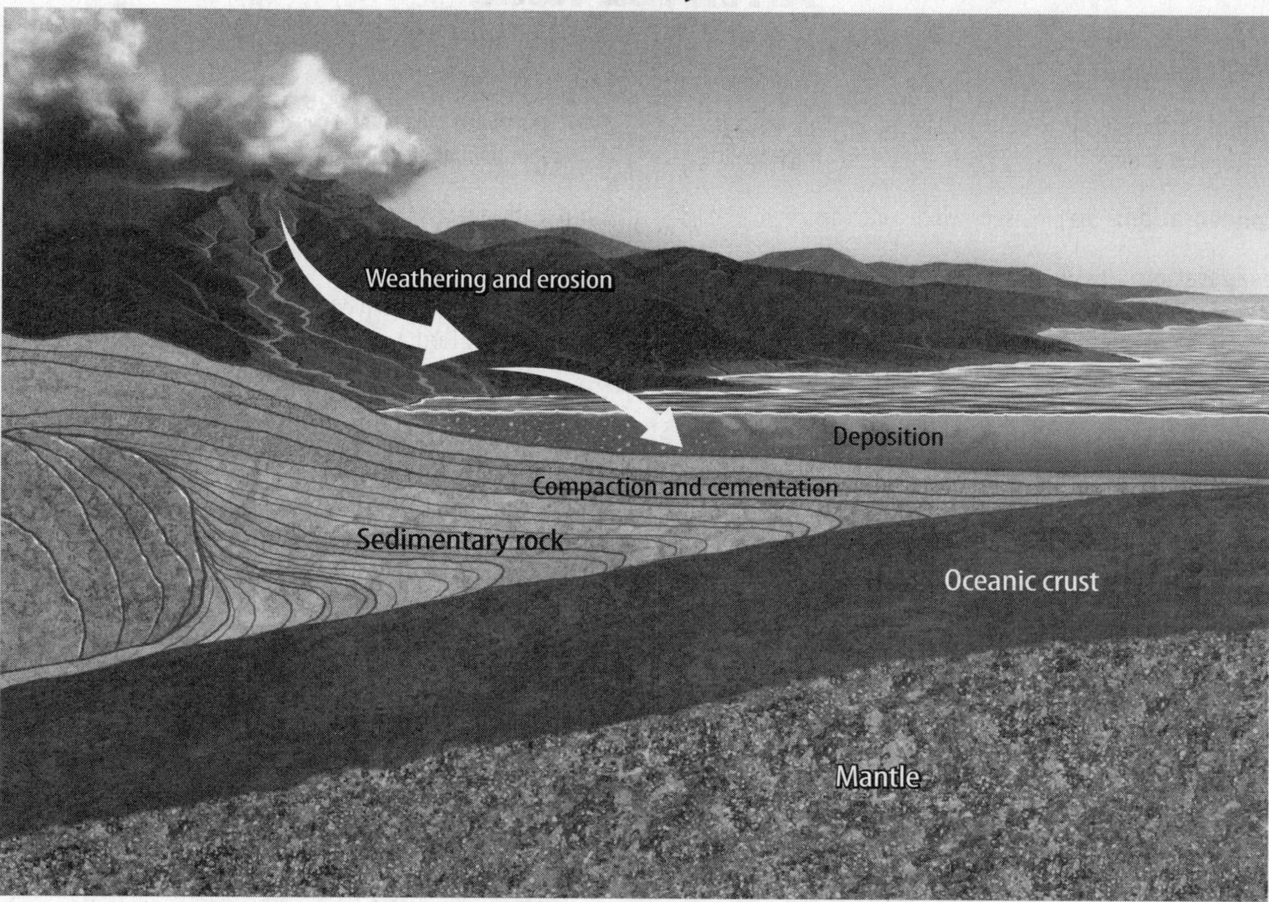

A Unified System You have read about different interacting subsystems of Earth in this chapter:

- The atmosphere is the layer of gases surrounding Earth.
- The hydrosphere is the liquid water found on Earth.
- The cryosphere is the frozen water on Earth.
- The geosphere is Earth's entire solid body.
- The biosphere consists of all living organisms on Earth.

But even though these systems are different, they function together as one unified system—planet Earth.

Think it Over

15. Consider Are you a part of the biosphere? Explain your answer.

____

____

____

____

____

After You Read

Mini Glossary

climate: the average weather pattern for a region over a long period of time

condensation: the process by which a gas changes to a liquid

evaporation: the process by which a liquid, such as water, changes into a gas

precipitation: moisture that falls from clouds to Earth's surface

rock cycle: the series of processes that transport and continually change rocks into different forms

transpiration: the process by which plants release water vapor through their leaves

uplift: the process that moves large bodies of Earth materials to higher elevations

water cycle: the continuous movement of water on, above, and below Earth's surface

weather: the state of the atmosphere at a certain time and place

1. Review the terms and their definitions in the Mini Glossary. Write a sentence contrasting evaporation and condensation.

____

____

2. In the graphic organizer below, identify the five factors scientists use to describe weather.

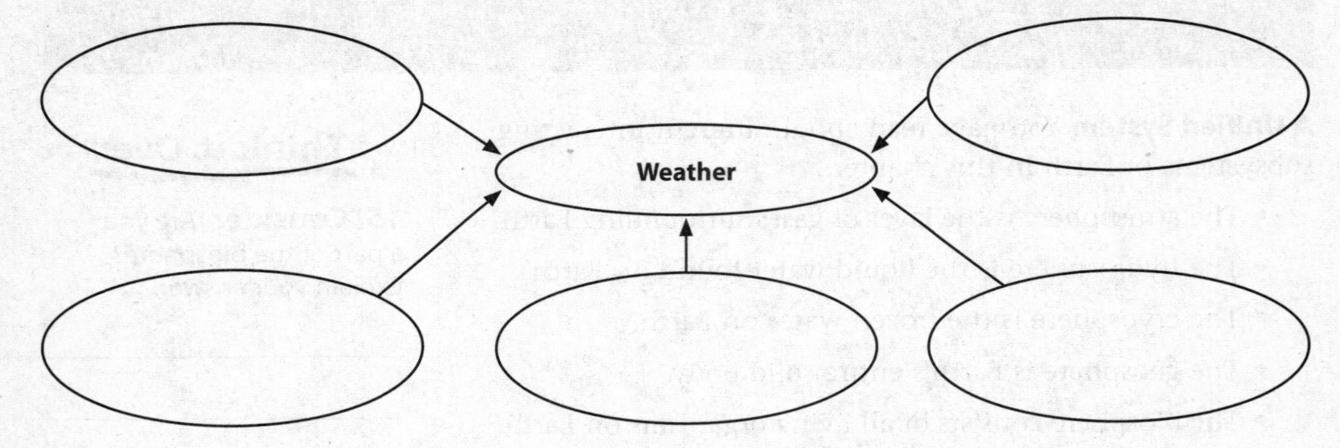

3. How do metamorphic rocks form? Where do they form?

____

____

What do you think NOW?

Reread the statements at the beginning of the lesson. Fill in the After column with an A if you agree with the statement or a D if you disagree. Did you change your mind?

Log on to ConnectED.mcgraw-hill.com and access your textbook to find this lesson's resources.

Earth's Dynamic Surface

Earth's Moving Surface

Before You Read

What do you think? Read the two statements below and decide whether you agree or disagree with them. Place an A in the Before column if you agree with the statement or a D if you disagree. After you've read this lesson, reread the statements to see if you have changed your mind.

Before	Statement	After
	1. Earth's surface is made up of tectonic plates.	
	2. Tectonic plate motion is too slow to measure.	

Key Concepts 

- What is the theory of plate tectonics?
- What are the differences between divergent, convergent, and transform plate boundaries?
- What causes tectonic plates to move on Earth's surface?

Read to Learn

Plate Tectonics

Earth's surface is not the same everywhere. Some regions have tall, rugged mountains. Some regions have flat plains. What causes different landforms? What processes shape and change Earth's surface?

During the 1960s, scientists developed a theory to explain many of the features on Earth's surface. The theory is called **plate tectonics** (tek TAH nihks) and states that *Earth's surface is broken into large, rigid pieces that move with respect to each other.* These pieces, or tectonic plates, move slowly over Earth's surface. You will read how tectonic plate motion forms volcanoes and mountains and causes earthquakes. The theory of plate tectonics was revolutionary because it explains much about how Earth changes.

Mark the Text

Building Vocabulary Skim this lesson and circle any words you do not know. If you still do not understand a word after reading the lesson, look it up in the dictionary. Keep a list of these words and definitions to refer to when you study other chapters.

Key Concept Check

1. Define What is the theory of plate tectonics?

____

____

____

____

What is a tectonic plate?

You might know that Earth is not a solid ball of rock. Earth is made of layers of material.

Lithosphere Earth's outermost layer is called the crust. *The crust and uppermost part of the mantle make up the* ***lithosphere*** (LIH thuh sfihr). The lithosphere forms a rigid shell on the outside of Earth. However, the lithosphere is broken into large pieces. These pieces are tectonic plates. The rocks in the lithosphere are strong and do not bend easily.

Visual Check

2. Identify On what layer of Earth do you live?

Think it Over

3. Summarize what you know about tectonic plates.

Visual Check

4. Determine How many large plates have continents on them?

Asthenosphere The figure on the right shows Earth's layers. *The partially melted portion of the mantle below the lithosphere is the* **asthenosphere** (as THEN uh sfihr). The asthenosphere is hotter than the lithosphere and can bend more easily. As you will read, the ability of the asthenosphere to bend relates to tectonic plate movement.

Earth's Layers

Crust
Oceanic crust
Continental crust
Lithosphere
Uppermost mantle
Asthenosphere
Mantle
Liquid outer core
Solid inner core

Major Tectonic Plates

Scientists have identified 15 large tectonic plates within Earth's crust, as shown in the figure below. Some plates are so large they support entire continents. Other plates are so small that a map of this scale cannot represent them. Earth's tectonic plates fit together like puzzle pieces. The plates are in constant motion across Earth's surface.

Many of you live on the North American Plate. To the east of it is the Eurasian Plate. To the west are two plates—a small plate called the Juan de Fuca Plate and the largest plate, the Pacific Plate.

Oceans completely cover some plates, such as the Juan de Fuca Plate. Other plates, such as the North American Plate, are made of both oceanic crust and continental crust.

Earth's Tectonic Plates

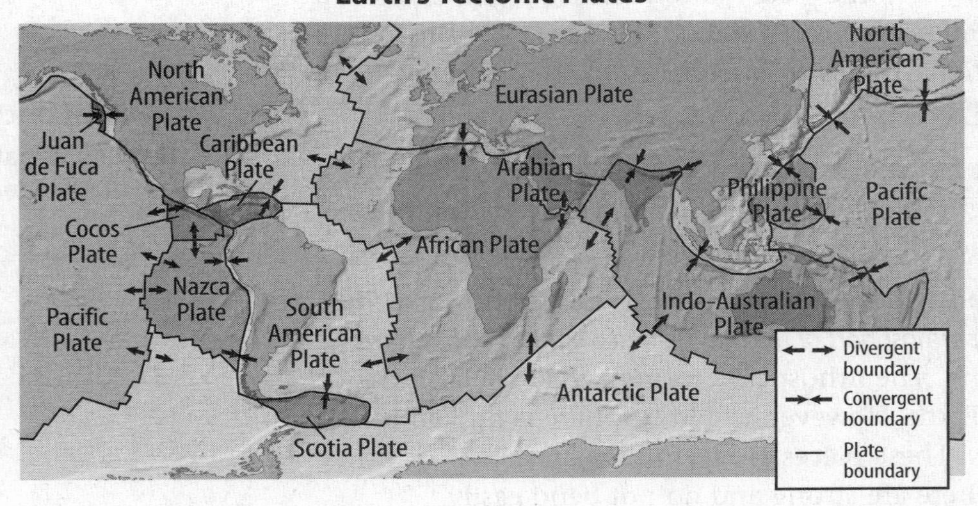

Relative Motion of Tectonic Plates

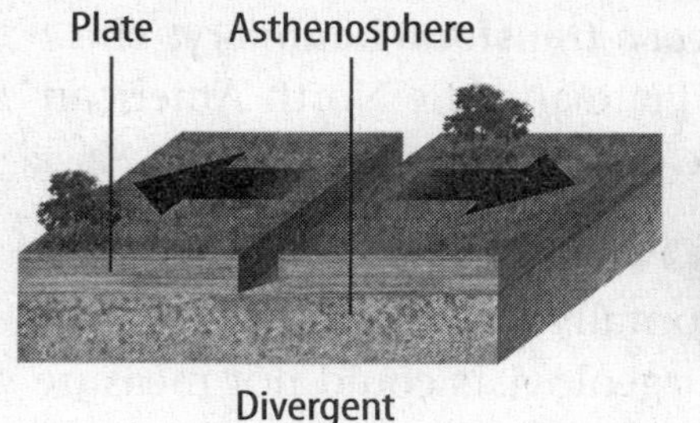

Divergent

Convergent

Transform

Plate Boundaries

How do scientists describe the movement of a tectonic plate? They describe a plate's relative motion—how it moves in relation to another plate. For example, the North American Plate is moving away from the Eurasian Plate, but it is also moving toward the Pacific Plate.

To better understand how plates move, place two pieces of paper side by side on your desk. Think about how you can make one sheet move relative to the other. You can push the sheets together so that one sheet goes over or under the other sheet. Or you might move them apart or slide them past each other. Tectonic plates move in similar ways, as illustrated in the figure above. As plates move relative to each other, different types of boundaries form where the plates meet. The type of boundary that forms depends on the relative motion of the plates.

Divergent Boundaries

A boundary where two plates move away from each other is called a **divergent boundary.** The boundary between the North American Plate and the Eurasian Plate is a divergent boundary. As plates move apart, new crust forms between them.

Convergent Boundaries

A boundary where two plates move toward each other is a **convergent boundary.** In some locations, one plate is pushed under the other plate and down into the mantle. That plate melts and becomes part of the mantle.

Subduction *is the process that occurs when one tectonic plate moves under another tectonic plate.* At a convergent boundary, the process of subduction forces one plate into the mantle. The Pacific Plate is subducting under the North American Plate at the convergent plate boundary near Alaska.

Visual Check

5. Describe the relative motion of plates at a convergent boundary.

____

____

____

FOLDABLES®

Make a horizontal two-tab book with an extended tab to summarize the causes and effects of tectonic plate movement.

Tectonic Plate Movement

Cause | Effect

Reading Check

6. Explain what happens to a plate that is pushed under another plate and down into the mantle.

____

____

____

____

 **Key Concept Check**

7. Differentiate What are the differences between divergent, convergent, and transform plate boundaries?

____

____

____

____

____

Math Skills

The plates along the Mid-Atlantic Ridge spread at an average rate of 2.5 cm/y. How long will it take the plates to spread 100 m? Use proportions to find the answer.

a. Convert the distances to the same unit.

100 cm = 1 m

2.5 cm = 0.025 m

b. Set up a proportion.

$$\frac{0.025 \text{ m}}{1 \text{ y}} = \frac{100 \text{ m}}{x \text{ y}}$$

c. Cross multiply and solve for *x*.

$0.025 \text{ m} \times x \text{ y} = 100 \text{ m} \times 1 \text{ y}$

d. Divide both sides by 0.025 m.

$$x \text{ y} = \frac{100 \text{ m/y}}{0.025 \text{ m}}$$

$$x = 4{,}000 \text{ y}$$

8. Use Proportions The Eurasian Plate travels at about 0.7 cm/y. How long would it take the plate to travel 1 km?

(1 km = 100,000 cm)

____

Transform Boundaries

Two plates slide past each other at a **transform boundary.** The boundary between the Pacific Plate and the North American Plate in California is an example of a transform boundary.

Measuring Plate Movement

Tectonic plates move horizontally over Earth's surface. The plates move so slowly that geologists could not measure their movement before the mid-twentieth century. However, during the 1970s, scientists and engineers developed new technologies that made it possible for them to measure how fast tectonic plates move. This technology has determined that North America is separating from Europe at an average rate of just 2.5 cm/y.

The position of any point on Earth's surface can be accurately measured using the network of satellites known as the Global Positioning System (GPS). GPS is a set of 24 satellites in orbit around Earth that send signals to help locate and track various moving objects. By tracking tectonic plate positions over several years, scientists can measure the speed and the direction of plate movement.

Even though plates move slowly, dramatic changes occur over long periods of time. North America and Europe once were part of a large continent called Pangaea (pan GEE uh). A divergent boundary formed between North America and Europe about 200 million years ago. The plates moved apart, and the Atlantic Ocean formed.

Why do tectonic plates move?

You have read that tectonic plates move over Earth's surface. As the plates move and interact with each other, they form convergent, divergent, and transform boundaries. But what causes plates to move?

Convection

Recall that density is the amount of matter per unit of volume. As the temperature of a fluid increases, the molecules in the fluid spread out. There is less matter in the same amount of volume. So, the fluid becomes less dense. However, fluids do not heat evenly. Some of a fluid can be warmer and less dense, while some is cooler and more dense. The warmer, less dense fluid rises, and the cooler, denser fluid sinks. *The circulation within fluids caused by differences in density and thermal energy is called* **convection.**

Convection Currents

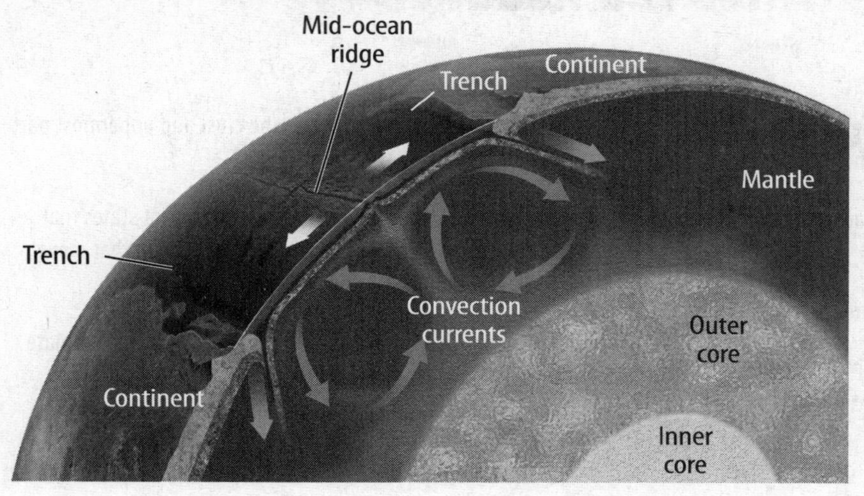

Visual Check

9. Interpret Name the type of boundary formed by convection currents at the mid-ocean ridge.

____

Convection also occurs in Earth's asthenosphere, the layer just below the lithosphere. Recall that rocks in the mantle are hot enough to bend easily. They can flow in a way similar to how fluids flow. Convection in the mantle can drag plates over the surface of Earth, as illustrated in the figure above. As the mantle convects, it pulls and pushes the tectonic plates.

Another process that causes plate movement is subduction. A plate at Earth's surface is colder and denser than the mantle below it. When two plates collide, one can subduct or sink into the hotter, less-dense mantle. When this happens, the sinking part of the plate pulls the rest of the plate along with it.

Key Concept Check

10. Explain What causes tectonic plates to move?

____

____

____

After You Read

Mini Glossary

asthenosphere (as THEN uh sfihr): the partially melted portion of the mantle below the lithosphere

convection: the circulation within fluids caused by differences in density and thermal energy

convergent boundary: a boundary where two plates move toward each other

divergent boundary: a boundary where two plates move away from each other

lithosphere (LIH thuh sfihr): the crust and uppermost part of the mantle

plate tectonics (tek TAH nihks): theory that states that Earth's surface is broken into large, rigid pieces that move with respect to each other

subduction: the process that occurs when one tectonic plate moves under another tectonic plate

transform boundary: a boundary where two plates slide past each other

1. Review the terms and their definitions in the Mini Glossary. Write a sentence that explains why convection currents form.

____

____

2. The diagram below illustrates the motion of two tectonic plates. Label it by identifying the type of plate boundary and the two processes that are at work.

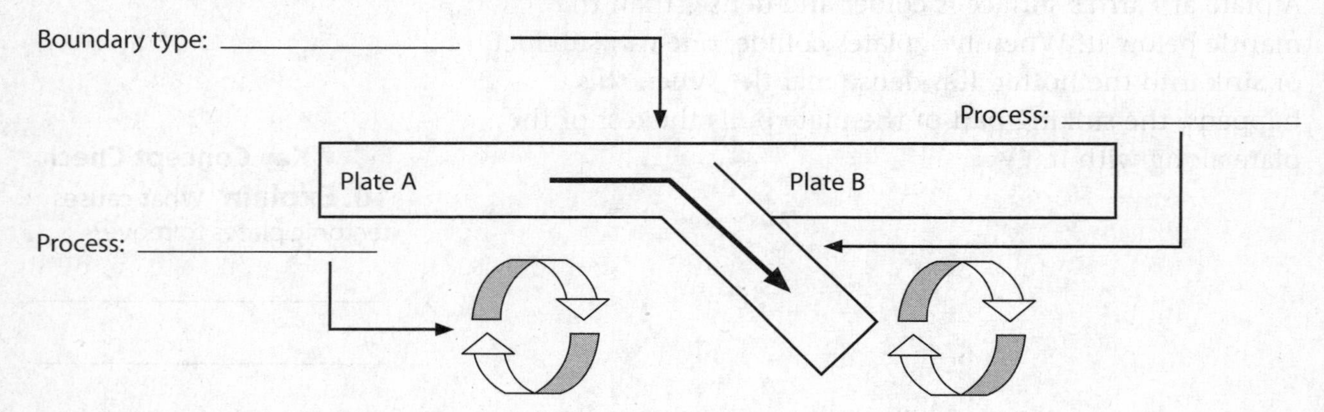

3. Explain the meaning of one of the words you circled as you skimmed the lesson.

____

____

What do you think NOW?

Reread the statements at the beginning of the lesson. Fill in the After column with an A if you agree with the statement or a D if you disagree. Did you change your mind?

Log on to ConnectED.mcgraw-hill.com and access your textbook to find this lesson's resources.

Earth's Dynamic Surface

Shaping Earth's Surface

Before You Read

What do you think? Read the two statements below and decide whether you agree or disagree with them. Place an A in the Before column if you agree with the statement or a D if you disagree. After you've read this lesson, reread the statements to see if you have changed your mind.

Before	Statement	After
	3. Most earthquakes occur near tectonic plate boundaries.	
	4. Volcanoes can erupt anywhere.	

Key Concepts 

- Where do most earthquakes occur?
- How are landforms related to plate tectonics?
- Where do most volcanoes form?
- How does plate movement form mountains?

Read to Learn

Earthquakes

An **earthquake** *is the vibrations caused by the rupture and sudden movement of rocks along a break or a crack in Earth's crust.* On Earth, Earthquakes occur every day. The strong shaking can damage both natural features and human-made structures.

Fault

Earthquakes can occur at faults. *A* **fault** *is a crack or a fracture in Earth's crust along which movement occurs.* One place where a fault can exist is at a plate boundary.

Tectonic plates do not continually slide past each other along faults. But, because of the convection currents beneath the tectonic plates, forces build up along faults. Eventually, these forces become so great that the rocks on either side of the fault move and slide along the fault. When this happens, the fault is said to rupture, and Earth's crust moves along the fault, causing an earthquake.

Where Earthquakes Occur

Most earthquakes occur at plate boundaries. Plate boundaries are long and do not rupture all at once. Instead, usually only small segments rupture. The amount of energy an earthquake releases determines the size of the earthquake. Scientists express the size of an earthquake in units of magnitude.

Study Coach

Create a Quiz Write a quiz question for each paragraph. Answer the question with information from the paragraph. Then work with a partner to quiz each other.

FOLDABLES®

Make a vertical three-tab Venn book to compare and contrast earthquakes and faults.

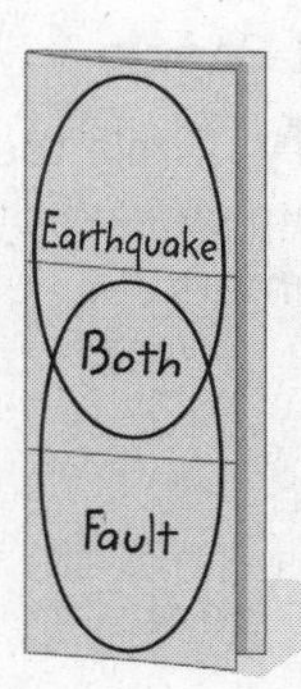

Interpreting Tables

1. Analyze How are fault length and magnitude related?

______

______

______

Earthquake Magnitudes			
Magnitude	**Average Number per Year**	**Typical Fault Length on Surface**	**Typical Movement on Fault**
3	>100,000	15 m	1 mm
4	15,000	100 m	5 mm
5	3,000	800 m	3 cm
6	100	6 km	20 cm
7	20	40 km	1 m
8	2	300 km	6 m

Earthquake magnitude can range from less than one to at least 9.9. As shown in the table above, small earthquakes occur more frequently than large ones. Still, more than a few major earthquakes occur each year.

A plate boundary is made up of more than one fault. The boundary covers a large region, and many smaller faults can branch out from the main fault. Faults can be many kilometers from the plate boundary. Earthquakes can occur on these remote faults, just as they do on faults at plate boundaries. Faults are largest where one plate subducts into the mantle. The strongest and most damaging earthquakes occur at these locations. Higher magnitude earthquakes occur when movement along faults covers large distances. In 2004, the convergent boundary between the Indian Plate and the Burma Plate (part of the Eurasian Plate) ruptured. That earthquake had a magnitude greater than 9.

Key Concept Check

2. Locate Where do most earthquakes occur?

______

______

How Earthquakes Change Earth's Surface

The movement of crust along faults can make mountains, valleys, and other landforms. Different types of movement occur at the three types of plate boundaries.

Transform Boundary Blocks of crust move horizontally past each other at a transform fault, as shown in the figure to the right. Both plate movement and earthquakes shift features that cross the fault, such as streams and roads. Transform faults also are called strike-slip faults.

Transform Boundary

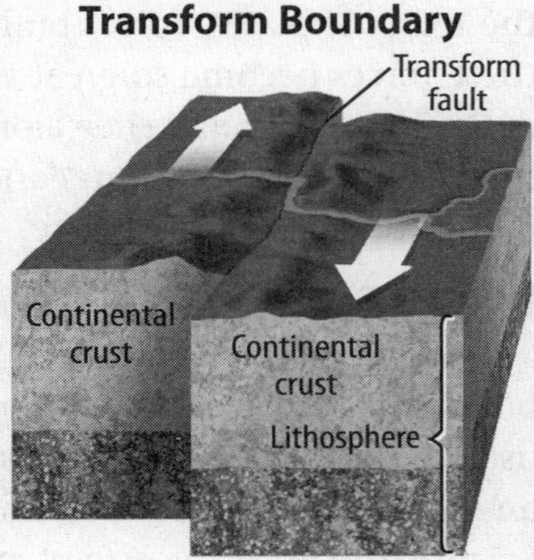

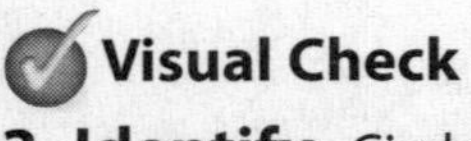

3. Identify Circle the shift in the stream caused by the plate movement.

Divergent Boundary Mid-ocean ridges form between oceanic plates, as illustrated to the right. Between continental plates, one side of the fault moves down relative to the other side of the fault. Normal faults form valleys at divergent boundaries.

Divergent Boundary

Visual Check

4. Name What landforms occur at divergent boundaries between oceanic plates?

____

____

Convergent Boundary—Subduction Zone The figure at left below shows what happens to plates at a convergent boundary where subduction occurs. The plate that does not subduct deforms and crumples as the two plates push toward each other. As the mantle near the subducted plate melts, magma rises and forms a volcanic arc on the plate that does not subduct.

Convergent Boundary—No Subduction What happens at a convergent boundary when subduction does not occur? The edges of both tectonic plates crumple and deform, as shown in the figure below on the right. Because neither plate subducts, blocks of crust slide upward along a complex series of faults called reverse faults. This results in the formation of tall mountains.

Key Concept Check

5. Describe How are landforms related to plate tectonics?

____

____

____

Convergence and Subduction

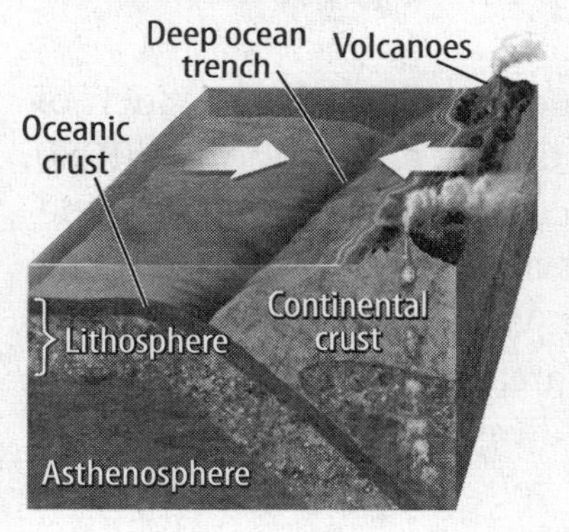

Convergence, No Subduction

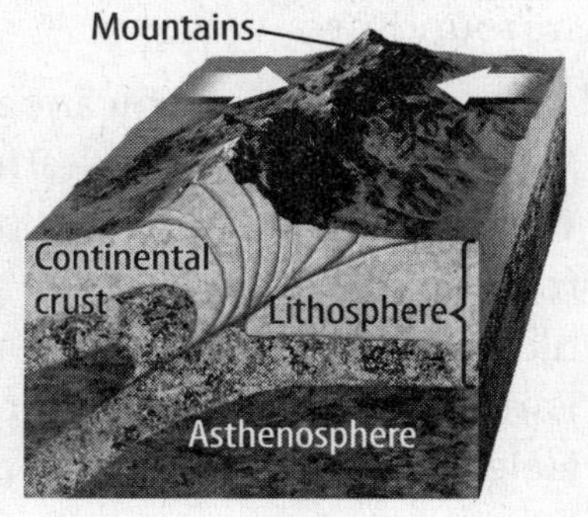

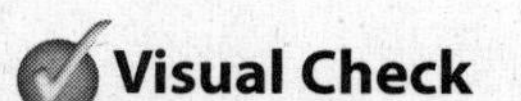

Visual Check

6. Point Out Circle the landforms that develop at a convergent boundary.

Volcanoes

The temperature inside Earth is high enough for rock to melt. Geologists call *molten rock stored beneath Earth's surface* **magma. Lava** *is magma that erupts onto Earth's surface.* **Volcanoes** *are vents in Earth's crust through which molten rock flows.* Volcanoes are common on Earth. During the last 10,000 years, more than 1,500 different volcanoes have erupted. Although they are common, volcanoes do not form everywhere.

Where Volcanoes Occur

Most volcanoes form at convergent plate boundaries. Recall that at some convergent boundaries, one plate subducts under another plate. Some rocks contain water within their structure. As the rocks subduct, heat and pressure drive the water out. This water can lower the melting temperature of the mantle.

Magma then rises toward the surface and forms volcanoes on the plate that does not subduct. A line of volcanoes forms parallel to the plate boundary directly above the plate that subducted. The volcanoes in Washington and Oregon, such as Mount Rainier, Mount St. Helens, and Mount Hood, formed above the subducting Juan de Fuca Plate.

Reading Check
7. Explain How does water released from subducting rocks affect the mantle?

Key Concept Check
8. Locate Where do most volcanoes form?

How Volcanoes Change Earth's Surface

Volcanoes are some of Earth's most distinctive landforms. Compared to other mountains, volcanoes can form quickly. Mountains can form over millions of years, but volcanoes can form in hundreds to thousands of years. Sometimes volcano formation happens even more quickly. In Mexico, Paricutín volcano grew 365 m above its surroundings within one year.

Volcanoes erupt in two ways. Sometimes, lava can flow over Earth's surface before cooling, hardening, and becoming solid rock. This is called a lava flow. Lava flows can be more than 10 km long and over time can cover large areas surrounding a volcano.

At other times, volcanoes can erupt explosively. Much of Mount St. Helens was destroyed during an eruption in 1980. This kind of eruption can produce tiny pieces of glass made from solidified lava. These pieces are called ash and can be blown high into the atmosphere. When the ash falls back to Earth's surface, it can cover vast areas. Ash from Mount St. Helens in Washington fell as far away as Minnesota and Oklahoma.

Reading Check
9. Describe How can volcanoes change Earth's surface?

Ocean Basins

You have read that lava erupts from volcanoes on land. However, not all lava flows on land. Recall that the land masses that make up North America and Europe began to separate 200 million years ago when a divergent plate boundary formed between them. What happened in the area between these land masses as it became larger?

Lava on the Ocean Floor

Lava erupts at both convergent plate boundaries and divergent plate boundaries. This lava hardens and forms new crust. At an oceanic divergent plate boundary, magma rises between two plates and forms new crust, as illustrated below. As the plates move apart, more lava rises. The lava fills in the space and forms more ocean crust. The seafloor between North America and Europe is made of ocean crust that formed after the continents began to spread apart.

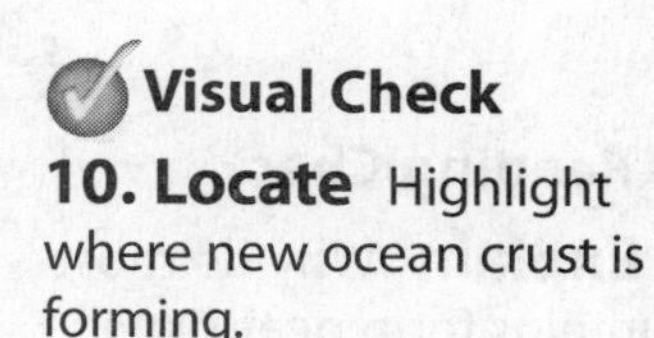

Visual Check

10. Locate Highlight where new ocean crust is forming.

Formation of Oceanic Crust

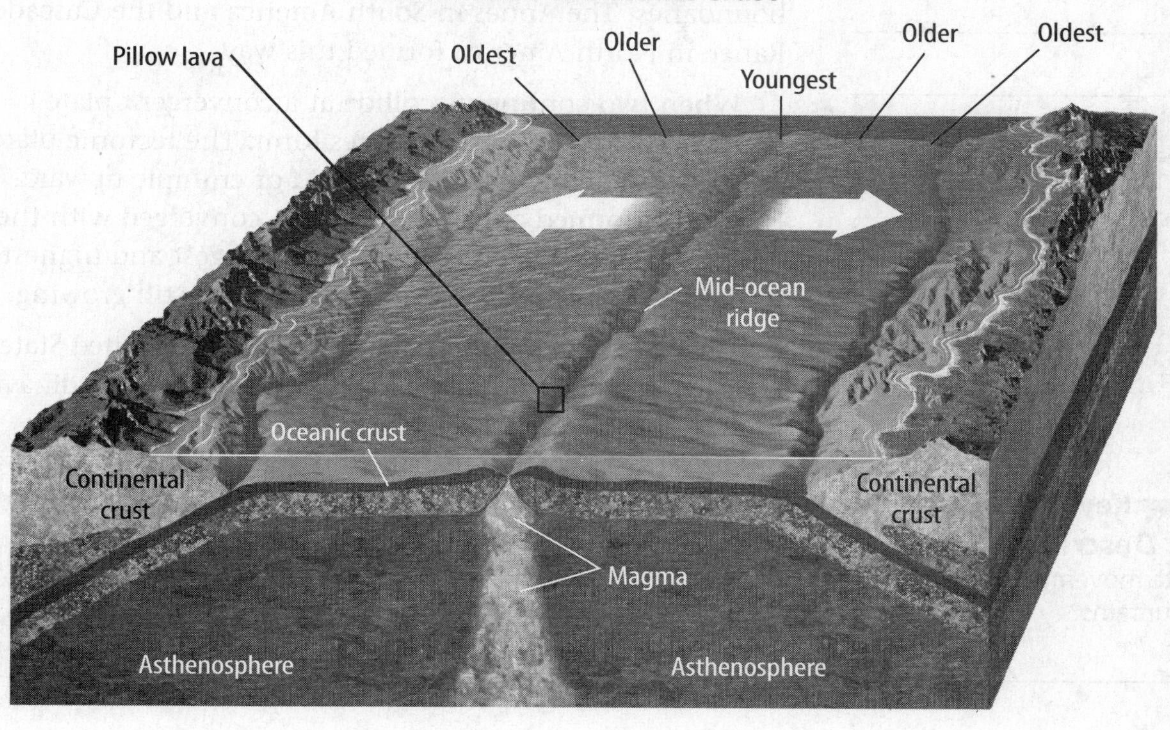

Mountains on the Ocean Floor

The ocean crust made at divergent plate boundaries is not flat. **Mid-ocean ridges** *are long, narrow mountains formed by magma at divergent boundaries.* The mid-ocean ridge in the Atlantic Ocean extends through the middle of the Atlantic Ocean, from near the North Pole to near the South Pole. Mid-ocean ridges usually have gentle slopes and are about 2 km high. ✓

Even though explosive volcanic eruptions usually occur near convergent plate boundaries, more lava erupts at divergent plate boundaries. Three-quarters of all lava erupts at mid-ocean ridges. As lava erupts under water, it hardens into flows and unique shapes, such as pillow lava.

Reading Check

11. Define What is a mid-ocean ridge?

Mountains at Convergent Boundaries

Mountains form when Earth's crust folds and crumples. Where do you think this happens? Recall that tectonic plates are rigid pieces of lithosphere. Collisions usually do not fold the centers of these rigid plates. Instead, folding and crumpling usually occur at the edges of plates. This is why most mountains form near plate boundaries.

Reading Check
12. Explain Why do most mountains form near plate boundaries?

Recall that volcanoes form at convergent plate boundaries where one plate subducts under another plate. These volcanoes form volcanic mountain chains along the plate boundaries. The Andes in South America and the Cascade Range in North America formed this way.

When two continents collide at a convergent plate boundary, large mountain ranges form. The tectonic plates are under extreme pressure and fold or crumple upward. The Himalayas formed as the Indian Plate converged with the Eurasian Plate. The Himalayas are the largest and highest mountain range in the world, and they are still growing.

The Appalachian Mountains in the eastern United States and the Caledonian mountains in Scotland and Scandinavia formed at the same convergent boundary. However, over millions of years, tectonic plate motion broke up Pangaea. The plates moved apart and the mountain chains separated.

Key Concept Check
13. Describe How does plate movement form mountains?

After You Read

Mini Glossary

earthquake: the vibrations caused by the rupture and sudden movement of rocks along a break or a crack in Earth's crust

fault: a crack or a fracture in Earth's crust along which movement occurs

lava: magma that erupts onto Earth's surface

magma: molten rock stored beneath Earth's surface

mid-ocean ridge: long, narrow mountains formed by magma at a divergent boundary

volcano: vent in Earth's crust through which molten rock flows

1. Review the terms and their definitions in the Mini Glossary. Write a sentence that explains how an earthquake and a fault are related.

2. For each effect on landforms described in the diagram, identify the type of plate boundary where the movement occurs. Choose from the list below.

convergent boundary, with subduction | **divergent boundary**
convergent boundary, without subduction | **transform boundary**

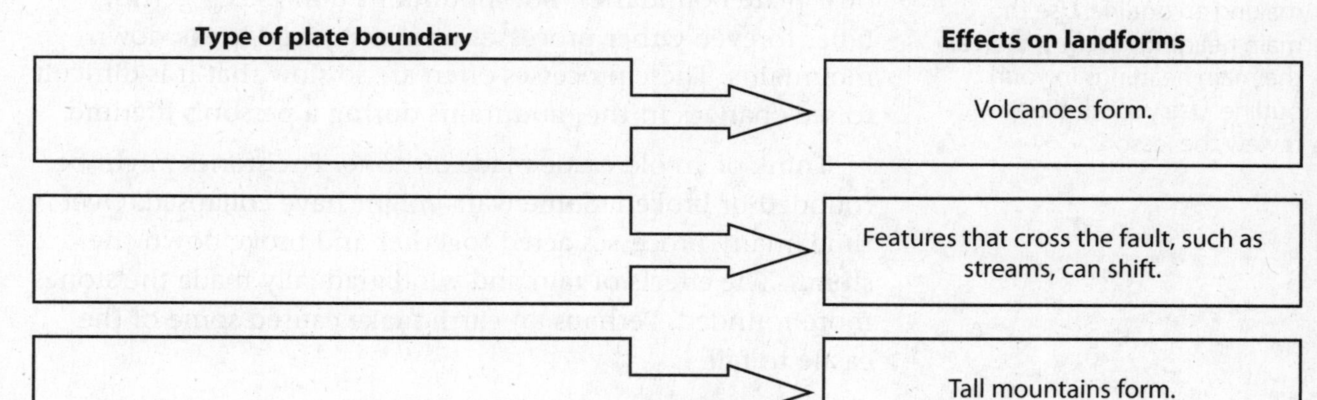

What do you think NOW?

Reread the statements at the beginning of the lesson. Fill in the After column with an A if you agree with the statement or a D if you disagree. Did you change your mind?

Log on to ConnectED.mcgraw-hill.com and access your textbook to find this lesson's resources.

CHAPTER 4
LESSON 3

Earth's Dynamic Surface

Changing Earth's Surface

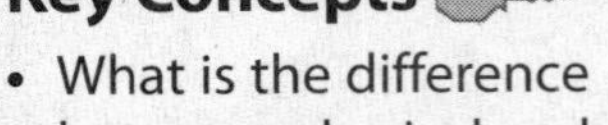

Key Concepts
- What is the difference between physical and chemical weathering?
- How do water, ice, and wind change Earth's surface?

Before You Read

What do you think? Read the two statements below and decide whether you agree or disagree with them. Place an A in the Before column if you agree with the statement or a D if you disagree. After you've read this lesson, reread the statements to see if you have changed your mind.

Before	Statement	After
	5. Wind erosion only occurs in the desert.	
	6. Rivers are the only cause of erosion.	

Study Coach

Make an Outline As you read, summarize the information in the lesson by making an outline. Use the main headings in the lesson as the main headings in your outline. Use your outline to review the lesson.

Read to Learn

Breaking Down Earth Materials

Tall mountains form as a result of movement along faults near plate boundaries. But mountains don't keep getting taller forever. Other processes wear away and break down mountains. These processes often are so slow that it is difficult to see changes in the mountains during a person's lifetime.

Think of an old castle made of stone. The stones might be rounded or broken. Some walls might have collapsed. Over time, many processes acted together and broke down the stones. The effects of rain and wind gradually made the stones more rounded. Perhaps an earthquake caused some of the castle to fall.

Mountains are similar. Over time, the processes that changed the stones of the castle can also change the rocks that make up mountains.

Weathering *refers to the mechanical and chemical processes that change Earth's surface over time.* Weathering can affect rocks in different ways. The processes of weathering can break, scrape, smooth, or chemically change rock.

Sediment *is the material formed from rocks broken down by weathering.* Weathering can produce sediment of different sizes. Sediment can be rock fragments, sand, silt, or clay. 

Reading Check

1. Relate How is weathering related to sediment?

Physical Weathering

The first step in making sediment is to make smaller pieces of rock from larger ones. **Physical weathering** *is the process of breaking down rock without changing the composition of the rock.* Several natural processes cause physical weathering. For example, if a boulder rolls off a cliff and breaks apart, physical weathering is occurring. Forces from plate motion, such as when faults rupture, also can cause rock to break.

Physical weathering also can occur because of changes in weather. Water can seep into rocks. If the temperature is low enough, the water can freeze. Unlike most liquids, water expands when it freezes. The force from the expanding ice pushes outward and, over time, can shatter rocks.

Plants and animals also can break rocks. The roots of plants can grow into cracks in rocks. The force from the growing roots can pry the rock open. Lichens are formed from two organisms that grow together. Some lichens can grow on rock surfaces. Acids produced by the lichens break down the rock.

Chemical Weathering

Some minerals can react with water, air, or substances in water and air, such as carbon dioxide (CO_2). **Chemical weathering** *is the process that changes the composition of rocks.*

Some minerals dissolve in water that is slightly acidic, such as rainwater. Limestone contains calcite, a mineral that dissolves in slightly acidic water.

Other minerals react with air and water to form new minerals. CO_2 in the atmosphere and in water can react with minerals, such as feldspar, to form clay. Some minerals contain iron, which reacts with oxygen in the atmosphere to form iron oxide, or rust. The red color of many rocks is caused by iron oxide.

Chemical weathering happens faster where water is abundant. It also happens faster in warm climates because chemical reactions happen faster at higher temperatures.

Chemical weathering and physical weathering affect each other. For example, physical weathering causes rock to break into smaller pieces. When rocks break, chemical weathering can occur on the newly exposed surfaces. Chemical weathering can weaken rocks. Weaker rocks break more easily.

FOLDABLES®

Make a vertical three-column chart book to organize your notes on weathering and erosion.

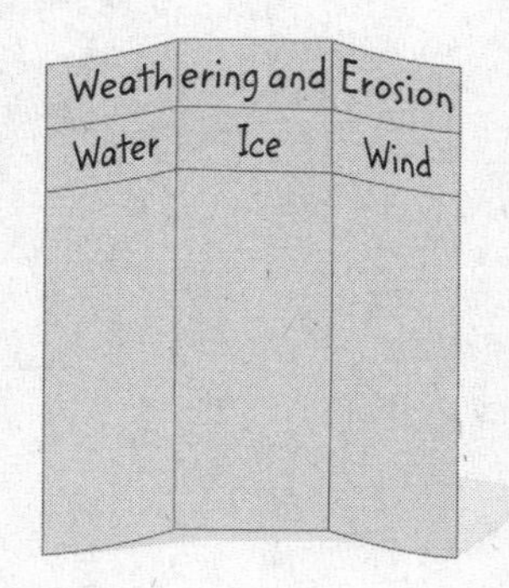

Reading Check

2. Name What processes can break down rock into smaller pieces?

Key Concept Check

3. Contrast What is the difference between physical and chemical weathering?

Moving Earth Materials

Mountains wear away for many reasons. Weathering produces smaller rocks, which move more easily than larger rocks. Chemical weathering dissolves minerals that make up these smaller rocks. But this slow weathering is not the only way mountains wear down. Processes can also remove rocks from the tops of mountains. Geologists use the term **erosion** to describe *the moving of weathered material, or sediment, from one location to another.* **Deposition** *is the laying down or settling of eroded material.* Together, erosion and deposition change the surface of Earth.

Gravity's Influence

Gravity causes material to move downhill. **Mass wasting** *is the downhill movement of a large mass of rocks or soil due to gravity.* If mountains are tall enough or slopes are steep enough, the force of gravity can create landslides, a type of mass wasting. In just a few moments, large amounts of rock and soil can come crashing downhill. Some landslides start from the tops of mountains and end at valley floors.

Erosion requires energy. During a landslide, gravity provides this energy. The energy of flowing water, wind, and moving ice also can move rocks and soil.

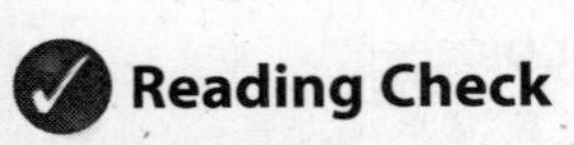

4. Identify What provides the forces that can cause rock to move downhill?

____

____

Water

Most erosion and transport of sediment occurs because of flowing water. Water flows fastest where the land is steep. Water also flows faster in larger rivers than in smaller ones. Large rivers cause the most erosion. The faster water flows, the larger the pieces of sediment it can carry. Rivers can even wear away solid rock and create land features such as the Grand Canyon and Niagara Falls.

5. Name What causes most erosion on Earth?

____

Water often slows as it flows downstream. Slowly flowing water has less energy and can carry less sediment. As water slows, the sediment in the water is deposited on the sides of the river. Sediment also is deposited when rivers enter oceans or lakes, creating land features called deltas.

REVIEW VOCABULARY

delta
triangular deposit of sediment that forms where a stream enters a large body of water

Wind

Sometimes wind is strong enough to cause erosion. In deserts, erosion by wind can be the most important process that changes landforms. Blowing wind can carry sand. Sand dunes and ripples are landforms formed by wind. Wind also can slowly weather and erode solid rocks, carving them into unusual shapes.

Ice

In cold climates, such as high mountains or near the North Pole and the South Pole, *large masses of ice, formed by snow accumulation on land, that move slowly across Earth's surface are called* **glaciers.** Gravity causes this ice to flow downhill. Sliding and flowing ice can weather the rocks over which the ice moves. This process creates sediment that glaciers carry away. Over time, glaciers can carve deep valleys.

When a glacier melts, it deposits the sediment it carried. Ice covered much of North America 20,000 years ago, as shown in the figure below. When this ice sheet melted, it left behind sediment—some as large as boulders. Rocks and smaller sediment were carried from northern Canada and deposited in the United States.

Glaciers Change Earth's Surface

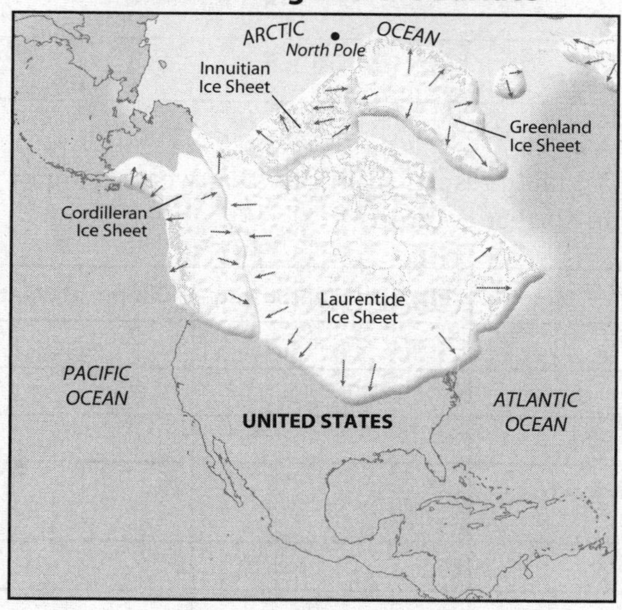

Earth's Changing Surface

Because of the theory of plate tectonics, there is an explainable connection between many of Earth's surface features and the processes that occur on Earth. The processes that move Earth material depend on climate, or the average weather in a region over a long period of time. Temperature, amount of precipitation, the pattern of winds, and circulation of the ocean affect climate. The location of continents affects ocean circulation. The locations of mountains affect wind patterns and precipitation. The processes that change the features made by plate movement are affected by plate movement itself.

Key Concept Check

6. Describe How do water, ice, and wind change Earth's surface?

Visual Check

7. Identify Name the ice sheet that once covered much of North America.

ACADEMIC VOCABULARY

processes
(noun) a series of actions or operations that lead to an end result

After You Read

Mini Glossary

chemical weathering: the process that changes the composition of rocks

deposition: the laying down or settling of eroded material

erosion: the moving of weathered material, or sediment, from one location to another

glacier: a large mass of ice, formed by snow accumulation on land, that moves slowly across Earth's surface

mass wasting: the downhill movement of a large mass of rocks or soil due to gravity

physical weathering: the process of breaking down rock without changing the composition of the rock

sediment: the material formed from rocks broken down by weathering

weathering: the mechanical and chemical processes that change Earth's surface over time

1. Review the terms and their definitions in the Mini Glossary. Write a sentence that gives an example of erosion and deposition.

____

____

2. Determine whether each example in the table results from physical weathering or chemical weathering. Write an *X* in the correct column.

Example	Physical Weathering	Chemical Weathering
Calcite in limestone dissolves in rainwater.		
A boulder rolls off a cliff and breaks apart.		
Plant roots in the cracks of a rock pry the rock open.		
A rock takes on a red color when iron in the rock reacts with oxygen and forms iron oxide.		
Clay forms from carbon dioxide and feldspar.		
A fault ruptures, causing rock to break.		
Water in a rock freezes and shatters the rock.		

3. Give an example of how physical weathering and chemical weathering work together.

____

____

What do you think NOW?

Reread the statements at the beginning of the lesson. Fill in the After column with an A if you agree with the statement or a D if you disagree. Did you change your mind?

Log on to ConnectED.mcgraw-hill.com and access your textbook to find this lesson's resources.

Natural Resources

Energy Resources

Before You Read

What do you think? Read the two statements below and decide whether you agree or disagree with them. Place an A in the Before column if you agree with the statement or a D if you disagree. After you've read this lesson, reread the statements to see if you have changed your mind.

Before	Statement	After
	1. Nonrenewable energy resources include fossil fuels and uranium.	
	2. Energy use in the United States is lower than in other countries.	

Key Concepts 

- What are the main sources of nonrenewable energy?
- What are the advantages and disadvantages of using nonrenewable energy resources?
- How can individuals help manage nonrenewable resources wisely?

Read to Learn

Sources of Energy

Think about all the times you use energy in one day. You use it for electricity, transportation, and other needs. That is one reason it is important to know where energy comes from and how much is available for humans to use.

Energy comes from nonrenewable and renewable sources. Most energy in the United States comes from nonrenewable resources. **Nonrenewable resources** *are resources that are used faster than they can be replaced by natural processes.* Fossil fuels, such as coal and oil, and uranium, which is used in nuclear reactions, are nonrenewable energy resources.

Renewable resources *are resources that can be replaced by natural processes in a relatively short amount of time.* The Sun's energy, also called solar energy, is a renewable energy resource. Others are wind, water, geothermal, and biomass. You will read more about renewable energy resources in Lesson 2.

Mark the Text

Sticky Notes As you read, use sticky notes to mark information that you do not understand. Read the text carefully a second time. If you still need help, write a list of questions to ask your teacher.

Nonrenewable Energy Resources

You might turn on a lamp so you can read, turn on a heater to stay warm, or ride the bus to school. In the United States, the energy you use to power lamps, heat houses, and run vehicles probably comes from nonrenewable energy resources, such as fossil fuels.

 **Key Concept Check**

1. Name What are the main nonrenewable energy resources?

____

____

Fossil Fuels

Coal, oil (also called petroleum), and natural gas are fossil fuels. They are nonrenewable because they form over millions of years.

Reading Check

2. Identify Where do the fossil fuels come from?

The fossil fuels used today formed from the remains of prehistoric organisms. The decayed remains of these organisms were buried by layers of sediment and changed chemically by extreme temperatures and pressure.

The type of fossil fuel that formed depended on three factors:

- the type of organic matter
- the temperature and pressure
- the length of time the organic matter was buried 

Reading Check

3. State the factors that determine which type of fossil fuel forms.

Coal Earth was very different 300 million years ago, when the coal used today began forming. Plants, such as ferns and trees, grew in prehistoric swamps. As shown in the figure below, the first step of coal formation occurred when those plants died.

Bacteria, extreme temperatures, and pressure acted on the plant remains over time. Eventually, a brownish material, called peat, formed.

Peat can be used as a fuel. However, peat contains moisture and produces a lot of smoke when it burns. As shown in the figure below, peat eventually can change into harder and harder types of coal. The hardest coal, anthracite, contains the most carbon per unit of volume and burns most efficiently.

Visual Check

4. Locate Circle the layer that is the hardest type of coal called anthracite.

Coal Formation

When plants in prehistoric swamps died, their remains built up. Over time, sediment covered the plant remains. Inland seas formed where the swamps once were.

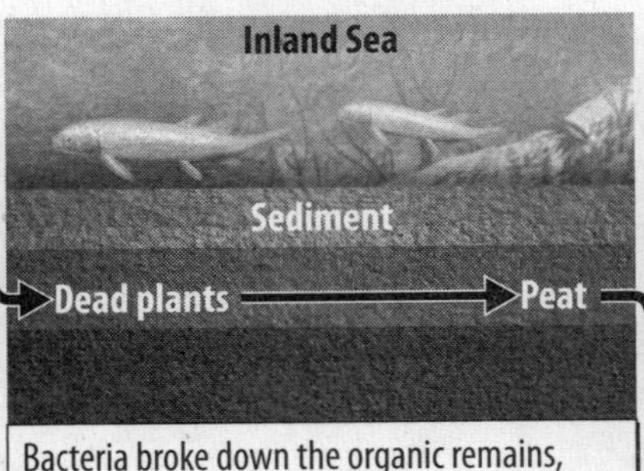

Bacteria broke down the organic remains, leaving behind mostly carbon. Extreme temperatures and pressure compressed the material and squeezed out gas and moisture. A brownish material, called peat, formed.

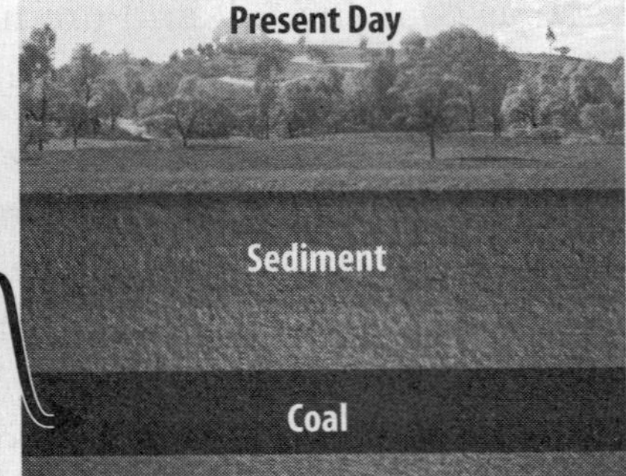

As additional layers of sediment covered and compacted the peat,over time it changed into successively harder types of coal.

Oil and Natural Gas Like coal, the oil and natural gas used today formed millions of years ago. The process that formed oil and natural gas is similar to the process that formed coal. However, oil and natural gas formation involves different types of organisms.

Scientists theorize that oil and natural gas formed from the remains of microscopic marine organisms called plankton. The plankton died and fell to the ocean floor. There, layers of sediment buried their remains. Bacteria decomposed the organic matter, and then pressure and extreme temperatures acted on the sediments. During this process, thick, liquid oil formed first. If the temperature and pressure were great enough, natural gas formed.

Most oil and natural gas formed where forces within Earth folded and tilted thick rock layers. Hundreds of meters of sediments and rock layers often covered oil and natural gas. However, oil and natural gas were less dense than these layers. As a result, oil and natural gas began to rise to the surface by passing through the pores, or small holes, in rocks.

As shown in the figure below, oil and natural gas eventually reached layers of rock through which they could not pass, or impermeable rock layers. Deposits of oil and natural gas formed under these impermeable rocks. The less-dense natural gas settled on top of the denser oil. ✓

Oil and Natural Gas

FOLDABLES®

Make a three-tab book to compare and contrast the use of fossil fuels and nuclear energy.

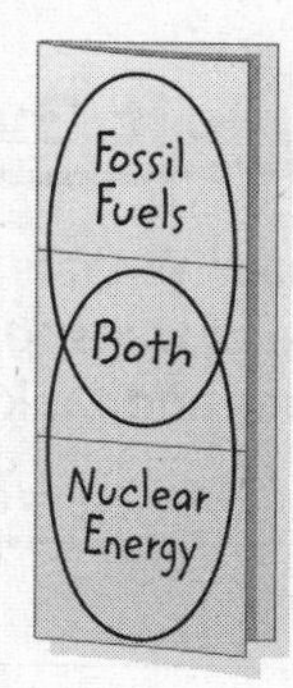

✓ Reading Check

5. Explain How is coal formation different from oil formation?

____

____

____

____

✓ Visual Check

6. Summarize What prevents oil and natural gas from rising to the surface?

____

____

____

Advantages of Fossil Fuels

Fossil fuels store chemical energy. Burning fossil fuels transforms this energy. The steps involved in changing chemical energy in fossil fuels into electric energy are fairly easy and direct. This process is one advantage of using these nonrenewable resources.

Fossil fuels are also relatively inexpensive and easy to transport. Coal is often transported by trains. Oil is transported by pipelines or large ships called tankers.

Disadvantages of Fossil Fuels

Although fossil fuels provide energy, using them has some disadvantages.

Limited Supply One disadvantage of using fossil fuels is that they are nonrenewable. No one knows for sure when supplies will be gone. Scientists estimate that at current rates of consumption, known reserves of oil will last only another 50 years.

Habitat Disruption In addition to being nonrenewable, the process of obtaining fossil fuels disturbs environments. Coal comes from underground mines or strip mines. Oil and natural gas come from wells drilled into Earth.

Mines in particular disturb habitats. Strip-mining involves removing layers of soil and rock to reach the coal. Forests can become fragmented, or broken into areas of trees that are no longer connected. Fragmentation can negatively affect birds and other organisms that live in forests.

Pollution Another disadvantage of using fossil fuels as an energy resource is pollution. For example, runoff from coal mines can pollute soil and water. Oil spills from tankers can harm organisms such as birds.

Pollution also occurs when fossil fuels are used. Burning fossil fuels releases chemicals into the atmosphere. These chemicals react in the presence of sunlight and produce a brownish haze.

This haze can cause respiratory problems, particularly in young children. The chemicals also can react with water in the atmosphere and make rain and snow more acidic. The acidic precipitation can change the chemistry of soil and water and harm living organisms.

Think it Over

7. Discuss What is the advantage of transporting oil by pipelines and tankers?

Reading Check

8. State How much longer are known oil reserves predicted to last?

Key Concept Check

9. Describe What is one advantage and one disadvantage of using fossil fuels?

Nuclear Energy

Atoms are too small to be seen with the unaided eye. Even though they are small, atoms can release large amounts of energy.

Energy released from atomic reactions is called **nuclear energy.** Stars release nuclear energy by fusing atoms. The type of nuclear energy used on Earth involves a different process.

Nuclear Fission Nuclear power plants, such as the one shown in the figure below, produce electricity using nuclear fission. This process splits atoms.

The nuclear fission process begins when uranium atoms are placed into fuel rods. Neutrons are aimed at the rods and hit the uranium atoms.

When hit, each atom splits and releases two to three neutrons and thermal energy. The released neutrons hit other atoms, causing a chain reaction of splitting atoms.

Countless atoms split and release large amounts of thermal energy. This energy heats water and changes it to steam.

The steam turns a turbine connected to a generator. As the generator spins, it produces electricity.

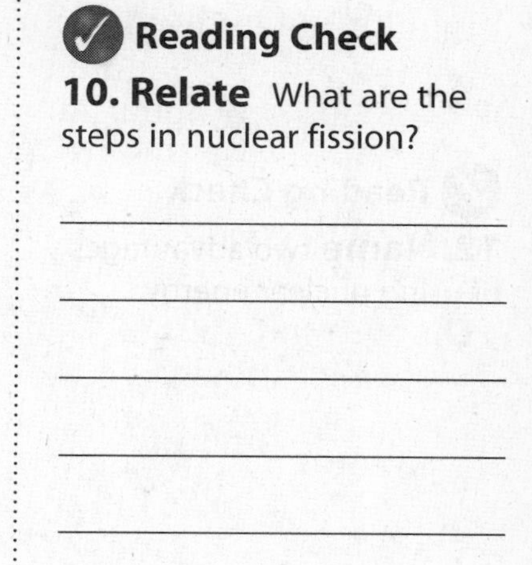

Reading Check

10. Relate What are the steps in nuclear fission?

Visual Check

11. Identify What does the nuclear power plant produce in the second step of the process?

Nuclear Energy

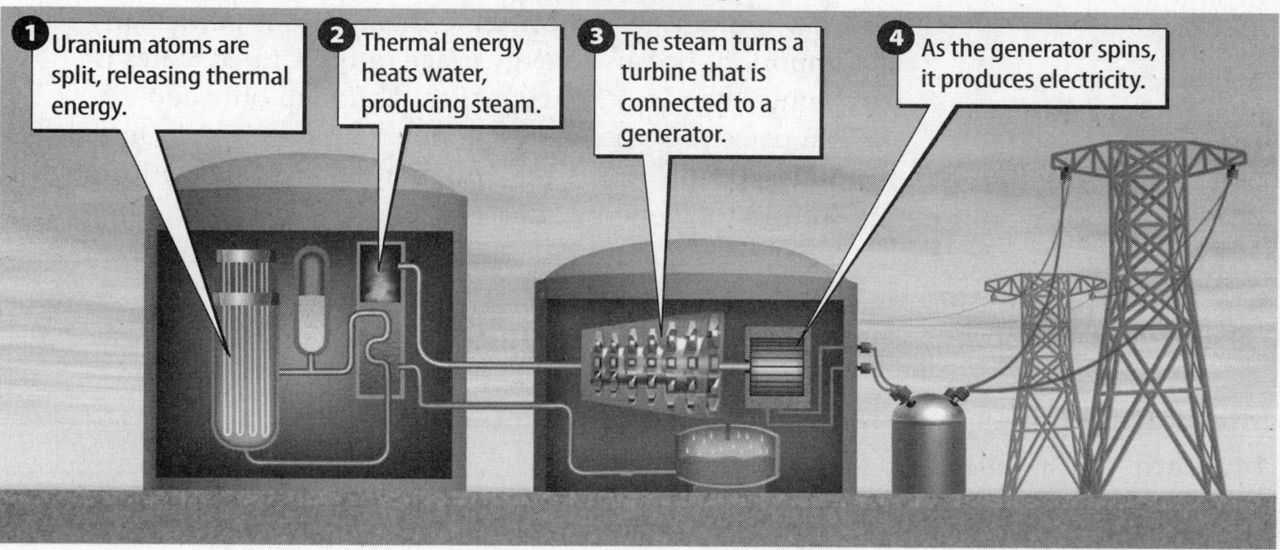

Advantages and Disadvantages of Nuclear Energy

One advantage of using nuclear energy is that a relatively small amount of uranium produces a large amount of energy. In addition, a well-run nuclear power plant does not pollute the air, the soil, or the water.

However, using nuclear energy has disadvantages. Nuclear power plants use a nonrenewable resource—uranium—for fuel. In addition, the chain reaction in the nuclear reactor must be monitored carefully. If not controlled, a nuclear reactor can release harmful radioactive substances into the environment.

The waste materials from nuclear power plants are highly radioactive and dangerous to living organisms. The waste materials remain dangerous for thousands of years. Storing them safely is important for the environment and for public health.

Reading Check

12. Name two advantages of using nuclear energy.

Reading Check

13. Explain Why is it important to control a chain reaction?

Managing Nonrenewable Energy Sources

As shown in the table below, fossil fuels and nuclear energy provide about 93 percent of U.S. energy. Because these sources eventually will be gone, we must understand how to manage and conserve them. This is particularly important because energy usage in the United States is higher than in other countries. Although only about 4.5 percent of the world's population lives in the United States, people in the U.S. use more than 22 percent of the world's total energy.

Interpreting Tables

14. State Which energy source is used most in the United States?

Sources of Energy Used in the United States in 2007

Source	Percentage
Oil	40
Natural gas	23
Coal	22
Nuclear power	8
Biomass	3.5
Hydroelectric	2.5
Solar, geothermal, and wind	1

Management Solutions

Mined land must be reclaimed. **Reclamation** *is a process in which mined land must be recovered with soil and replanted with vegetation.*

Laws also help ensure that mining and drilling take place in an environmentally safe manner. In the United States, the Clean Air Act limits the amount of pollutants that can be released into the air. In addition, the U.S. Atomic Energy Act and the Energy Policy Act include regulations that protect people from nuclear emissions.

ACADEMIC VOCABULARY
regulation
(noun) a rule dealing with procedures, such as safety

What You Can Do

Have you ever heard of vampire energy? Vampire energy is the energy used by appliances and other electronic equipment, such as microwave ovens, washing machines, televisions, and computers. These items use electricity 24 hours a day.

Reading Check

15. Define What is vampire energy?

Even when turned off, these appliances still consume energy. They consume about 5 percent of the energy used each year. You can conserve energy by unplugging DVD players, printers, and other appliances when they are not in use.

You also can walk or ride your bike to help conserve energy. And, you can use renewable energy resources, which you will read about in the next lesson.

Key Concept Check

16. Describe How can you help manage nonrenewable resources wisely?

After You Read

Mini Glossary

nonrenewable resource: a resource that is used faster than it can be replaced by natural processes

nuclear energy: energy released from atomic reactions

reclamation: a process in which mined land must be recovered with soil and replanted with vegetation

renewable resource: a resource that can be replaced by natural processes in a relatively short amount of time

1. Review the terms and their definitions in the Mini Glossary. Write a sentence that describes the importance of reclamation.

2. Fill in the diagram below to show the disadvantages of using fossil fuels.

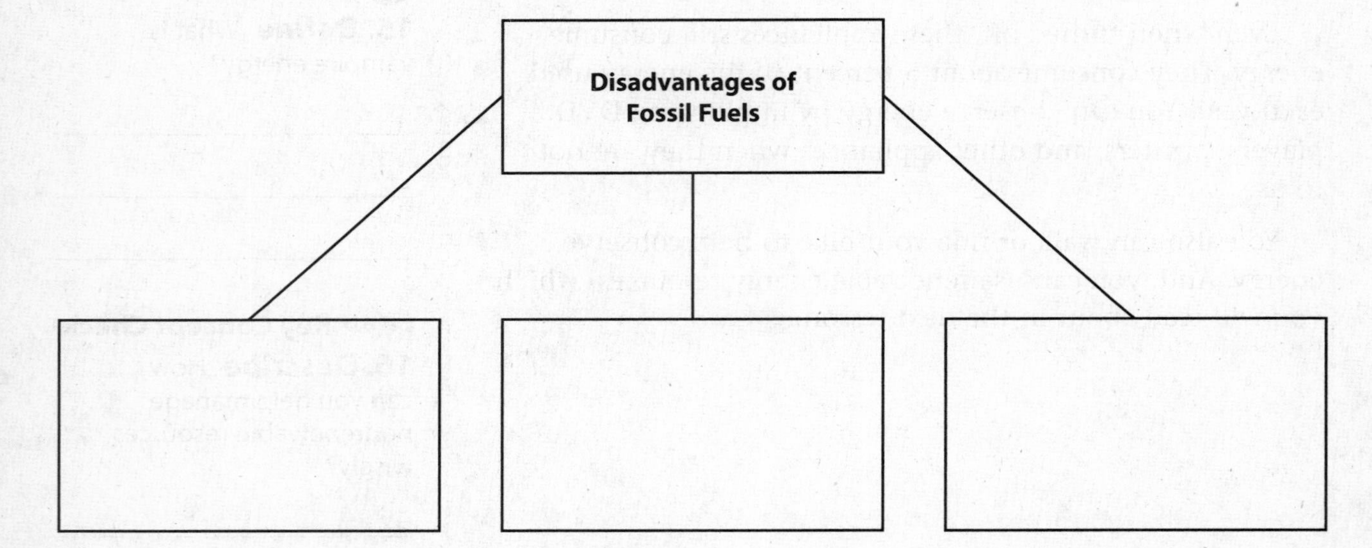

3. Distinguish between nonrenewable and renewable resources.

What do you think NOW?

Reread the statements at the beginning of the lesson. Fill in the After column with an A if you agree with the statement or a D if you disagree. Did you change your mind?

Log on to ConnectED.mcgraw-hill.com and access your textbook to find this lesson's resources.

Natural Resources

Renewable Energy Resources

Before You Read

What do you think? Read the two statements below and decide whether you agree or disagree with them. Place an A in the Before column if you agree with the statement or a D if you disagree. After you've read this lesson, reread the statements to see if you have changed your mind.

Before	Statement	After
	3. Renewable energy resources do not pollute the environment.	
	4. Burning organic material can produce electricity.	

Key Concepts 

- What are the main sources of renewable energy?
- What are the advantages and disadvantages of using renewable energy resources?
- What can individuals do to encourage the use of renewable energy resources?

Read to Learn

Renewable Energy Resources

The Sun and wind are renewable resources. Renewable resources come from natural processes that have been happening for billions of years and will continue to happen.

Solar Energy

Solar energy *is energy from the Sun.* Solar cells, such as those in watches and calculators, capture light energy and transform it to electric energy. Solar power plants can generate electricity for large areas. They transform energy in sunlight, which then turns turbines connected to generators. Some people use active solar energy in their homes. Solar panels gather and store solar energy to heat water and homes. Passive solar energy uses design elements that capture energy in sunlight. An example of passive solar energy is windows on the south side of a house that can let in sunlight to help heat a room.

Study Coach

Make an Outline Summarize the information in the lesson by making an outline. Use the main headings in the lesson as the main headings in your outline. Use your outline to review the lesson.

Wind Energy

Have you ever dropped papers outside and had them scattered by the wind? If so, you experienced wind energy. This renewable resource has been used since ancient times to sail boats and to turn windmills. Today, wind turbines can produce electricity on a large scale. *A group of wind turbines that produce electricity is called a* **wind farm.** ✓

Reading Check

1. Explain How is wind energy a renewable resource?

____

____

____

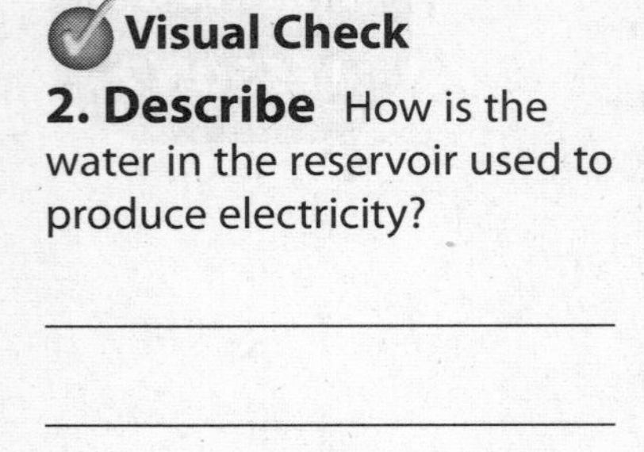

Visual Check

2. Describe How is the water in the reservoir used to produce electricity?

____

____

____

Water Energy

Like wind energy, flowing water has been used as an energy source since ancient times. Today, water energy produces electricity using different methods, such as hydroelectric power and tidal power.

Hydroelectric Power *Electricity produced by flowing water is called* **hydroelectric power.** To produce hydroelectric power, humans build a dam across a powerful river. The figure below shows how flowing water is used to produce electricity.

Hydroelectric Plant

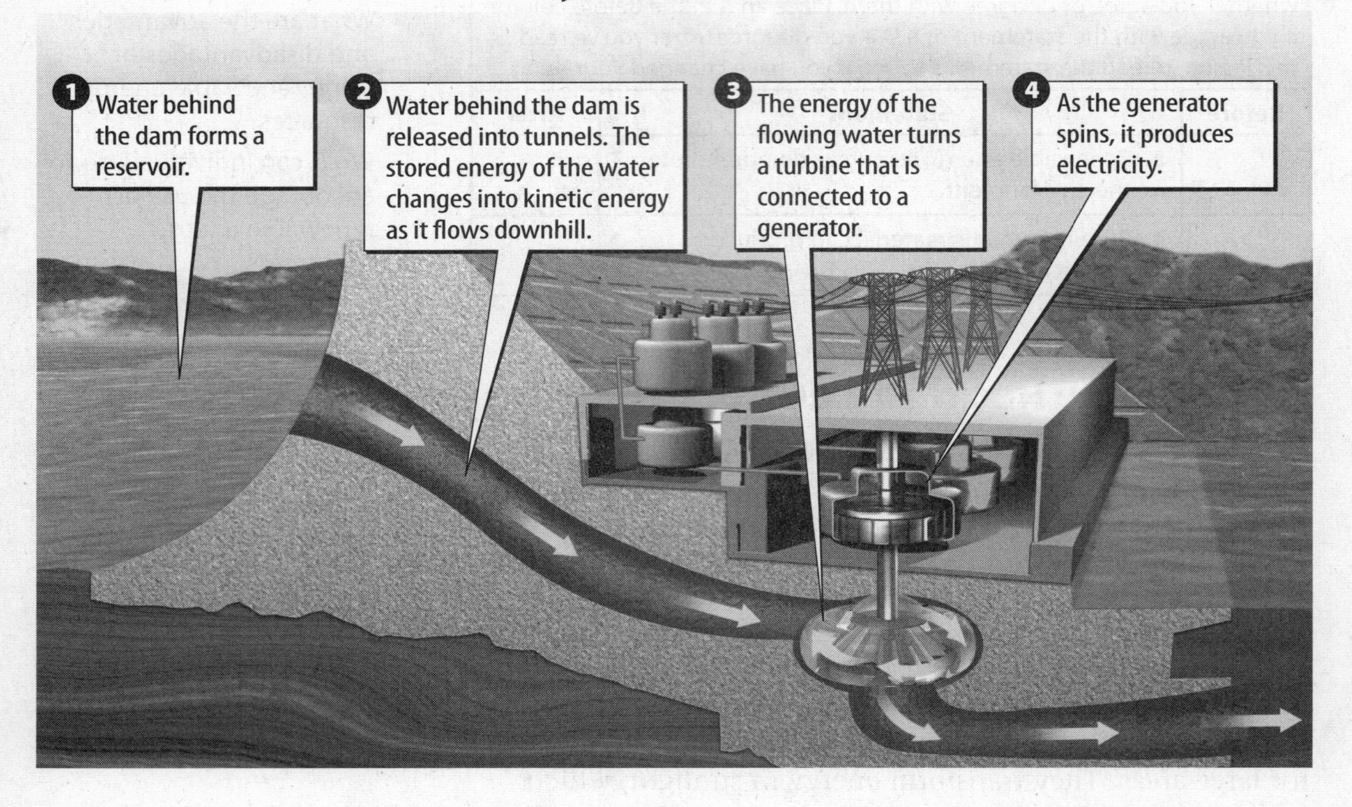

In a hydroelectric power plant, stored energy of the water changes into kinetic energy as it flows downhill. The energy of the flowing water turns a turbine that is connected to a generator. The generator produces electricity as it spins.

Tidal Power Coastal areas that have great differences between high and low tides can be a source of tidal power. Water flows across turbines as the tide comes in during high tides and as it goes out during low tides. The flowing water turns turbines connected to generators that produce electricity.

Reading Check

3. State Which areas can produce tidal power?

____

____

____

Geothermal Energy

Earth's core is nearly as hot as the Sun's surface. This thermal energy flows outward to Earth's surface. *Thermal energy from Earth's interior is called* **geothermal energy.**

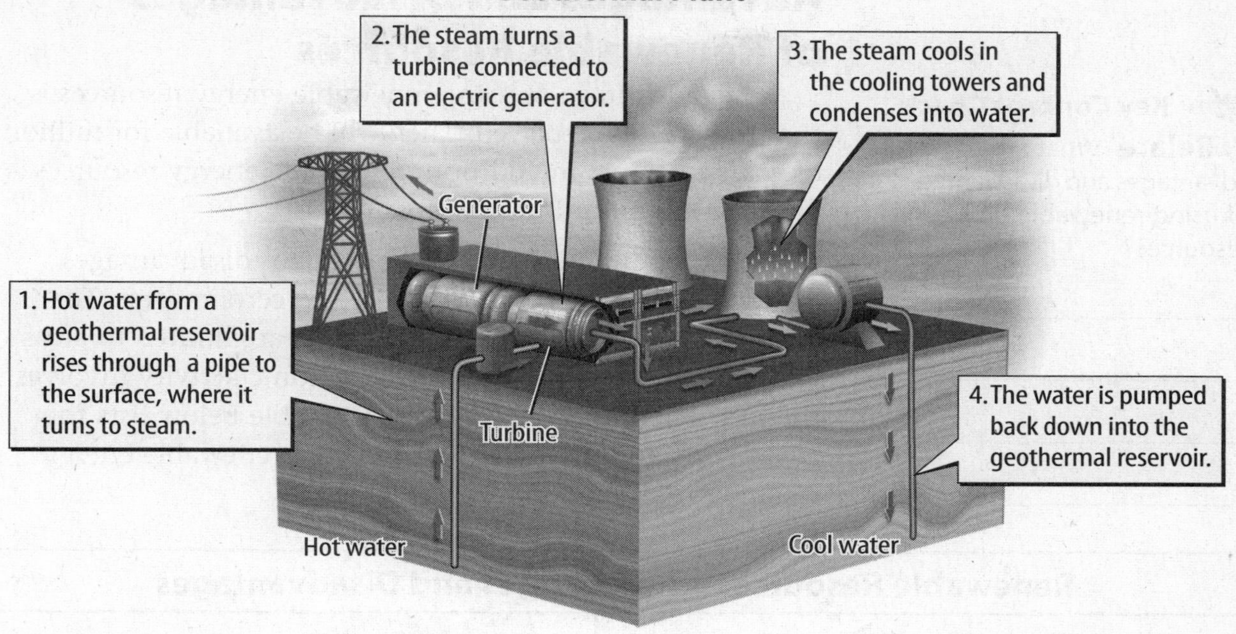

Geothermal energy can be used to heat homes and generate electricity in power plants, such as the one shown in the figure above. People drill wells to reach hot, dry rocks or bodies of magma. The thermal energy from the hot rocks or magma heats water that makes steam. The steam turns turbines connected to generators that produce electricity.

Biomass Energy

Since humans first lit fires for warmth and cooking, biomass has been an energy source. **Biomass energy** *is energy produced by burning organic matter, such as wood, food scraps, and alcohol.*

Wood is the most widely used biomass. Industrial wood scraps and organic materials, such as grass clippings and food scraps, are burned to generate electricity on a large scale.

Biomass also can be converted into fuels for vehicles. Ethanol is made from sugars in plants such as corn. Ethanol often is blended with gasoline. This reduces the amount of oil used to make the gasoline. Adding ethanol to gasoline also reduces the amount of carbon monoxide and other pollutants released by vehicles.

Another renewable fuel, biodiesel, is made from vegetable oils and fats. It emits few pollutants and is the fastest-growing renewable fuel in the United States.

Visual Check

4. Summarize How does steam create electricity?

Reading Check

5. State What is the most widely used biomass?

Key Concept Check

6. Describe What are the main sources of renewable energy?

Advantages and Disadvantages of Renewable Resources

Key Concept Check
7. Relate What are some advantages and disadvantages of using renewable energy resources?

A big advantage of using renewable energy resources is that they can be replaced. They will be available for millions of years to come. In addition, renewable energy resources produce less pollution than fossil fuels.

Using renewable resources has certain disadvantages, however. Some are costly or limited to certain areas. For example, large-scale geothermal plants are limited to areas with tectonic activity. Recall that tectonic activity involves the movement of Earth's plates. The table below lists the advantages and disadvantages of using renewable energy resources.

Renewable Resources—Advantages and Disadvantages

Renewable Resource	Advantages	Disadvantages
Solar energy	• nonpolluting • available in the United States	• less energy produced on cloudy days • no energy produced at night • high cost of solar cells • requires a large surface area to collect and produce energy on a large scale
Wind energy	• nonpolluting • relatively inexpensive • available in the United States	• large-scale use limited to areas with strong, steady winds • best sites for wind farms are far from urban areas and transmission lines • potential impact on bird populations
Water energy	• nonpolluting • available in the United States	• large-scale use limited to areas with fast-flowing rivers or great tidal differences • negative impact on aquatic ecosystems • production of electricity affected by long periods of little or no rainfall
Geothermal energy	• produces little pollution • available in the United States	• large-scale use limited to tectonically active areas • habitat destruction from drilling to build a power plant
Biomass energy	• reduces amount of organic material discarded in landfills • available in the United States	• air pollution results from burning some forms of biomass • less energy efficient than fossil fuels; costly to transport

Interpreting Tables
8. State the advantages and disadvantages of biomass energy.

Managing Renewable Energy Resources

Renewable energy currently meets only 7 percent of U.S. energy needs. Most renewable energy comes from biomass. Solar energy, wind energy, and geothermal energy meet only a small percentage of U.S. energy needs.

Sources of Renewable Energy Resources Used in the United States	
Energy Resource	**Percent**
Biomass	53%
Hydroelectric	36%
Wind	5%
Geothermal	5%
Solar	1%

The table above shows the percentages of renewable energy resources used in the United States. Some states are passing laws that require the state's power companies to produce a percentage of electricity using renewable resources. Management of renewable resources often focuses on encouraging their use.

Management Solutions

The U.S. government has begun programs to encourage use of renewable resources. In 2009, billions of dollars were granted to the U.S. Department of Energy's Office of Efficiency and Renewable Energy for renewable energy research and programs that reduce the use of fossil fuels.

What You Can Do

You might be too young to own a house or a car, but you can help educate others about renewable energy resources. You can talk with your family about ways to use renewable energy at home. You can participate in a renewable energy fair at school. As a consumer, you also can make a difference by buying products that are made using renewable energy resources.

Interpreting Tables

9. Interpret Wind accounts for what percentage of the renewable energy resources used in the United States?

____

FOLDABLES®

Make a vertical, five-tab book to identify the advantages and disadvantages of alternative fuels.

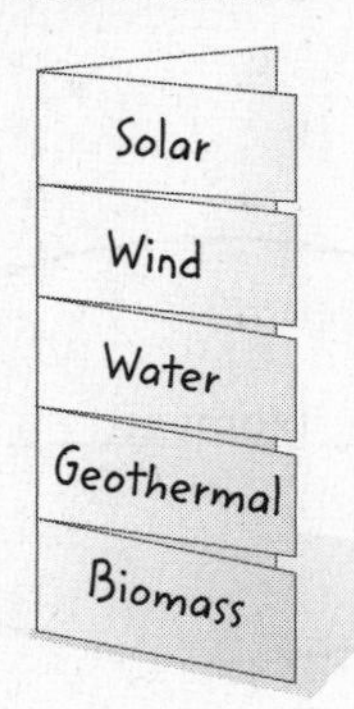

Key Concept Check

10. Describe What can you do to encourage the use of renewable energy resources?

____

____

____

____

After You Read

Mini Glossary

biomass energy: energy produced by burning organic matter, such as wood, food scraps, and alcohol

geothermal energy: thermal energy from Earth's interior

hydroelectric power: electricity produced by flowing water

solar energy: energy from the Sun

wind farm: a group of wind turbines that produce electricity

1. Review the terms and their definitions in the Mini Glossary. Write a sentence discussing a disadvantage of solar energy.

____

____

2. Use what you have learned about renewable energy sources to complete the diagram.

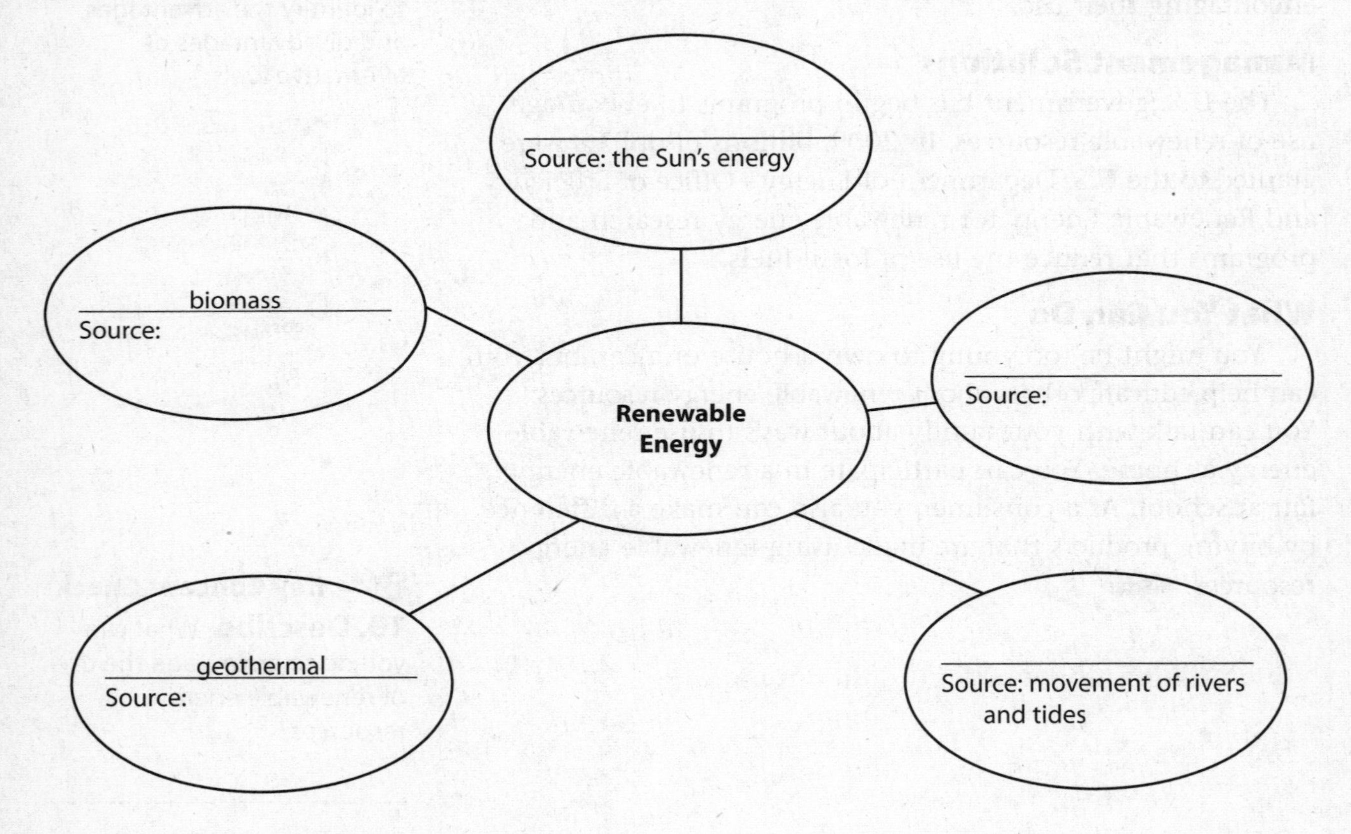

What do you think NOW?

Reread the statements at the beginning of the lesson. Fill in the After column with an A if you agree with the statement or a D if you disagree. Did you change your mind?

Log on to ConnectED.mcgraw-hill.com and access your textbook to find this lesson's resources.

Natural Resources

Land Resources

Before You Read

What do you think? Read the two statements below and decide whether you agree or disagree with them. Place an A in the Before column if you agree with the statement or a D if you disagree. After you've read this lesson, reread the statements to see if you have changed your mind.

Before	Statement	After
	5. Cities cover most of the land in the United States.	
	6. Minerals form over millions of years.	

Key Concepts 

- Why is land considered a resource?
- What are the advantages and disadvantages of using land as a resource?
- How can individuals help manage land resources wisely?

Read to Learn

Land as a Resource

A natural resource is something from Earth that living organisms use to meet their needs. People use soil for growing crops, harvest wood from forests, and mine minerals from the land. In each of these cases, people use land as a natural resource to meet their needs.

Living Space

No matter where you live, you and all living organisms use land for living space. Living space includes natural habitats, as well as the land on which buildings and streets are built. As shown in the table below, cities make up only a small percentage of land use in the United States. Most land is used for agriculture, grasslands, and forests.

Land Use in the United States

Type of Use	Percent
Agriculture	20%
Grassland and pasture	26%
Forest	29%
Miscellaneous	22%
Urban	3%

Mark the Text

Building Vocabulary As you read, underline the words and phrases that you do not understand. When you finish reading, discuss these words and phrases with another student or your teacher.

Key Concept Check

1. Explain Why is land considered a resource?

____

____

____

Interpreting Tables

2. Evaluate What is the largest category of land use in the United States?

____

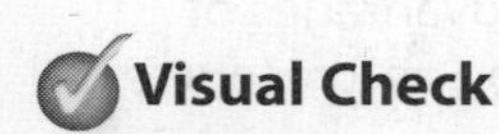

Make a horizontal two-tab book to record your notes about renewable and nonrenewable land resources.

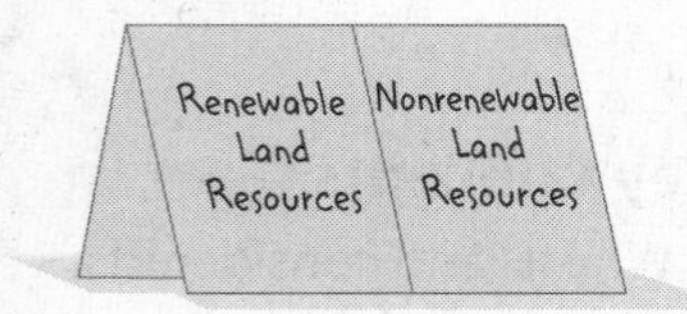

Forests and Agriculture

Forests covered much of the eastern United States in 1650. By 1920, many of the forests had been cut. Forests have regrown, but the new trees are not as tall. Also, the forests are not as complex as the original forests. Trees are cut to get wood for fuel, paper products, and wood products and to clear land for development and agriculture.

Today, about one-fifth of the land in the United States is used for growing crops, and about one-fourth is used for grazing livestock. Though the amount of land used for agriculture has decreased, crop production has increased in some areas because of advances in farming techniques.

Visual Check

3. Identify a mineral resource used in the production of paint.

____

____

Mineral Resources

Coal, an energy resource, is mined from the land. Certain minerals also are mined to make products you use every day. These minerals often are called ores. **Ores** *are deposits of minerals that are large enough to be mined for a profit.* The house in the figure below identifies some common items and the mineral resources they are made from.

Mineral Resources

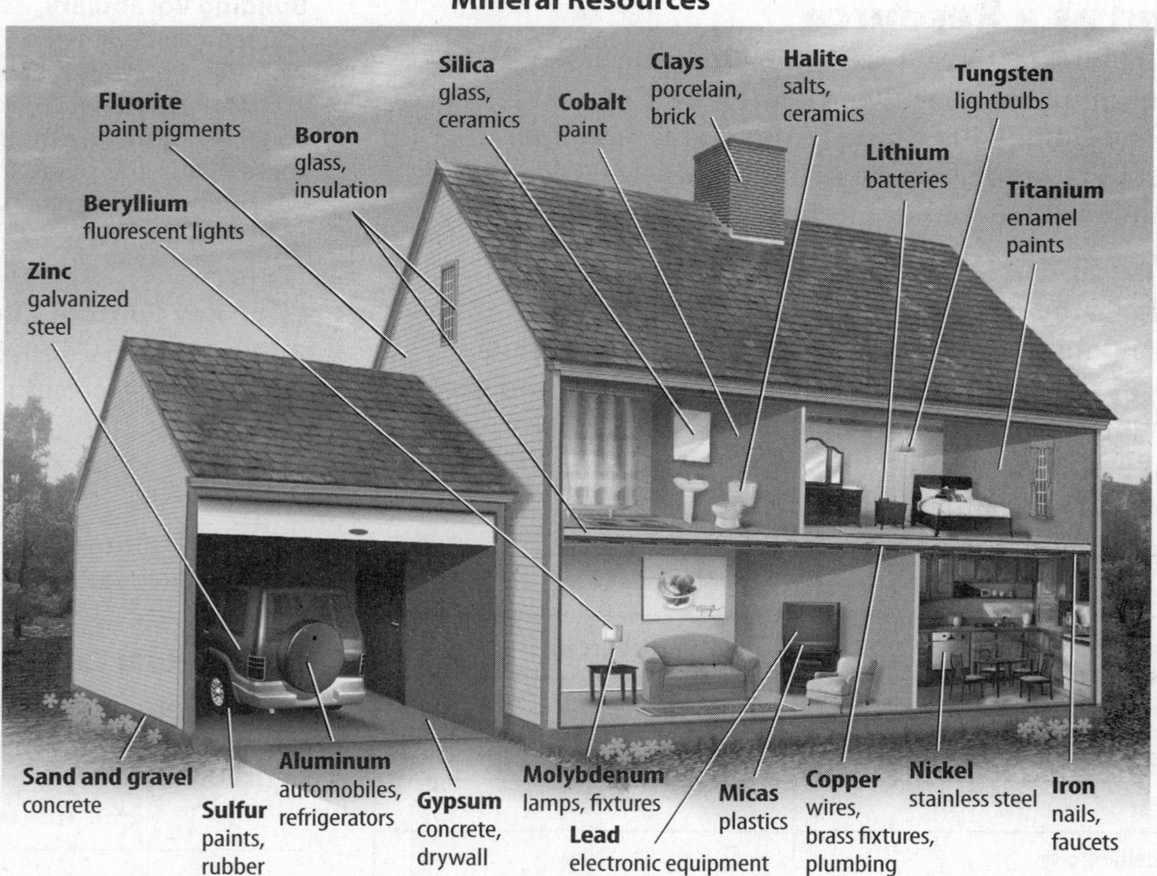

Metallic Mineral Resources Ores such as bauxite and hematite are metallic mineral resources. They are used to make metal products. The aluminum in automobiles and refrigerators comes from bauxite. The iron in nails and faucets comes from hematite.

Nonmetallic Mineral Resources Some mineral resources come from nonmetallic mineral resources, such as sand, gravel, gypsum, and halite. Other nonmetallic mineral resources are mined. The sulfur used in paints and rubber and the fluorite used in paint pigments are also nonmetallic mineral resources. ✓

Advantages and Disadvantages of Using Land Resources

Land resources such as soil and forests are widely available and easy to access. In addition, crops and trees are renewable. This means they can be replanted and grown in a relatively short amount of time. These are advantages of using land resources. Some land resources, however, are nonrenewable. It can take millions of years for minerals to form. This is one disadvantage of using land resources. Other disadvantages include deforestation and pollution.

Deforestation

Humans sometimes cut forests to clear land for grazing, farming, and other uses. **Deforestation** *is the cutting of large areas of forests for human activities*. It leads to soil erosion and loss of animal habitats. In tropical rain forests—complex ecosystems that can take hundreds of years to replace—deforestation is a serious problem.

Deforestation also can affect global climates. Trees remove carbon dioxide from the atmosphere during photosynthesis. When large areas of trees are cut down, less photosynthesis occurs. More carbon dioxide remains in the atmosphere. Carbon dioxide helps trap thermal energy within Earth's atmosphere. Increased concentrations of carbon dioxide can cause Earth's average surface temperatures to increase. ✓

Pollution

Recall that runoff from coal mines can affect soil and water quality. The same is true of mineral mines. Runoff that contains chemicals from these mines can pollute soil and water. In addition, chemical fertilizers are used on farmland to improve crop growth. Runoff containing fertilizers can pollute rivers, soil, and underground water supplies.

Reading Check

4. Identify two products in the house in the figure made from nonmetallic resources.

____

____

Reading Check

5. Name three negative results of deforestation.

____

____

____

Key Concept Check

6. Describe What are some advantages and disadvantages of using land resources?

____

____

____

____

Managing Land Resources

Because some land uses involve renewable resources but others do not, managing land resources is complex. For example, a tree is renewable. But forests can be nonrenewable because some can take hundreds of years to fully regrow. In addition, the amount of land is limited, so there is competition for space. Those who manage land resources must balance all of these issues.

Management Solutions

SCIENCE USE V. COMMON USE
preserve
Science Use to keep safe from injury, harm, or destruction
Common Use to can, pickle, or save something for future use

One way governments can manage forests and other unique ecosystems is by preserving them. On preserved land, logging and development is banned or strictly controlled. Large areas of forests cannot be cut. Instead, loggers cut selected trees and then plant new trees to replace the ones they cut.

Land mined for mineral resources also must be preserved. On public and private lands, mined land must be restored according to government regulations.

Land used for farming and grazing can be managed to conserve soil and improve crop yield. Crop stalks that remain in the field after harvesting help protect the soil from erosion. Farming techniques that use organic materials instead of synthetic fertilizers are also being used.

What You Can Do

You can help conserve land resources by recycling products made from land resources. You can use yard waste and vegetable scraps to make rich compost for gardening, reducing the need to use synthetic fertilizers. Compost is a mix of decayed organic material, bacteria and other organisms, and small amounts of water. Assisting with a community garden is one way you can help manage land resources wisely.

Key Concept Check
7. Relate What can you do to help manage land resources wisely?

After You Read

Mini Glossary

deforestation: the cutting of large areas of forests for human activities

ore: a deposit of minerals that is large enough to be mined for a profit

1. Review the terms and their definitions in the Mini Glossary. Write a sentence describing why ores are important.

2. Fill in the diagram below to show the effects of deforestation.

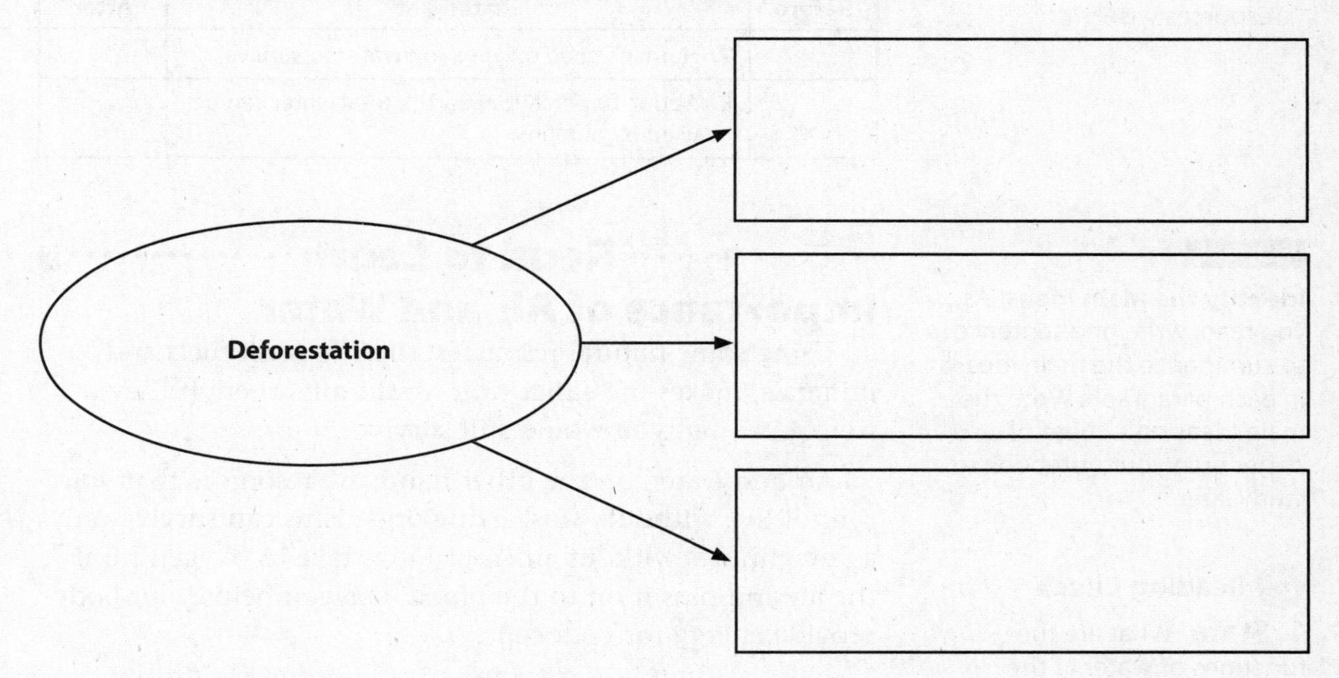

3. In the space below, define a word or phrase that you were better able to understand after discussing it with another person.

What do you think NOW?

Reread the statements at the beginning of the lesson. Fill in the After column with an A if you agree with the statement or a D if you disagree. Did you change your mind?

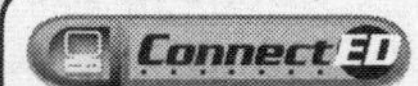

Log on to ConnectED.mcgraw-hill.com and access your textbook to find this lesson's resources.

Natural Resources

Air and Water Resources

Key Concepts
- Why is it important to manage air and water resources wisely?
- How can individuals help manage air and water resources wisely?

Before You Read

What do you think? Read the two statements below and decide whether you agree or disagree with them. Place an A in the Before column if you agree with the statement or a D if you disagree. After you've read this lesson, reread the statements to see if you have changed your mind.

Before	Statement	After
	7. Humans need oxygen and water to survive.	
	8. About 10 percent of Earth's total water can be used by humans.	

Study Coach

Identify the Main Ideas As you read, write one sentence to summarize the main idea in each paragraph. Write the main ideas on a sheet of paper or in your notebook to study later.

Read to Learn

Importance of Air and Water

Using some natural resources, such as fossil fuels and minerals, makes life easier. You would miss them if they were gone, but you would still survive.

Air and water, on the other hand, are resources that you cannot live without. Most living organisms can survive only a few minutes without air. Your lungs take in oxygen from the air and pass it on to the blood. Oxygen helps your body provide energy for your cells.

Water is needed for many life functions. Water is the main component of blood. Water also helps protect body tissues, helps maintain body temperature, and plays a role in many chemical reactions, such as the digestion of food.

In addition to drinking water, people use water for other purposes that you will learn about later in this lesson, including agriculture, transportation, and recreation.

Reading Check

1. State What are the functions of water in the human body?

______

______

______

Air

Most living organisms need air to survive. The polluted air described in the figure at the top of the next page can harm humans and other living organisms.

Air pollution is produced when fossil fuels burn in homes, vehicles, and power plants. It also can be caused by natural events, such as volcanic eruptions or forest fires.

Reading Check

2. Explain What natural events can cause air pollution?

______

______

______

Smog Formation

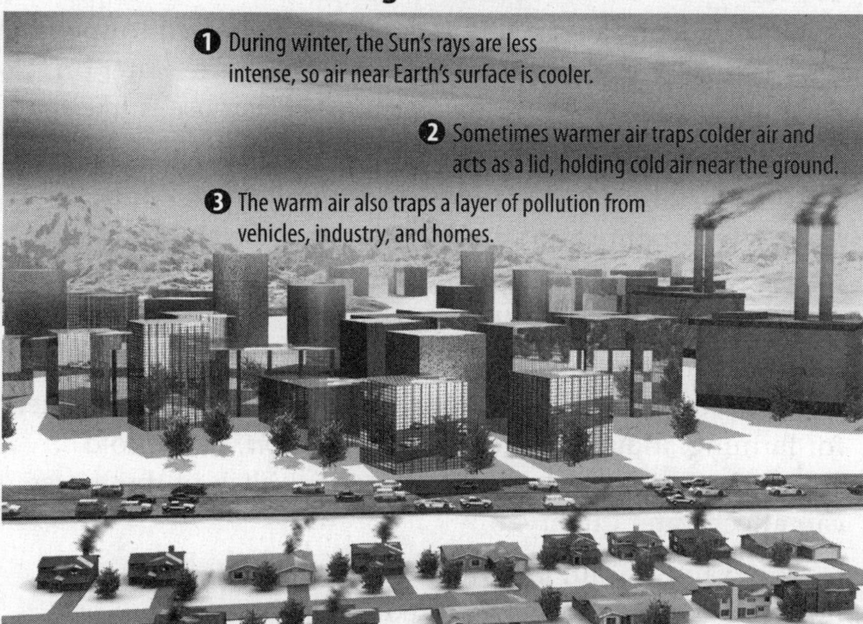

Smog Burning fossil fuels releases not only energy, but also substances such as nitrogen compounds. **Photochemical smog** *is a brownish haze produced when nitrogen compounds and other pollutants in the air react in the presence of sunlight*.

Smog can irritate your respiratory system. In some individuals, it can increase the chance of asthma attacks. Smog can be particularly harmful when it remains in an area for several days. Pollution becomes trapped under a layer of warm air, as shown in the figure above.

Acid Precipitation Nitrogen and sulfur compounds released when fossil fuels burn can react with water in the atmosphere and produce acid precipitation. **Acid precipitation** *is precipitation that has a pH less than 5.6.*

When acid precipitation falls into lakes, it can harm fish and other organisms. It also can pollute soil and kill trees and other plants. Acid precipitation can even damage buildings and statues made of some types of rocks.

Natural Events Forest fires and volcanic eruptions release gases, ash, and dust into the air. Dust and ash from one volcanic eruption can spread around the world. Materials from forest fires and volcanic eruptions can cause health problems similar to those caused by smog.

Visual Check

3. Identify Where does the pollution that forms smog come from?

____

____

____

Math Skills

The carbon monoxide (CO) level in Seattle air went from 7.8 parts per million (ppm) in 1990 to 1.8 ppm in 2007. What was the percent change in CO levels?

a. Subtract the starting value from the final value.

$1.8 \text{ ppm} - 7.8 \text{ ppm} = -6.0 \text{ ppm}$

b. Divide the difference by the starting value.

$-6.0 \text{ ppm}/7.8 \text{ ppm} = -0.769 \text{ ppm}$

c. Multiply by 100 and add a % sign.

$-0.769 \times 100 = -76.9\%$

It decreased by 76.9%.

4. Use Percentages Between 1900 and 2000, the ozone (O_3) levels in New York City went from 0.098 ppm to 0.086 ppm. What was the percent change in ozone levels?

____

Water

Suppose you saved $100, but you were allowed to spend only 90 cents. You might be frustrated! If all the water on Earth were your $100, freshwater that we can use is like that 90 cents you can spend.

FOLDABLES

Make a horizontal two-tab book to discuss the importance of air and water.

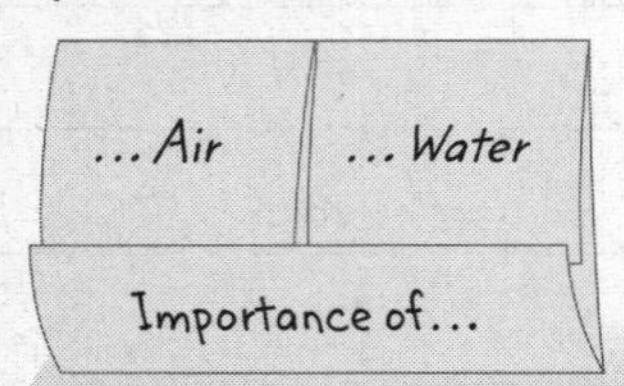

Most water on Earth is salt water. Only 3 percent is freshwater, and most of that is frozen in glaciers. That leaves just a small part, 0.9 percent, of the total amount of water on Earth for humans to use.

This relatively small supply of freshwater must meet many needs. In addition to drinking water, people use water for farming, industry, electricity production, household activities, transportation, and recreation. Each of these uses can affect water quality.

For example, rain and water used to irrigate fields can mix with fertilizers. This polluted water then can run off into rivers and groundwater, reducing the quality of these water supplies.

Let's look at another example. Water used in industry often is heated to high temperatures. This hot water can harm aquatic organisms when it is returned to the environment.

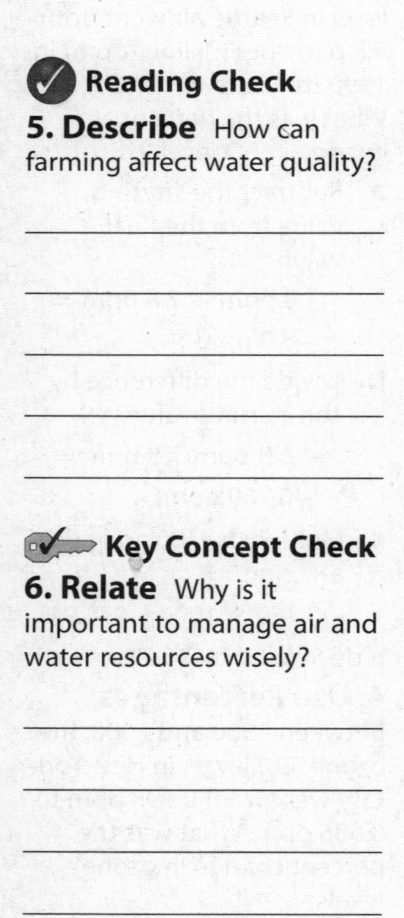

Reading Check

5. Describe How can farming affect water quality?

____

____

____

____

____

Key Concept Check

6. Relate Why is it important to manage air and water resources wisely?

____

____

____

____

Managing Air and Water Resources

Animals and plants do not use natural resources to produce electricity or to raise crops. But they do use air and water. Those who manage these important resources must consider human needs and the needs of other living organisms.

Management Solutions

Legislation is an effective way to reduce air and water pollution. The regulations of the U.S. Clean Air Act, passed in 1970, limit the amount of certain pollutants that can be released into the air. The graph at the top of the next page shows how levels of sulfur compounds have decreased since the act became law.

Similar laws are now in place to maintain water quality. The U.S. Clean Water Act legislates the reduction of water pollution. The Safe Drinking Water Act legislates the protection of drinking water supplies. By reducing pollution, these laws help ensure that all living organisms have access to clean air and water.

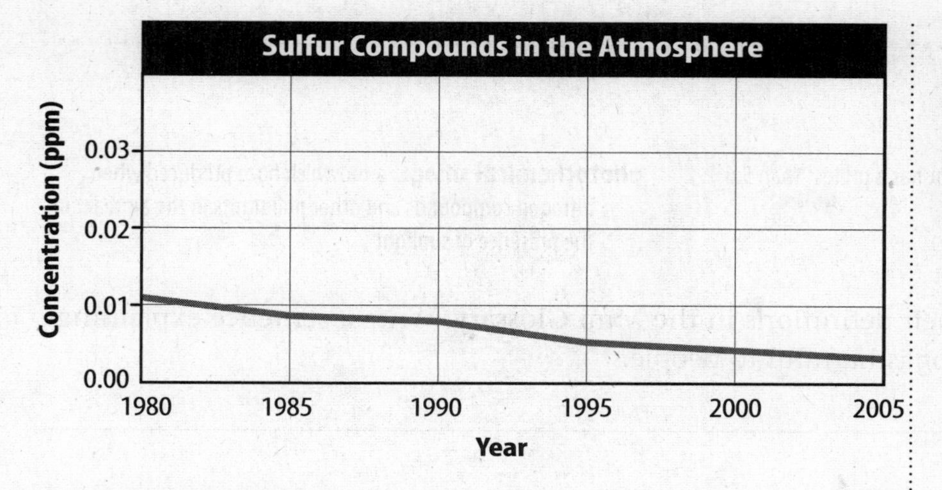

Visual Check

7. Describe the trend of the concentration of sulfur compounds in the atmosphere from 1980 to 2005.

What You Can Do

You have learned that reducing fossil fuel use and improving energy efficiency can reduce air pollution. You can make sure your home is energy efficient by keeping the filters in the air conditioner or the furnace clean and by using energy-saving lightbulbs.

You can help reduce water pollution by properly disposing of harmful chemicals so that less pollution runs off into rivers and streams. You can volunteer to help clean up litter from a local stream. You also can conserve water so enough of this resource remains for you and other living organisms in the future.

Key Concept Check

8. Apply How can individuals help manage air and water resources wisely?

After You Read

Mini Glossary

acid precipitation: precipitation that has a pH less than 5.6

photochemical smog: a brownish haze produced when nitrogen compounds and other pollutants in the air react in the presence of sunlight

1. Review the terms and their definitions in the Mini Glossary. Write a sentence explaining why photochemical smog is harmful to people.

______

______

2. One use of water is for household activities. Complete the diagram by writing the other uses in the empty circles.

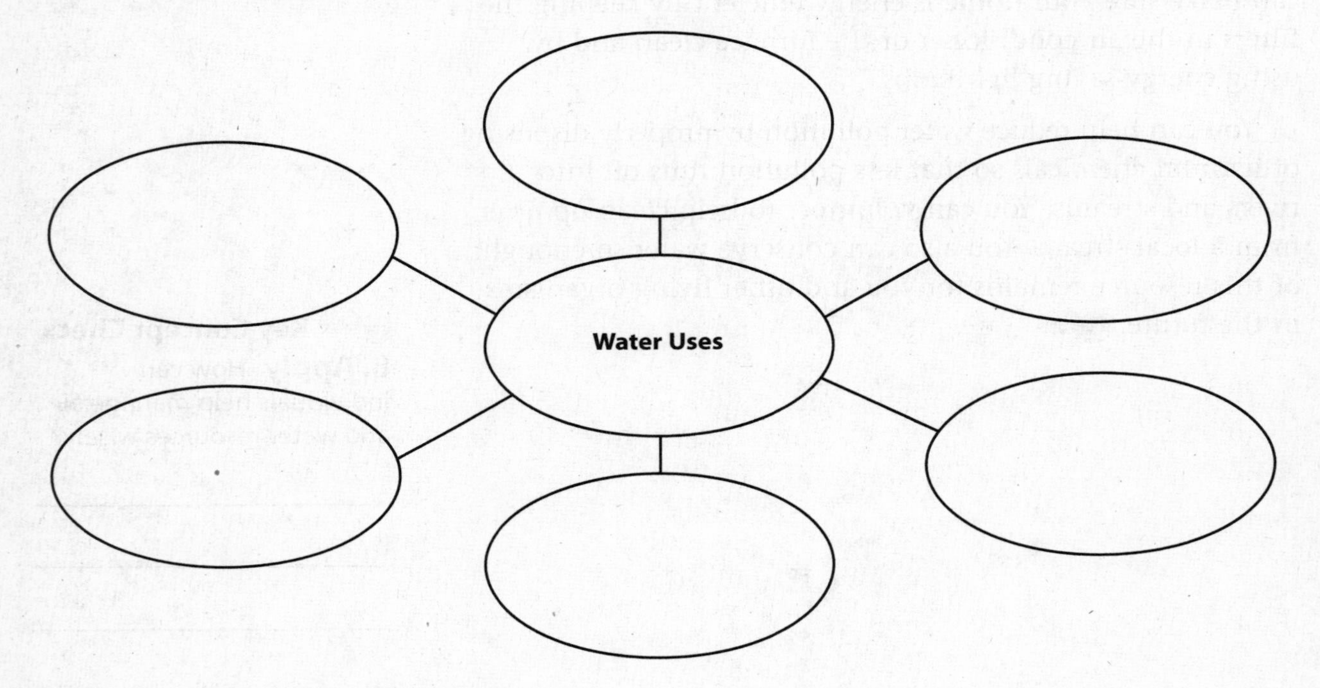

3. Describe the effects of acid precipitation.

______

______

What do you think NOW?

Reread the statements at the beginning of the lesson. Fill in the After column with an A if you agree with the statement or a D if you disagree. Did you change your mind?

Log on to ConnectED.mcgraw-hill.com and access your textbook to find this lesson's resources.

Life's Classification and Structure

Classifying Living Things

Before You Read

What do you think? Read the three statements below and decide whether you agree or disagree with them. Place an A in the Before column if you agree with the statement or a D if you disagree. After you've read this lesson, reread the statements to see if you have changed your mind.

Before	Statement	After
	1. All living things are made of cells.	
	2. A group of organs that work together and perform a function is called a tissue.	
	3. Living things are classified based on similar characteristics.	

Key Concepts 

- What are living things?
- What do living things need?
- How are living things classified?

Read to Learn

What are living things?

It might be easy to tell whether a bird, a tree, or a person is alive. But it is harder to tell whether some organisms are even living things. For example, think about moldy bread. Is the bread a living thing? What about the mold on the bread? All living things have six characteristics in common:

- Living things are made of cells.
- Living things are organized.
- Living things grow and develop.
- Living things respond to their environment.
- Living things reproduce.
- Living things use energy.

In the moldy bread example, the bread is not living, but the molds growing on the bread are living things. Mold is a type of fungus. If you looked at the mold using a microscope, you would see that the mold is made of cells. Mold cells respond to their environment by growing and reproducing. The mold cells obtain energy, which they need to grow, from the bread.

Study Coach

Create a Quiz Create a 5-question quiz about classifying living things. Exchange quizzes with a partner. After taking the quizzes, discuss your answers.

Key Concept Check

1. Identify What are living things?

____

____

____

____

Living things are organized.

Marching bands are made up of rows of people playing different instruments. Some rows are made up of people playing flutes, and other rows are filled with drummers. Although marching bands are organized into different rows, all band members work together to play a song.

Like marching bands, living things are organized. Some living things are more complex than others, but all organisms are made of cells. In all cells, macromolecules are organized into different structures that help cells function.

REVIEW VOCABULARY
macromolecule
substance in a cell that forms by joining many small molecules together

You might recall that cells have four macromolecules—nucleic acids, lipids, proteins, and carbohydrates. Nucleic acids, such as DNA, store information. Lipids are the main component of cell membranes and provide structure. Some proteins are enzymes, and others provide structure. Carbohydrates are used for energy. ✓

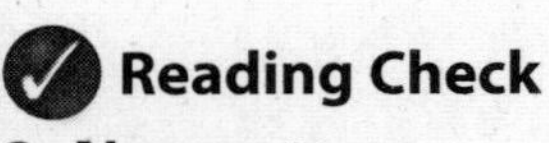

2. Name the four macromolecules in cells.

____

____

____

Unicellular Organisms Some living things are unicellular, which means they are made up of only one cell. In fact, most living things on Earth are unicellular organisms. Unicellular organisms are the oldest forms of life.

There are many groups of unicellular organisms, each with unique characteristics. Bacteria, amoebas (uh MEE buhz), and paramecia (per uh MEE see ah) are unicellular organisms. A unicellular organism has everything needed to obtain and use energy, reproduce, and grow.

ACADEMIC VOCABULARY
unique
(adjective) without an equal, distinctive

Some unicellular organisms are tiny and cannot be seen without a microscope. Other unicellular organisms are large. For example, the plasmodial (plaz MOH dee ul) slime mold is a huge cell formed by many cells that joined together and formed one cell.

Multicellular Organisms Soccer teams are made up of many types of players, including goalkeepers, forwards, and fullbacks. Each team member has a specific job, but all the team members work together when playing a game.

Many living things are made of more than one cell and are called multicellular organisms. Like the different types of players on a soccer team, multicellular organisms have different types of cells that carry out specialized functions. One example is the ladybug. A ladybug has cells that form wings and other cells that form eyes.

Organization in Multicellular Organisms Multicellular organisms have different levels of organization. Groups of cells that work together and perform a specific function are called tissues. Tissues that work together and carry out a specific function are called organs. Organs that work together and perform a specific function are called organ systems. Organ systems work together and perform all the functions an organism needs to survive. ✓

Living things grow, develop, and reproduce.

During their lifetimes, living things grow, or increase in size. For a unicellular organism, the size of its cell increases. For a multicellular organism, the number of its cells increases.

Development Living things also develop, or change, during their lifetimes. For some organisms, it is easy to see the changes that happen as they grow and develop. As shown in the figure below, ladybug larva grow into pupae (PYEW pee; singular, pupa), an intermediate stage, before developing into adults. ✓

Ladybug Larva and Pupa

Pupa

Reading Check

3. Select Tissues are ____ that work together and perform a specific function. (Circle the correct answer.)

a. organs
b. organ symptoms
c. groups of cells

Reading Check

4. Describe How does a multicellular organism grow?

Visual Check

5. Differentiate What differences do you see between the two stages?

Reproduction Once an organism is an adult, it can reproduce asexually or sexually and form new organisms. Unicellular organisms, such as bacteria, reproduce asexually when one cell divides and forms two new organisms. Some multicellular organisms also can reproduce asexually; one parent organism produces offspring when body cells replicate and divide.

Sexual reproduction occurs when the reproductive cells of one or two parent organisms join and form a new organism. Multicellular organisms, such as humans and other mammals, reproduce sexually. Some organisms such as yeast can reproduce both asexually and sexually. ✓

Reading Check
6. Relate What are the two ways reproduction can occur?

______

______

Living things use energy.

All living things need energy to survive. Some organisms are able to convert light energy to chemical energy that is used for many cellular processes. *Organisms that convert light energy to usable energy are called* **autotrophs** (AW tuh trohfs).

Many autotrophs use energy from light and convert carbon dioxide and water into carbohydrates, or sugars. Autotrophs use the carbohydrates for energy. Plants and algae are autotrophs.

Other autotrophs are called chemoautotrophs (kee moh AW tuh trohfs). Chemoautotrophs use energy released by chemical reactions of inorganic substances such as sulfur and ammonia. Many chemoautotrophs are bacteria that live in extreme environments such as deep in the ocean or in hot sulfur springs. ✓

Reading Check
7. Explain How do some autotrophs use energy from light?

______

______

______

Heterotrophs (HE tuh roh trohfs) *are organisms that obtain energy from other organisms*. Heterotrophs eat autotrophs or other heterotrophs to obtain energy. Animals and fungi are examples of heterotrophs.

Living things respond to stimuli.

All living things sense their environments. When an organism detects a change in its external environment, it will respond to that change. A change in an organism's environment is called a stimulus (STIHM yuh lus; plural, stimuli). ✓

Reading Check
8. Define What is a stimulus?

______

______

______

Responding to a stimulus might help an organism protect itself. For example, an octopus responds to a stimulus such as a predator by releasing a black liquid. The black ink hides the octopus while it escapes. In many organisms, nerve cells detect the environment, process the information, and coordinate a response.

What do living things need?

You just read that all living things need energy to survive. Some organisms obtain energy from food. What else do living things need to survive?

Living things also need water and a place to live. Organisms live in environments specific to their needs where they are protected, can obtain food and water, and can be sheltered.

A Place to Live

Living things are everywhere. Organisms live in the soil, in lakes, and in caves. Some living things live on or in other organisms. For example, bacteria live in your intestines and on other body surfaces. *A specific environment where an organism lives is its* **habitat.**

Most organisms can survive in only a few habitats. A land iguana lives in warm, tropical environments and would not survive in cold places such as the Arctic.

Food and Water

Living things also need food and water. Food is used for energy. Water is essential for survival. In Lesson 2, you will read about how water is in all cells and helps them function.

The type of food that an organism eats depends on the habitat in which it lives. Marine iguanas live near the ocean and eat algae. Land iguanas live in hot, dry areas and eat cactus fruits and leaves. The food is processed to obtain energy. Plants and some bacteria use energy from light and produce chemical energy for use in cells.

How are living things classified?

You might have a notebook with different sections. Each section might contain notes from a different class. This organizes information and makes it easy to find notes on different subjects.

Scientists use a classification system to group organisms with similar traits. Classifying living things makes it easier to organize organisms and to recall how they are similar and how they differ.

FOLDABLES®

Make a vertical three-column chart book to organize your notes about living things, their needs, and classification criteria.

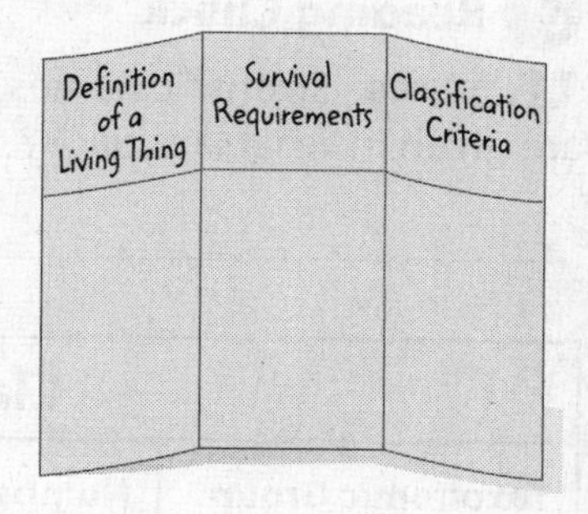

Think it Over

9. Consider Why would a land iguana have difficulty surviving in the Arctic?

Key Concept Check

10. Express What do living things need?

Naming Living Things

Scientists use a system called binomial nomenclature (bi NOH mee ul • NOH mun klay chur) to name living things. **Binomial nomenclature** *is a naming system that gives each living thing a two-word scientific name.*

More than 300 years ago, a scientist named Carolus Linnaeus created the binomial nomenclature system. All scientific names are in Latin. *Homo sapiens* is the scientific name for humans. As the table below shows, the scientific name for an Eastern chipmunk is *Tamias striatus.*

Reading Check

11. State Scientific names are given in what language?

Classification of the Eastern Chipmunk

Taxonomic Group	Number of Species	Examples
Domain Eukarya	about 4–10 million	
Kingdom Animalia	about 2 million	
Phylum Chordata	about 50,000	
Class Mammalia	about 5,000	
Order Rodentia	about 2,300	
Family Sciuridae	299	
Genus *Tamias*	25	
Species *Tamias striatus*	1	

Interpreting Tables

12. Point Out What taxonomic group has about 50,000 species?

Classification Systems

Linnaeus also classified organisms based on their behavior and appearance. Today, the branch of science that classifies living things is called taxonomy.

A group of organisms is called a **taxon** (plural, taxa). There are many taxa, as shown in the table above. Recall that all living things share similar traits. However, not all living things are exactly the same.

Taxonomy

Using taxonomy, scientists divide all living things on Earth into three groups called domains. Domains are divided into kingdoms, and then phyla (FI luh; singular, phylum), classes, orders, families, genera (singular, genus), and species. A species is made of all organisms that can mate with one another and produce offspring that can reproduce. The first word in an organism's scientific name is the organism's genus (JEE nus), and the second word might describe a distinguishing characteristic of the organism. For example, dogs belong to the genus *Canis*. The *Canis* genus also includes wolves, coyotes, and jackals.

Recall that Linnaeus used similar physical traits to group organisms. Today, scientists also look for other similarities, such as how an organism reproduces, how it processes energy, and the types of genes it has.

Dichotomous Keys

A dichotomous (di KAH tah mus) key is a tool used to identify an organism based on its characteristics. Dichotomous keys contain descriptions of traits that are compared when classifying an organism. Dichotomous keys are organized in steps. Each step might ask a yes or a no question and have two answer choices. Which question is answered next depends on the answer to the previous question. Based on the features, a choice is made that best describes the organism.

Math Skills 

A ratio expresses the relationship between two or more things. Ratios can be written

3 to 5, 3:5, or $\frac{3}{5}$.

Reduce ratios to their simplest form. For example, of about 3 million species in the animal kingdom, about 50,000 are mammals. What is the ratio of mammals to animals?

Write the ratio as a fraction.

$$\frac{50,000}{3,000,000}$$

Reduce the fraction to the simplest form.

$$\frac{50,000}{3,000,000} = \frac{5}{300} = \frac{1}{60}$$

(or 1:60 or 1 to 60)

13. Use Ratios Of the 5,000 species of mammals, 250 species are carnivores. What is the ratio of carnivores to mammals? Write the ratio in all three ways.

Key Concept Check

14. Describe How are living things classified?

After You Read

Mini Glossary

autotroph (AW tuh trohf): an organism that converts light energy to usable energy

binomial nomenclature (bi NOH mee ul • NOH mun klay chur): a naming system that gives each living thing a two-word scientific name

habitat: a specific environment where an organism lives

heterotroph (HE tuh roh trohf): an organism that obtains energy from other organisms

taxon: a group of organisms

1. Review the terms and their definitions in the Mini Glossary. Write a sentence that compares autotrophs to heterotrophs.

2. Use the following graphic organizer to summarize the characteristics of living things.

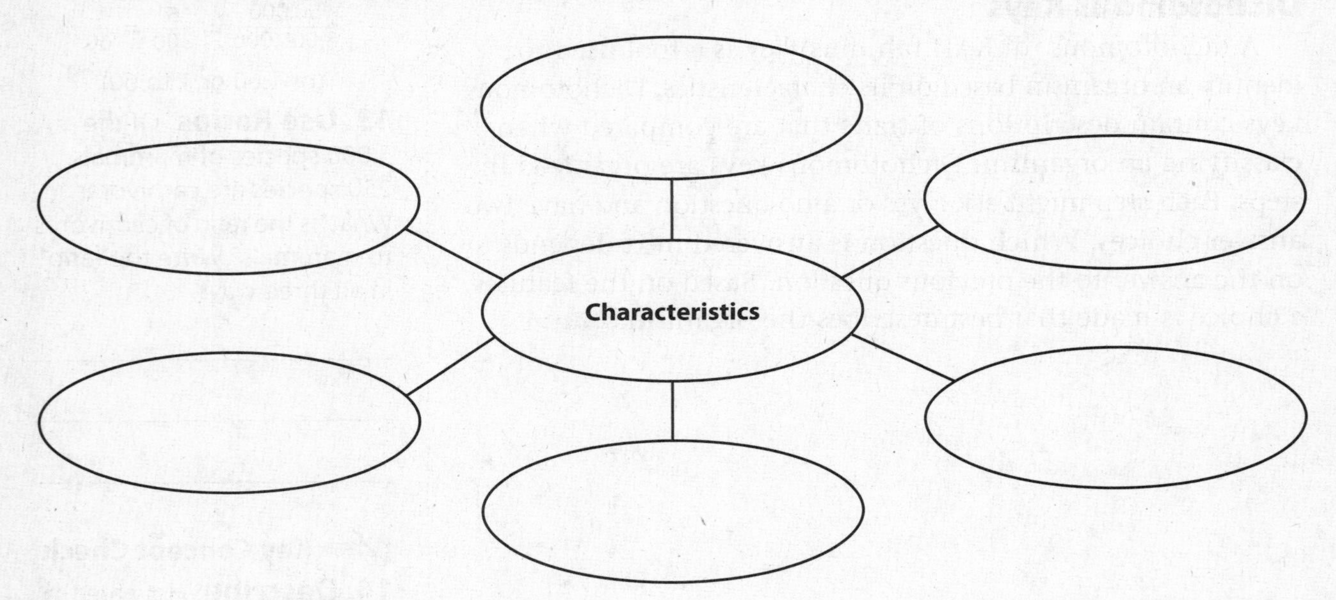

3. How did creating a quiz and answering another quiz help you learn about classifying living things? Explain what you learned.

What do you think NOW?

Reread the statements at the beginning of the lesson. Fill in the After column with an A if you agree with the statement or a D if you disagree. Did you change your mind?

Log on to ConnectED.mcgraw-hill.com and access your textbook to find this lesson's resources.

Life's Classification and Structure

Cells

Before You Read

What do you think? Read the three statements below and decide whether you agree or disagree with them. Place an A in the Before column if you agree with the statement or a D if you disagree. After you've read this lesson, reread the statements to see if you have changed your mind.

Before	Statement	After
	4. *Cell wall* is a term used to describe the cell membrane.	
	5. Prokaryotic cells contain a nucleus.	
	6. Plants use chloroplasts to process energy.	

Key Concepts 
- What is a cell made of?
- How do the parts of a cell enable it to survive?

Read to Learn

What are cells?

What is one thing all living organisms have in common? All organisms have one or more cells. Cells are the basic units of organisms. Most organisms have only one cell. Some organisms have many cells. Humans have about 100 trillion cells! Most cells are so small that they can be seen only with a microscope. Microscopes are used to view details of small objects or things that are too small to be seen by the unaided eye.

Scientists first used microscopes to look at cells more than 300 years ago. Cells can be different shapes and sizes. Nerve cells are long and slender. Many female reproductive cells, or eggs, are large and round.

Mark the Text

Identify the Main Ideas Write a phrase beside each paragraph that summarizes the main point of the paragraph. Use the phrases to review the lesson.

Reading Check

1. Explain Why is a microscope needed to view most cells?

______

______

______

What are cells made of?

Recall that all cells are made of four macromolecules—nucleic acids, lipids, proteins, and carbohydrates. Cells also have many other characteristics. For example, all cells are surrounded by an outer structure called a cell membrane. The cell membrane keeps substances such as macromolecules inside the cell. It also helps protect cells by keeping harmful substances from entering. About 70 percent of the inside of a cell is water. Because many of the substances inside a cell are dissolved in water, they move easily within the cell.

Key Concept Check

2. Relate What is a cell made of?

______

______

______

Types of Cells

There are two main types of cells, as shown in the figure below. Structures in the two types of cells are organized differently. **Prokaryotic** (pro kayr ee AH tihk) **cells** *do not have a nucleus or other membrane-bound organelles.* Organelles are structures in cells that carry out specific functions. The few organelles in prokaryotic cells are not surrounded by membranes. Organisms with prokaryotic cells are called prokaryotes. Most prokaryotes are unicellular organisms, such as bacteria.

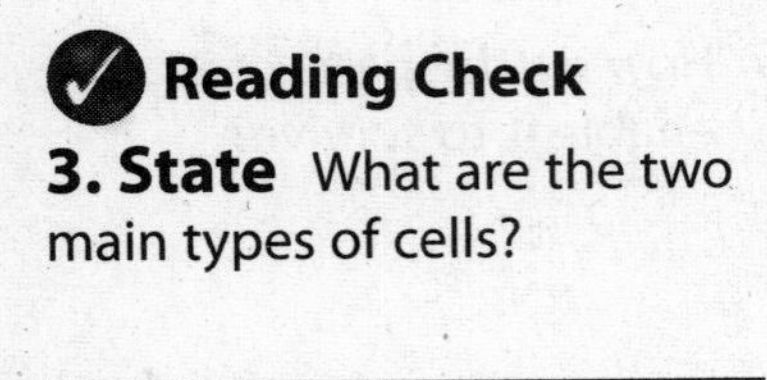

Reading Check

3. State What are the two main types of cells?

Eukaryotic (yew ker ee AH tihk) **cells** *have a nucleus and other membrane-bound organelles.* Most multicellular organisms and some unicellular organisms are eukaryotes. The eukaryotic cell shown in the figure below contains many structures that are not in a prokaryotic cell. In eukaryotes, membranes surround most of the organelles, including the nucleus.

Prokaryotic and Eukaryotic Cells

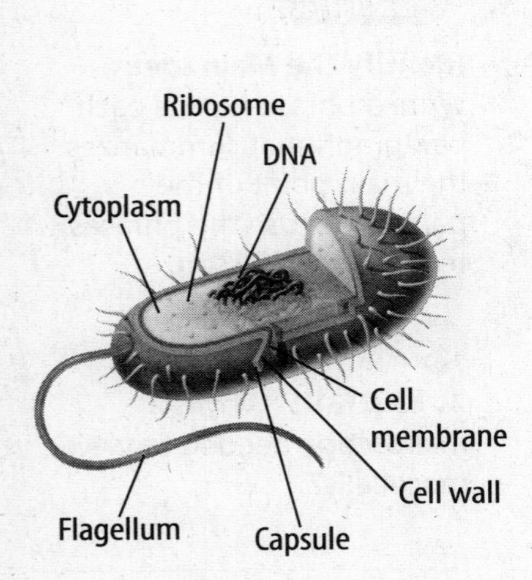

Prokaryotic Cell

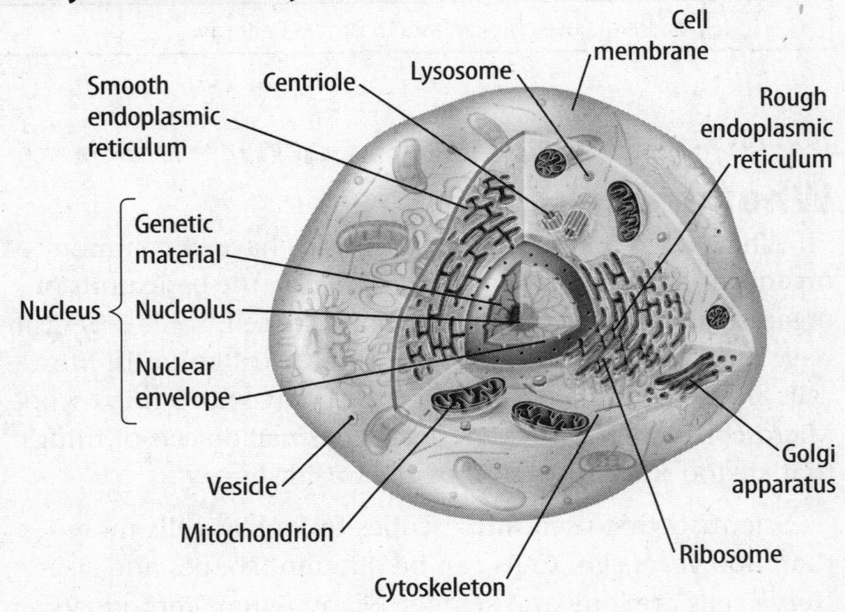

Eukaryotic Cell

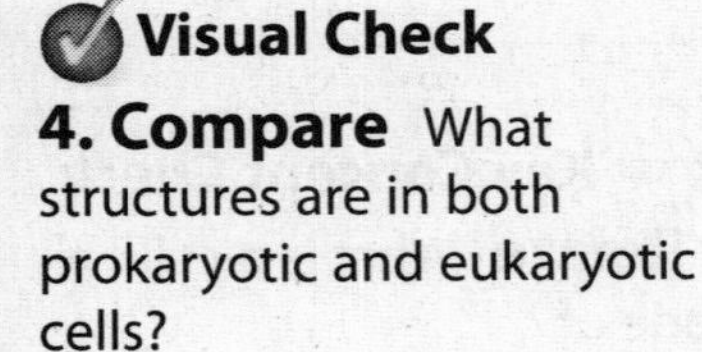

Visual Check

4. Compare What structures are in both prokaryotic and eukaryotic cells?

The Outside of a Cell

As you have just read, the cell membrane surrounds a cell. Much like a fence surrounds a school, the cell membrane helps keep substances inside a cell separate from the substances outside a cell. A more rigid layer, called a cell wall, also surrounds some cells.

Cell Membrane

The cell membrane is made of lipids and proteins. Recall that lipids and proteins are macromolecules that help cells function. Lipids in the cell membrane protect the inside of a cell from the external environment. Proteins in the cell membrane transport substances between a cell's environment and the inside of the cell. Proteins in the cell membrane also communicate with other cells and organisms and sense changes in the cell's environment.

Cell Wall

In addition to a cell membrane, some cells have a cell wall, as shown in the figure below. The cell wall is a strong, rigid layer outside the cell membrane. Cells in plants, fungi, and many types of bacteria have cell walls. They provide structure and help protect the cell from the outside environment. Most cell walls are made from different types of carbohydrates.

Reading Check

5. Summarize the major components of cell membranes.

Cell Membrane and Cell Wall

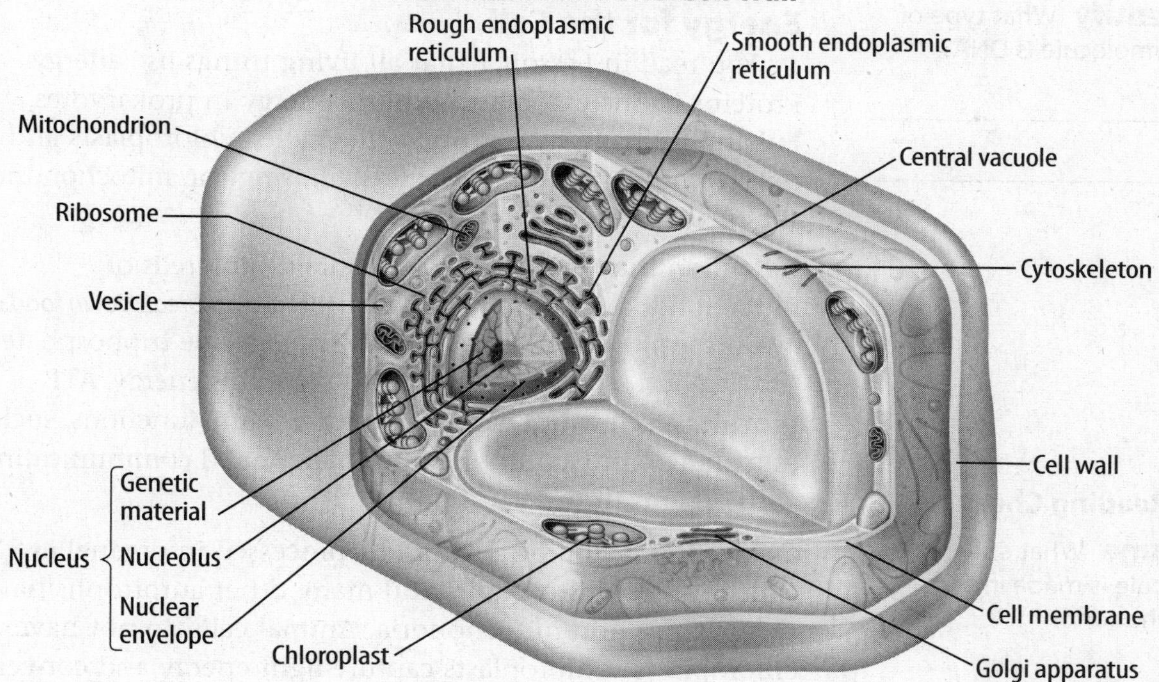

The Inside of a Cell

Recall that the inside of a cell is mainly water. Many substances used for communication, energy, and growth dissolve in water. This makes it easier for the substances to move around inside a cell. Water also gives cells their shapes and helps keep the structures inside a cell organized. The organelles inside a cell perform specific functions. They control cell activities, provide energy, transport materials, and store materials.

Visual Check

6. Locate Is the cell wall found inside or outside the cell membrane?

FOLDABLES

Make an envelope book and use the inside to illustrate a cell. Use the tabs to label and describe the cellular structures.

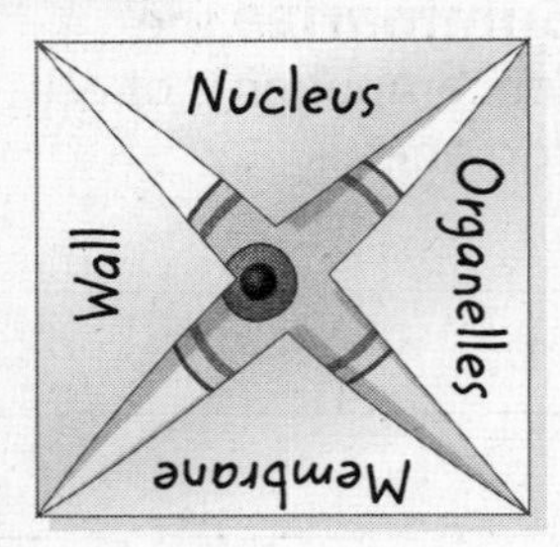

Cytoplasm

The liquid part of a cell inside the cell membrane is called the **cytoplasm.** It contains water, macromolecules, and other substances. In eukaryotic cells, the organelles are located in the cytoplasm. Proteins in the cytoplasm provide structure and help organelles and other substances move around.

Controlling Cell Activities

Genetic material, called DNA, stores the information that controls all of a cell's activities. DNA is a type of macromolecule called a nucleic acid. The DNA transfers its information to another nucleic acid called RNA. RNA gives cells instructions about which proteins need to be made.

In prokaryotic cells, DNA is in the cytoplasm. In eukaryotic cells, DNA is stored in an organelle called the nucleus. A membrane, called the nuclear membrane, surrounds the nucleus. Tiny holes in the nuclear membrane let certain substances move between the nucleus and the cytoplasm.

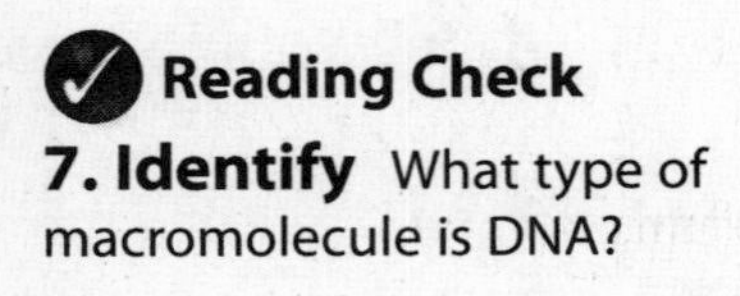

Reading Check

7. Identify What type of macromolecule is DNA?

____

____

Energy for the Cell

You read in Lesson 1 that all living things use energy. Proteins in the cytoplasm process energy in prokaryotes. Eukaryotes have special organelles called chloroplasts and mitochondria (mi tuh KAHN dree uh; singular, mitochondrion) that process energy.

Mitochondria Most eukaryotes contain hundreds of mitochondria. **Mitochondria** *are organelles that break down food and release energy*. Molecules called ATP—adenosine triphosphate (uh DEN uh seen • tri FAHS fayt)—store this energy. ATP provides a cell with energy to perform many functions, such as making proteins, storing information, and communicating with other cells.

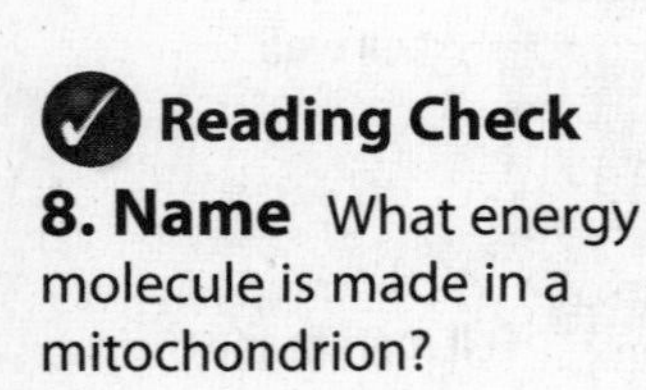

Reading Check

8. Name What energy molecule is made in a mitochondrion?

____

____

Chloroplasts Energy also can be processed in organelles called chloroplasts. Plants and many other autotrophs have chloroplasts and mitochondria. Animal cells do not have chloroplasts. Chloroplasts capture light energy and convert it into chemical energy in a process called photosynthesis. Chloroplasts contain many structures that capture light energy. Like the reactions that occur in mitochondria, ATP molecules are produced during photosynthesis. However, photosynthesis also produces carbohydrates such as glucose that also are used to store energy.

Protein Production

You just read that cells use protein for many functions. These proteins are made on the surface of ribosomes that are in the cytoplasm of prokaryotic and eukaryotic cells. In eukaryotic cells, some ribosomes are attached to an organelle called the endoplasmic reticulum (en duh PLAZ mihk • rih TIHK yuh lum). Endoplasmic reticulum is made of folded membranes. The proteins can be processed and can move inside the cell through the endoplasmic reticulum.

Cell Storage

What happens to the molecules that are made in a cell? An organelle called the Golgi (GAWL jee) apparatus packages proteins into tiny organelles called vesicles. Vesicles transport proteins around a cell. Organelles called vacuoles store other molecules. A vacuole is usually the largest organelle in a plant cell. In plant cells, vacuoles also store water and provide support. In contrast to the vacuoles in all plant cells, only some animal and bacterial cells contain vacuoles. The vacuoles in animal and bacterial cells are smaller than the ones in plant cells.

Reading Check

9. Express Where are ribosomes found in eukaryotic cells?

____

____

____

Key Concept Check

10. Describe How do the parts of a cell enable it to survive?

____

____

____

____

After You Read

Mini Glossary

cytoplasm: the liquid part of a cell inside the cell membrane

eukaryotic (yew ker ee AH tihk) cell: a cell that has a nucleus and other membrane-bound organelles

mitochondrion (mi tuh KAHN dree un): an organelle that breaks down food and releases energy

prokaryotic (pro kayr ee AH tihk) cell: a cell that does not have a nucleus or other membrane-bound organelles

1. Review the terms and their definitions in the Mini Glossary. Write a sentence discussing what is found in cytoplasm.

2. Identify the organelle that performs each function described in the table and write its letter in the left column.

a. chloroplast
b. cell wall
c. Golgi apparatus
d. nucleus
e. vacuole
f. vesicle

Organelle	Function
	stores DNA in eukaryotic cells
	transports proteins around a cell
	stores water and provides support in plants
	processes energy in plant cells
	packages proteins in vesicles
	protects the cell from the outside environment

3. Explain how DNA and RNA work together.

What do you think NOW?

Reread the statements at the beginning of the lesson. Fill in the After column with an A if you agree with the statement or a D if you disagree. Did you change your mind?

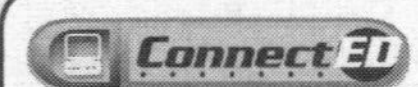

Log on to ConnectED.mcgraw-hill.com and access your textbook to find this lesson's resources.

Inheritance and Adaptations

Inheritance and Traits

Before You Read

What do you think? Read the three statements below and decide whether you agree or disagree with them. Place an A in the Before column if you agree with the statement or a D if you disagree. After you've read this lesson, reread the statements to see if you have changed your mind.

Before	Statement	After
	1. Genes are made of chromosomes.	
	2. A mutation is a permanent change in a gene.	
	3. The environment cannot affect an inherited trait.	

Key Concepts 
- What is inheritance?
- What is the role of genes in inheritance?
- How do environmental factors influence traits?
- How do mutations influence traits?

Read to Learn

What is inheritance?

You probably look like your parents or grandparents. If you have brothers or sisters, they probably look like your parents and grandparents, too. All of you might have some of the same characteristics, such as being tall or having brown eyes. *A distinguishing characteristic of an organism is a* **trait.** During reproduction, many traits are passed from one generation to the next. *The passing of traits from generation to generation is* **inheritance.** Inheritance is the reason offspring look like their parents, their grandparents, and even their distant ancestors.

Every organism has a set of inherited traits. A parrot, for example, has feathers, wings, claws, and a hooked beak. It gets these traits from its parents. Its body structure is also inherited. All of these traits can be passed to its offspring.

Not all traits are inherited. If a parrot lost a claw in an accident, its offspring would not be born missing a claw. Similarly, the parrot's offspring would not be born knowing how to do tricks that the parrot learned to do. Losing a claw and learning tricks are examples of acquired traits. An acquired trait is a trait that an organism acquires or develops during its lifetime.

Mark the Text

Identify the Main Ideas Write a phrase beside each paragraph that summarizes the main point of the paragraph. Use the phrases to review the lesson.

Key Concept Check
1. Define What is inheritance?

____

____

____

Think it Over

2. Analyze Why would asexual reproduction result in offspring that are identical to their parents?

____

____

____

____

Key Concept Check

3. Relate How are traits and genes related?

____

____

____

4. Describe how DNA would look if it were stretched out.

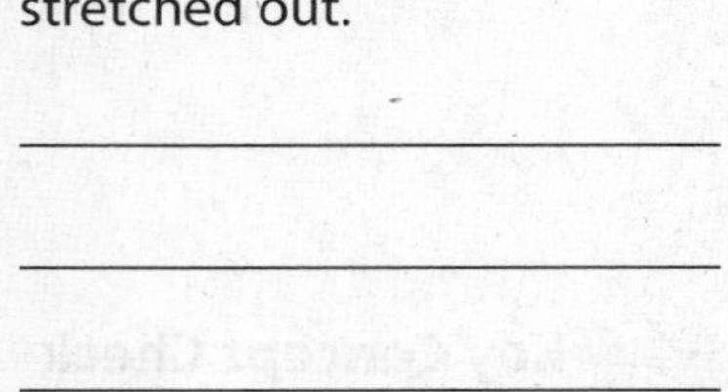

Inheritance and DNA

Organisms pass inherited traits to their offspring. Some organisms, such as amoebas, bacteria, and some plants, pass traits to their offspring by cell division and mitosis. This process is called asexual reproduction. Asexual reproduction produces offspring that are identical to the original organism. Many other organisms, including humans, reproduce sexually. This process produces offspring that are similar—but not identical—to the parent or parents.

DNA and Genes

Sexual reproduction requires DNA from a sperm cell and an egg cell. DNA, shown below, is a molecule inside a cell's nucleus. It looks like a twisted zipper. Genes are distinct segments of DNA. *A* ***gene*** *is a section of DNA that has genetic information for one trait.* Genes carry this information in a unique sequence within DNA, much as words have information given by the unique sequence of their letters.

DNA is long. If you stretched out the DNA in one of your cells, it would be almost 2 m long. DNA fits into a cell's nucleus because it is tightly coiled with proteins to form chromosomes. A chromosome is a structure made of long chains of DNA.

DNA Molecule

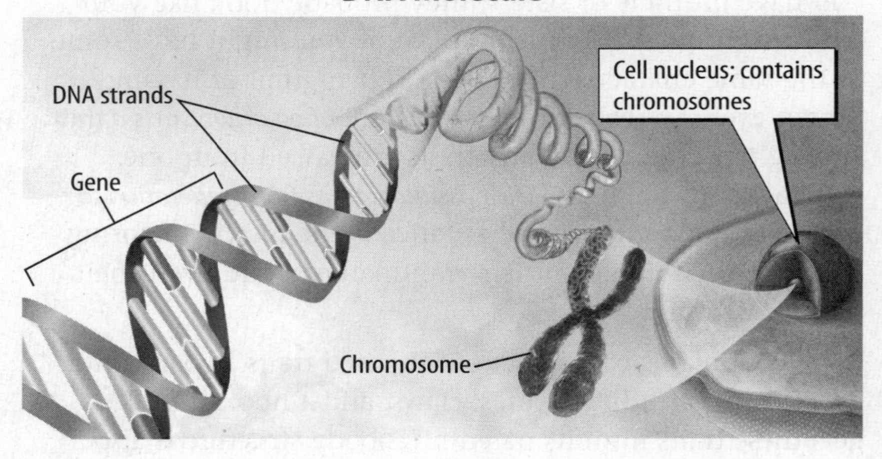

Chromosomes

Different species can have different numbers of chromosomes in a cell. In most species, chromosomes come in pairs. Humans have 23 pairs of chromosomes in each body cell. Each pair contains one chromosome from the father and one chromosome from the mother. Reproductive cells—called sperm and eggs—each contain only single chromosomes. Along each of these chromosomes lies hundreds or thousands of genes.

Combining Genes

You probably look like your parents in many ways, but you are not identical to them. For instance, you might have your father's blue eyes but your mother's brown hair. In sexual reproduction, an egg cell and a sperm cell each contribute one gene for a trait. Each gene for a single trait is called an allele (uh LEEL). How the alleles are sorted and combined into offspring during sexual reproduction is mostly a matter of chance.

FOLDABLES

Make a horizontal shutterfold book to organize your notes on factors that influence traits.

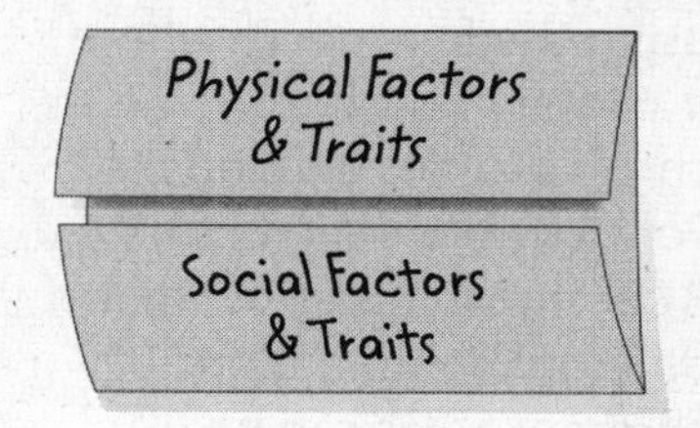

Meiosis

Much of the randomness of sexual reproduction occurs during meiosis. Meiosis is the process during which sperm and egg cells form. During meiosis, the chromosomes in existing egg and sperm cells replicate and divide, as shown in the figure below. Then each egg cell and each sperm cell splits into four separate cells. Each of these four cells has half the original number of chromosomes—23 individual chromosomes in human egg cells and sperm cells. Each sperm cell and egg cell contains a unique combination of genes on each chromosome.

Visual Check

5. Solve If four cells undergo meiosis, how many reproductive cells result?

____

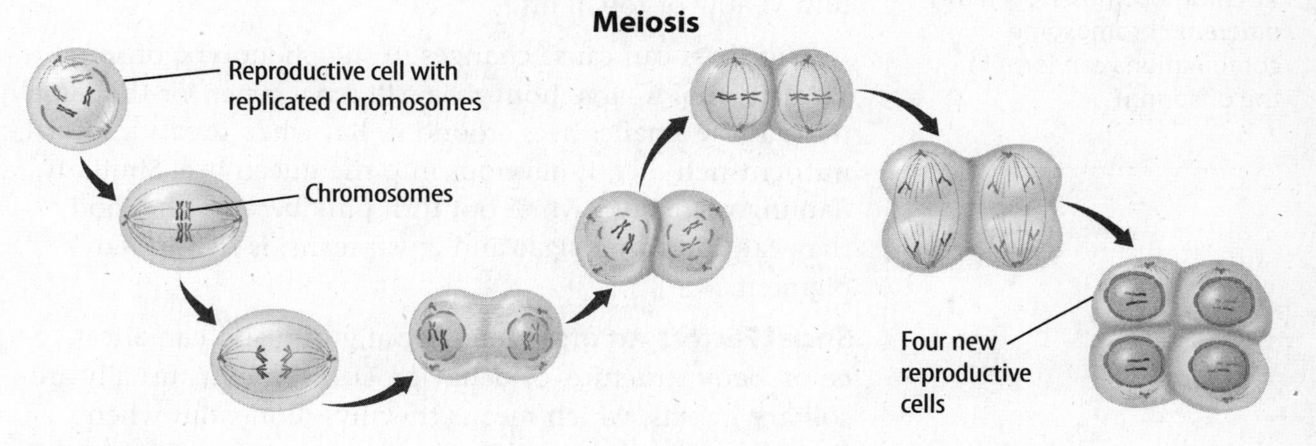

Fertilization

During fertilization, a sperm and an egg unite. When this happens, the egg cell chromosomes and the sperm cell chromosomes combine and form an offspring with a full set of paired chromosomes. Because each sperm cell and egg cell is unique, the resulting offspring also is unique. In humans, there are many possible gene arrangements from the joining together of sperm and egg chromosomes. So many arrangements are possible that if a mother and a father could have billions of offspring, with each offspring produced from a different fertilized egg, no two offspring would be alike. 

Reading Check

6. Infer Why is each sperm cell and egg cell produced by meiosis unique?

____

____

____

Influencing Traits

An organism's complete set of genes is its **genotype** (JEE nuh tipe). Once inherited, this genotype does not change. However, an organism's environment can influence traits expressed by the genotype. If the environment changes, it can affect the expression of a trait in an individual organism.

Phenotype and the Environment

Inherited traits are part of an organism's phenotype (FEE nuh tipe). *The* **phenotype** *of a trait is how the trait appears, or is expressed.* Phenotypes result from the interaction of an organism's genes and its environment. An organism's environment changes all the time. Light, temperature, moisture, nutrients, and social factors are not constant. These factors influence organisms in different ways. For example, light levels have a strong effect on plant phenotype. Plants that grow tall in full sunlight might not grow as tall in low light.

Physical Factors Many physical factors other than light can influence phenotype. For example, low levels of nutrients in soils, such as nitrogen or iron, might cause a plant's leaves to turn yellow or to fall off.

Nutrients can cause changes in the phenotype of some animals, too. A large honeybee will have genes for the same traits as the smaller bees around it. But when it eats a special, nutrient-rich diet, it develops into the queen bee. Similarly, flamingos are born white but turn pink because the food they eat, including algae and crustaceans, is rich in red pigment.

Social Factors An organism's social group also can affect color, body structure, or behavior. Desert locusts usually are solitary insects, which means they live alone. But when these locusts are in a large group, they apply pressure on each other's legs. This causes the locusts to change color, from green to yellowish-brown, and to swarm.

Flamingos are another type of animal that is influenced by social factors. Through studies conducted in zoos, scientists have learned that the large social group in which flamingos live is important because it triggers breeding among them. A flock must have at least 20 flamingos for breeding to take place in zoos. Studies have shown that adding more birds to the flock leads to increased breeding success. In the wild, flamingos live in flocks of up to 10,000 birds.

Math Skills

You can calculate the probability, or likelihood, of chromosome combinations using the formula 2^n where n = total number of chromosomes divided by 2. For example, fruit flies have 8 chromosomes. How many different combinations of chromosomes can be produced in the offspring?

a. Divide the number of chromosomes by 2.

$$\frac{8 \text{ chromosomes}}{2} = 4$$

b. Replace n in the formula with the answer to step *a* and calculate.

$$2^4 = 2 \times 2 \times 2 \times 2 = 16$$

7. Use Probability
The common housefly has 12 chromosomes. How many different chromosome combinations can form in the offspring?

____

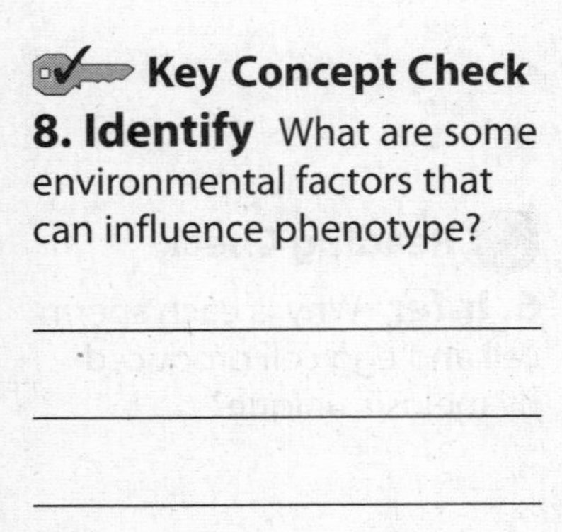

Key Concept Check
8. Identify What are some environmental factors that can influence phenotype?

____

____

____

____

Phenotype and Mutations

When an organism's phenotype changes in response to its environment, the organism's genes are not affected. Thus, the change cannot be passed on to the next generation. The only way that a trait can change so that it can be passed to the next generation is by mutation, or changing an organism's genes.

Random Changes *A* ***mutation*** *is a permanent change in the sequence of DNA in a gene.* A mutation is an error in the DNA's arrangement in a gene. Have you ever made an error when you were typing or texting? For example, you might use one letter instead of another in a word. This could change the meaning of the word. Similarly, a mutation can change the trait for which the gene holds information.

Although all genes can mutate, only mutated genes in egg or sperm cells are inherited. Some mutations in egg or sperm cells occur when an organism is exposed to harsh chemicals or severe radiation. But most mutations occur randomly. For example, lobsters are usually green-brown in color. Due to a random mutation in an egg or a sperm cell, about 1 in 5 million lobsters is naturally blue. The feather color of a penguin's stomach is usually white. Stomach feather color that is completely black is the result of a mutation.

Effects of Mutations Many mutations have no effect on an organism. They neither help nor hurt it. But some mutations change an organism's genes—and its traits—so much that they can affect an organism's ability to survive in its environment. Some mutations are harmful to an organism, but other mutations might help it survive. In Lesson 2, you will read how mutations that benefit an organism can spread to an entire population.

Reading Check

9. Point Out Why can't changes in phenotype be passed on to the next generation?

____

____

____

____

ACADEMIC VOCABULARY

random

(adjective) without a definite aim, rule, or method

Key Concept Check

10. Explain How can a mutation influence a trait?

____

____

____

After You Read

Mini Glossary

gene: a section of DNA that has genetic information for one trait

genotype (JEE nuh tipe): an organism's complete set of genes

inheritance: the passing of traits from generation to generation

mutation: a permanent change in the sequence of DNA in a gene

phenotype (FEE nuh tipe): how a trait appears, or is expressed

trait: a distinguishing characteristic of an organism

1. Review the terms and their definitions in the Mini Glossary. Write a sentence explaining the link between phenotype and environment.

2. In the table below, write *inherited* in the space next to each inherited trait. Write *acquired* next to those traits that are not inherited.

Inherited or Acquired?	Trait
	a. A bear cub has one brown eye and one gray eye.
	b. A dog performs a variety of special tricks.
	c. A fish changes color to match the pigment in its food.
	d. An athlete is six-and-a-half-feet tall.
	e. Your friend loses a finger in an accident.
	f. A deer's large ears help it to hear faint sounds.

3. How did summarizing each paragraph with a phrase help you review this lesson?

What do you think NOW?

Reread the statements at the beginning of the lesson. Fill in the After column with an A if you agree with the statement or a D if you disagree. Did you change your mind?

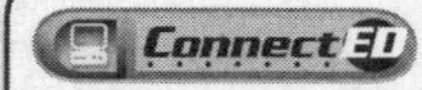

Log on to ConnectED.mcgraw-hill.com and access your textbook to find this lesson's resources.

Inheritance and Adaptations

Adaptations in Species

Before You Read

What do you think? Read the three statements below and decide whether you agree or disagree with them. Place an A in the Before column if you agree with the statement or a D if you disagree. After you've read this lesson, reread the statements to see if you have changed your mind.

Before	Statement	After
	4. Mutations are a source of variation.	
	5. All species on Earth are uniquely adapted to their environments.	
	6. Plants have adaptations for movement.	

Key Concepts 

- How do mutations cause variations?
- How does natural selection lead to adaptations in species?
- What are some ways adaptations help species survive in their environments?

Read to Learn

What is adaptation?

It is easy to see the differences among people, but what about the differences among plants or animals? Are all robins alike? What about sunflower seeds? Sexual reproduction ensures that offspring are different from their parents. This is true of all species that reproduce sexually. This includes robins, sunflowers, and giraffes. All giraffes are members of the same species, yet each one has a slightly different pattern of spots on its coat. *Slight differences in inherited traits among individual members of a species are* ***variations.***

Variations occur through mutations. A mutation might harm an organism's chances of survival. However, many mutations, such as those that cause the unique pattern of spots on a giraffe, cause no harm. Still other mutations can benefit an organism. They produce traits that help an organism survive.

Giraffes have different spot patterns depending on the genes they inherit. However, no matter what the pattern they inherit looks like, every giraffe has spots. The spots help the giraffes blend in with their environment—the grasslands of Africa. As a result, predators of giraffes, such as lions and hyenas, cannot see them as easily. The spotted coat of giraffes is an adaptation. *An* ***adaptation*** *is an inherited trait that helps a species survive in its environment.*

Study Coach

Preview Headings Before you read the lesson, preview all the headings. Make a chart and write a question for each heading beginning with *What* or *How*. As you read, write the answers to your questions.

Key Concept Check

1. Relate How are mutations related to variations?

How Adaptations Occur

Giraffe spots probably first came from a mutation in one giraffe many generations ago. The mutation produced a variation that helped that giraffe survive. In time, the mutated gene became common to the entire population. How did this happen?

Key Concept Check
2. Describe How does a variation become an adaptation?

____

____

____

____

____

Natural Selection

Natural selection *is the process by which organisms with variations that help them survive in their environment live longer, compete better, and reproduce more than those that do not have the variation.* If a variation helps an organism survive or compete better in its environment, the organism with that variation lives longer. Because it lives longer, it has more offspring that can inherit the variation. Over many generations, more and more offspring inherit the variation. Eventually, most of the population has the variation, and it becomes an adaptation, as shown in the figure below.

Visual Check
3. Point Out What change in the environment led to natural selection of brown beetles?

____

____

____

Because mutations are random and occur continually, so do new variations. The variations that become adaptations depend on the environment. Over time, all environments change. Huge volcanic eruptions can change a climate rapidly. The movement of continents causes slow, gradual changes. When an environment changes, the members of a population either survive or die off. The repeated dying off of entire populations can lead to the extinction of a species.

Natural Selection Leads to Adaptation

❶ Variation in Traits In this population of beetles, some are yellow and some are brown. The color does not affect the ability of the beetles to survive in their environment.

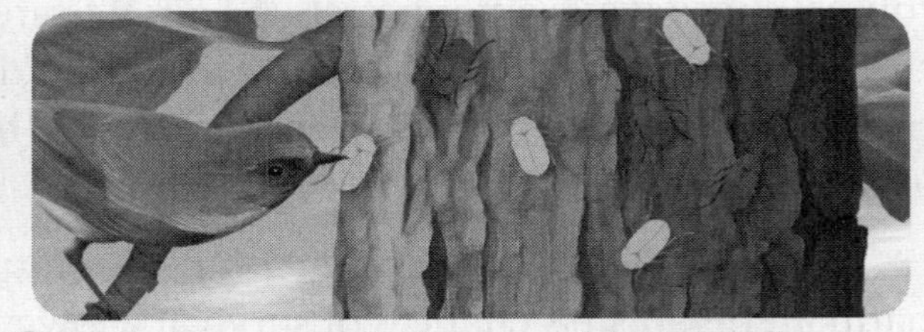

❷ Organisms Compete A new predator eats yellow beetles more often because it sees the yellow beetles more easily than the brown beetles.

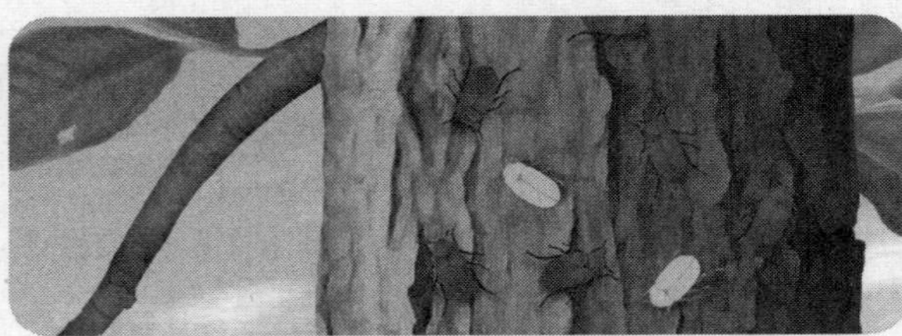

❸ Traits are Inherited The yellow beetles do not live as long as the brown beetles, and—since color is inherited—fewer yellow beetles hatch.

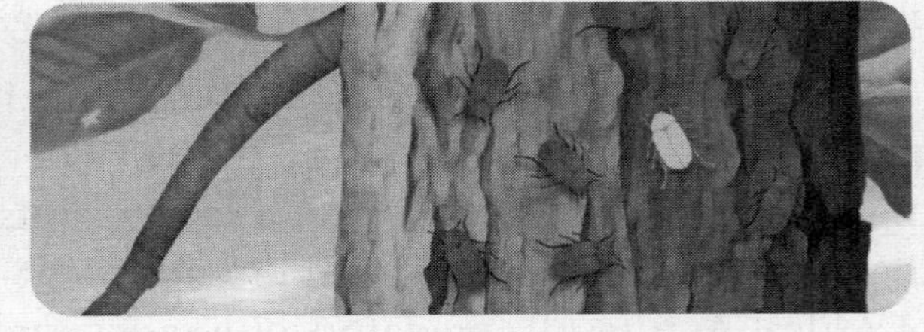

❹ Adaptation over Time Nearly all individuals in a population are brown. The color brown has become an adaptation that helps the beetles avoid predators in that environment.

Selective Breeding

Natural selection happens over so many generations that it usually cannot be seen. It is easier to observe a different type of selection. When humans breed organisms for food or for use as pets, they are selecting variations that occur naturally in populations. *The selection and breeding of organisms with desired traits is* **selective breeding**. Selective breeding is similar to natural selection except that humans, instead of nature, do the selecting. By breeding organisms with desired traits, humans change traits just as natural selection does. Cows with increased levels of milk production and roses of unique colors are products of selective breeding. ✓

Types of Adaptations

Through natural selection or selective breeding, each species is uniquely adapted to its environment. Chickens are adapted to life in a henhouse just as giraffes are adapted to life in the grasslands. Adaptations enable species to avoid predators, maintain homeostasis, find and eat food, and move. Adaptations can be structural, behavioral, or functional. The table below gives examples of each.

Types of Adaptations

Type of Adaptation	Description	Example
Structural	a physical trait—such as color, shape, or internal structure—that increases survival	The color and shape of an insect's eyes are structural adaptations.
Behavioral	a behavior or action—such as migration, hibernation, hunting at night, or playing dead—that increases survival	Some animals, like snakes, play dead—a behavioral adaptation that fools predators.
Functional	a biochemical change—such as hibernating, shedding, or spitting—that enables a species to increase survival or maintain homeostasis	Spraying venom, as a cobra might do, is a functional adaptation.

Maintaining Homeostasis

The ability of an organism to keep its internal conditions within certain limits is homeostasis. Sweating on a hot day is an adaptation that helps you maintain your internal body temperature when external temperatures increase. All species have adaptations that help them survive temporary changes in their environments. Species also have adaptations specific to their environments. Plants living in deserts store water in their leaves. Fish have gills that remove oxygen from water.

Reading Check

4. Contrast How is selective breeding different from natural selection?

____

____

____

Think it Over

5. Categorize What category of adaptation is a giraffe's long neck?

____

Interpreting Tables

6. Name What are some examples of behavioral adaptations?

____

____

____

____

![FOLDABLES]

Make a four-tab book to organize your notes on benefits of adaptations.

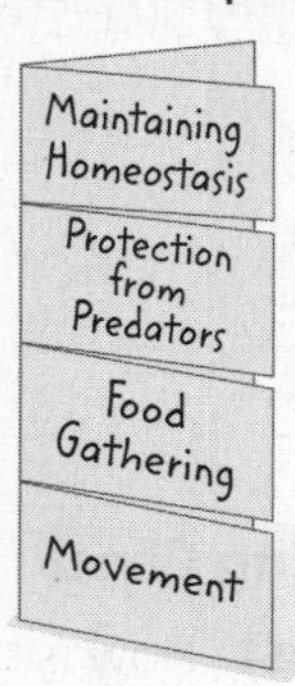

Key Concept Check

7. Name Give an example of how adaptations help species survive.

Interpreting Tables

8. Describe How is a condor's beak adapted for the food it eats?

Protection from Predators

Species also have adaptations that protect them from predators. For example, sharp quills protect porcupines. Sometimes, through natural selection, variations are selected that make an organism resemble something else. Camouflage and mimicry are two adaptations that cause animals to look like something else.

Camouflage (KAM uh flahj) *is an adaptation that enables a species to blend in with its environment.* A stonefish is a fish that looks like a rock. This makes it less visible to predators. **Mimicry** (MIH mih kree) *is an adaptation in which one species looks like another species.* The scarlet kingsnake is a nonpoisonous snake that looks like, or mimics, the poisonous coral snake. Predators often avoid the kingsnake because they cannot tell the two snakes apart.

Food Gathering

As you have just read, camouflage and mimicry protect species from predators. These same adaptations also can help species find food. A camouflaged stonefish is hidden not only from predators but also from its prey. Many other kinds of adaptations help species gather and eat food. An anteater has a long nose and a long, sticky tongue for gathering ants.

Beak Shape Though all birds have wings, beaks, and feathers, each bird species is adapted to a different environment. Each species uses its beak in a different way to gather food. Each of the birds described in the table below has a beak that helps it gather a different type of food.

Beak Adaptations

Bird	Beak Shape and Use
Woodpeckers	Woodpeckers use their long, thin beaks to search for insects in tree bark.
Parrots	Parrots have strong beaks that help them crack nuts and seeds.
Condors	Condors use their long, powerful beaks to tear the flesh from dead organisms.

Food Storage in Plants Some plants also have adaptations that enable them to store food. Potatoes, onions, and tulips have modified underground stems that are very thick. These thick stems store food for the plants.

Hunters and Their Prey As predators develop adaptations for hunting their prey, the species they hunt develop adaptations for avoiding them. A cheetah is a fast runner. But so are the gazelles it chases as prey. Over time, cheetahs might become even faster due to chance variations and natural selection. But faster gazelles also might arise from the same process. In this way, species adapt to each other.

Movement

Cheetahs and gazelles have long, powerful legs adapted to running fast. Legs, wings, flippers, fins, and even tails are adaptations that help different species move. Movement helps species search for food, avoid predators, and escape unpleasant stimuli. Even some plants have adaptations for movement. Their leaves turn to face the Sun as it moves across the sky.

Think it Over

9. Infer How would having leaves that follow the Sun help a plant survive?

After You Read

Mini Glossary

adaptation: an inheritied trait that helps a species survive in its environment

camouflage (KAM uh flahj): an adaptation that enables a species to blend in with its environment

mimicry (MIH mih kree): an adaptation in which one species looks like another species

natural selection: the process by which organisms with variations that help them survive in their environment live longer, compete better, and reproduce more than those that do not have the variation

selective breeding: the selection and breeding of organisms with desired traits

variation: a slight difference in inherited traits among individual members of a species

1. Review the terms and their definitions in the Mini Glossary. Write a sentence that describes how natural selection and selective breeding are alike.

____

____

2. Use what you have learned about adaptations to fill in the blanks below.

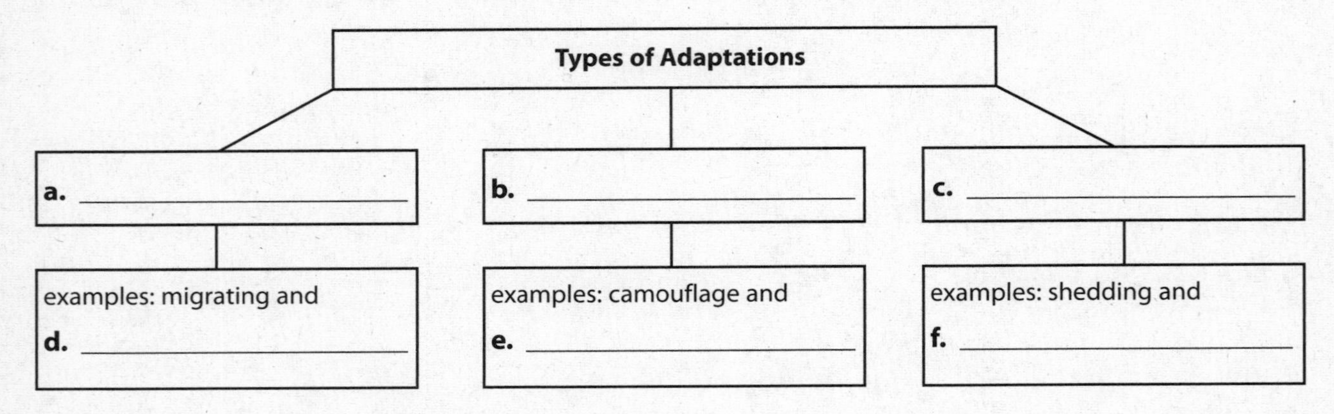

3. How can variation and natural selection cause species to change over time?

____

____

____

____

What do you think NOW?

Reread the statements at the beginning of the lesson. Fill in the After column with an A if you agree with the statement or a D if you disagree. Did you change your mind?

Log on to ConnectED.mcgraw-hill.com and access your textbook to find this lesson's resources.

Introduction to Plants

Plant Diversity

Before You Read

What do you think? Read the two statements below and decide whether you agree or disagree with them. Place an A in the Before column if you agree with the statement or a D if you disagree. After you've read this lesson, reread the statements to see if you have changed your mind.

Before	Statement	After
	1. Humans could survive without plants.	
	2. Plant cells contain the same organelles as animal cells.	

Key Concepts 

- How do a plant's structures ensure its survival?
- How are the different plant types alike and different?

Read to Learn

What is a plant?

Humans depend on plants for food, oxygen, building materials, and many other things. Plants have functions that make them a vital part of the world.

Study Coach

Make an Outline Summarize the information in this lesson by making an outline. Use the main headings in the lesson as the main headings in your outline. Use your outline to review the lesson.

Plant Cells

Animal cells and plant cells are similar. They have many of the same organelles. However, a plant cell also contains chloroplasts, which are organelles that make food. Also, a rigid cell wall surrounds a plant cell. It helps protect and support the cell. A plant cell also contains a large central vacuole, as shown below.

Plant Structures and Functions

Most plants have roots, stems, and leaves, which have specialized transport tissues. Some plants, such as mosses, don't have specialized transport tissues. Instead, they have root-, stem-, and leaflike structures that perform similar functions.

Plant Cell

Mitochondrion
Central vacuole
Nucleus
Chloroplast
Cell membrane
Cell wall

Visual Check

1. Draw a circle around the names of organelles in the plant cell that are unique to plant cells.

Roots There are many different types of roots. Some plants have a larger main root, called a taproot, with smaller roots growing from it. The stems and branches of some plants grow additional roots above ground, called prop roots, that help support the plant. Other plants have fibrous root systems that consist of many small, branching roots. Roots anchor a plant in the soil and enable it to grow upright and not be blown away by wind or carried away by water.

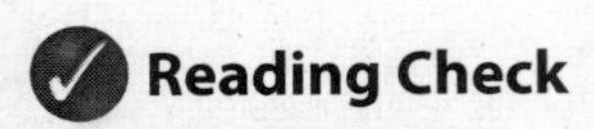

Reading Check

2. Name the three types of roots.

____

____

____

Roots also absorb water and minerals from the soil, which plants need for cellular processes. Some roots store food such as sugar and starch. Plants that survive from one growing season to the next use this stored food for growing leaves at the beginning of the next season.

Some plants, such as mosses and hornworts, have rootlike structures called rhizoids. **Rhizoids** *are structures that anchor a plant without transport tissue to a surface.* Scientists do not consider rhizoids roots because they do not have the transport tissues that roots have.

FOLDABLES®

Make a vertical three-tab book and use it to organize your notes on the structures of a plant.

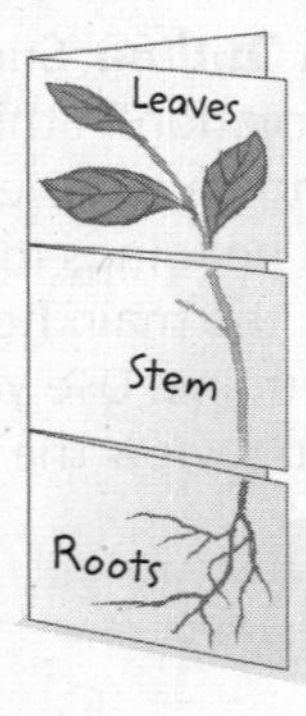

Stems Have you ever leaned against a tree? If so, you were leaning on a plant stem. Stems help support the leaves, and in some cases flowers, of a plant. There are two main types of stems, as shown in the figure at right. A woody stem is like the one you might have leaned against, and a herbaceous (hur BAY shus) stem is flexible and green, such as the stem of a morning glory vine.

Woody and Herbaceous Stems

Visual Check

3. Examine As the figure shows, a woody stem ____. (Circle the correct answer.)

a. is flexible

b. is rigid

c. has green leaves and flowers

Stems have tissues that help carry water and the minerals absorbed by the roots to a plant's leaves. These tissues also transport sugar to a plant's roots. This sugar is produced in chloroplasts during photosynthesis. (Photosynthesis is the process by which cells convert light energy into food energy.) In some plants, such as cacti, stems store water that the plants use during dry periods. Other plants, such as potatoes, have underground stems that store food.

Plant Leaves

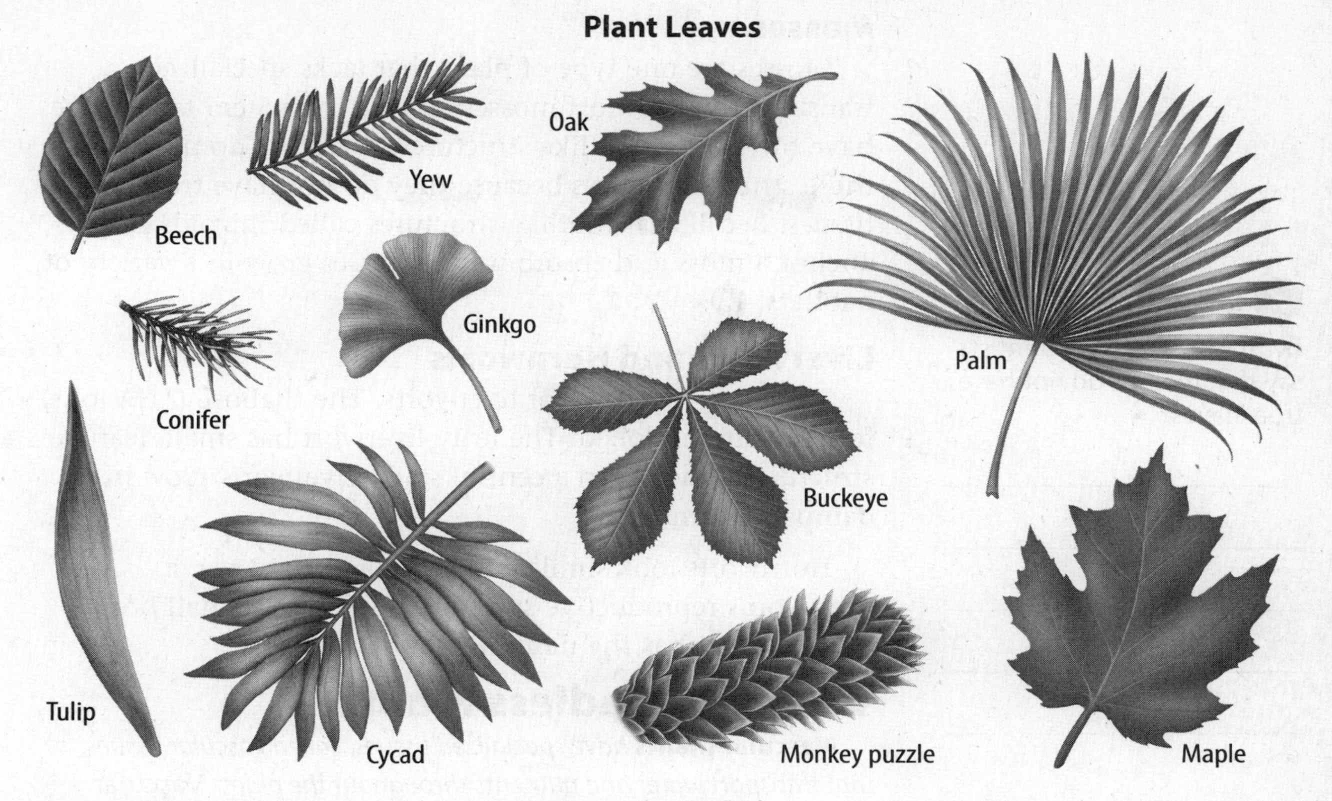

Leaves In most plants, leaves are the major sites for photosynthesis. Some cells in a leaf contain many chloroplasts, where light energy is converted into the chemical energy stored in sugar during photosynthesis. There are many different sizes and shapes of leaves, but all leaves have vascular tissue. A variety of leaves is shown in the figure above.

In addition to making food, leaves also are involved in the exchange of gases with their environment. *The* **stomata** (singular, stoma) *are small openings in the surfaces of most plant leaves.* Water vapor, carbon dioxide, and oxygen can pass into and out of a leaf through stomata.

Nonvascular Plants

Have you ever noticed tiny green plants growing on the bark of a tree? These plants might be one of several types of nonvascular plants. *Plants that lack specialized tissues for transporting water and nutrients are* **nonvascular plants.**

You might recall that animals are grouped into phyla and plants are grouped into divisions. The divisions of nonvascular plants include mosses, liverworts, and hornworts.

Visual Check

4. Circle one type of leaf that grows in your community.

ACADEMIC VOCABULARY
major
(adjective) greater in number, quantity, or extent

Key Concept Check

5. Explain How do plant structures such as roots, stems, and leaves ensure a plant's survival?

____
____
____
____

Mosses

Mosses are one type of plant that lacks specialized transport tissues. Most mosses are less than 5 cm tall. Mosses have tiny, green, leaflike structures. Scientists do not call these structures leaves because they do not have transport tissues. Recall that rootlike structures called rhizoids help anchor a moss and absorb water. Mosses grow in a variety of habitats.

Reading Check

6. State Why do scientists say that mosses do not have true leaves?

Liverworts and Hornworts

There are two types of liverworts. The thallose (THA lohs) form is flat and lobed. The leafy liverwort has small, leaflike structures attached to a central stalk. Liverworts grow in damp environments.

Hornworts look similar to liverworts. However, a hornwort's reproductive structure resembles a small horn. That's why it has the name *hornwort.*

Vascular Seedless Plants

Vascular plants *have specialized tissues, called vascular tissues, that transport water and nutrients throughout the plant.* Vascular plants are divided into two groups—those that produce seeds and those that do not. Ferns, horsetails, and club mosses do not produce seeds.

Reading Check

7. Identify What are the two types of vascular plants?

Ferns

Fossil evidence indicates that millions of years ago, ferns grew as large as today's trees. Present-day ferns are much smaller. Many grow to only 50 cm or less. A fern leaf is called a frond. A frond can have an intricate shape. A fern's fronds grow from an underground stem called a rhizome (RI zohm). Ferns usually grow in shady locations.

Horsetails

Horsetails get their name from a stage of their life cycle that looks like a horse's tail. Horsetails also are called scouring rushes due to an abrasive mineral called silica in the stems. This abrasiveness made the horsetail plant useful to early settlers for cleaning pots and pans.

Reading Check

8. Explain How did horsetails get their name?

Club Mosses

This group of plants gets its name from its reproductive structure that resembles a club. The plants often look like small pine trees. Club mosses were abundant in ancient forests. Much of the fossil fuel we use today comes from the remains of these forests. Present-day club mosses grow in diverse locations that include tropical and arctic habitats.

Vascular Seed Plants

Probably the plants that you are most familiar with are vascular seed plants. They are the most common type of plants. Grasses, flowering shrubs, and trees are vascular seed plants. Vascular seed plants range from tiny aquatic plants that are less than 1 mm across to towering redwood trees. All share one important characteristic—they produce seeds. Scientists further organize vascular seed plants into two groups—those that produce flowers and those that do not.

Nonflowering Seed Plants

For many plants, seeds are inside or on the surface of fruits. However, some plants produce seeds without fruits. **Gymnosperms** *are plants that produce seeds that are not part of a fruit.* Gymnosperms are a diverse group. The most common gymnosperms are conifers. Conifers are usually evergreen, meaning they stay green all year. They have needlelike or scalelike leaves, and most produce cones. The seeds are part of the cones. Conifers can grow in diverse habitats from near-arctic regions to tropical areas. Conifers also have many commercial uses. Lumber, paper products, and turpentine are products made from conifers.

Flowering Seed Plants

How many flowering plants can you name? More than 260,000 species of flowering plants exist! **Angiosperms** *are plants that produce flowers and develop fruits.* Squash, rosemary, pear trees, poppies, wisteria vines, palm trees, grass, cacti, and lily pads are just some of the different varieties of flowering seed plants.

Flowering plants have many adaptations that enable them to survive in most habitats on Earth. Their specialized vascular tissues carry water and nutrients throughout the plants. Plants that live in dry areas have special adaptations that help prevent water loss.

Perhaps the most amazing characteristic of flowering seed plants is the diversity of their flowers. The flowers attract insects and birds of all kinds. Some flowers are specialized so wind or water can aid in reproduction. As you will read in the next lesson, flowers play a key role in plant reproduction.

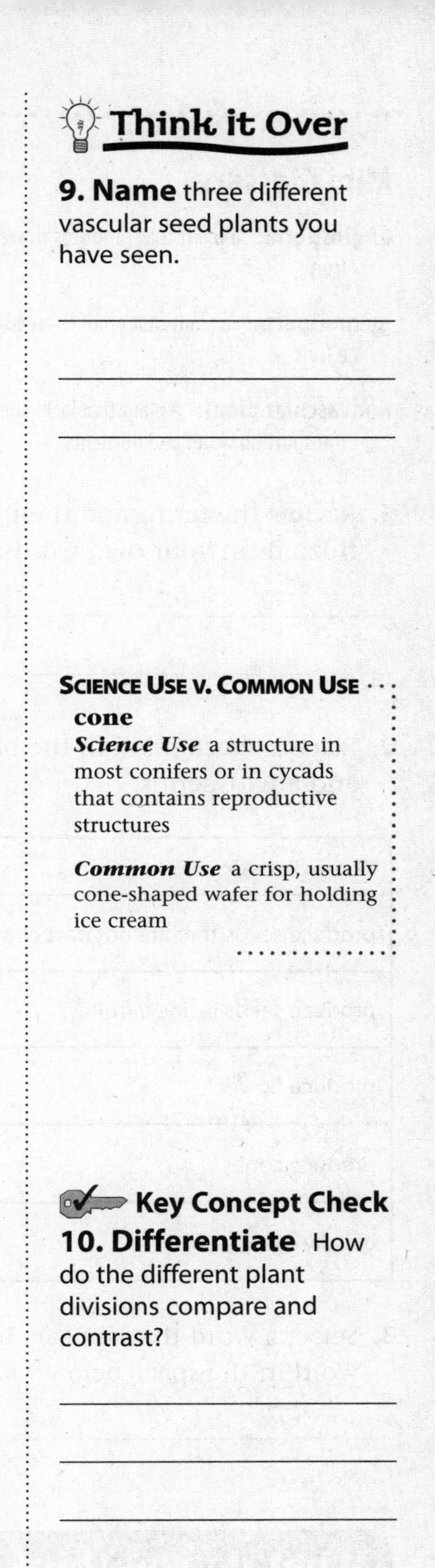

Think it Over

9. Name three different vascular seed plants you have seen.

SCIENCE USE V. COMMON USE

cone

Science Use a structure in most conifers or in cycads that contains reproductive structures

Common Use a crisp, usually cone-shaped wafer for holding ice cream

Key Concept Check

10. Differentiate How do the different plant divisions compare and contrast?

After You Read

Mini Glossary

angiosperm: a plant that produces flowers and develops fruit

gymnosperm: a plant that produces seeds that are not part of a fruit

nonvascular plant: a plant that lacks specialized tissues for transporting water and nutrients

rhizoid: a structure that anchors a plant without transport tissue to a surface

stoma: a small opening in the surfaces of most plant leaves

vascular plant: a plant that has specialized tissues, called vascular tissues, that transport water and nutrients throughout the plant

1. Review the terms and their definitions in the Mini Glossary. Write a sentence defining rhizoids in your own words.

2. Place a check mark in the blank columns to identify the characteristics of gymnosperms and angiosperms.

Characteristics	Gymnosperms	Angiosperms
produce seeds that are not part of a fruit		
produce seeds inside a fruit		
produce flowers		
produce cones		
usually evergreen		

3. Select a word that appears in a main heading of the outline you created. Define that word in the space below.

What do you think NOW?

Reread the statements at the beginning of the lesson. Fill in the After column with an A if you agree with the statement or a D if you disagree. Did you change your mind?

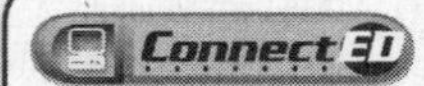

Log on to ConnectED.mcgraw-hill.com and access your textbook to find this lesson's resources.

Introduction to Plants

Plant Reproduction

Before You Read

What do you think? Read the two statements below and decide whether you agree or disagree with them. Place an A in the Before column if you agree with the statement or a D if you disagree. After you've read this lesson, reread the statements to see if you have changed your mind.

Before	Statement	After
	3. Plants can reproduce both sexually and asexually.	
	4. All plants have a two-stage life cycle.	

Key Concepts 

- How do asexual and sexual reproduction in plants compare and contrast?
- What are the differences between the life cycles of seedless and seed plants?

Read to Learn

Asexual Reproduction

Some types of plants produce new plants from a leaf, a stem, or another plant part instead of a seed. Asexual reproduction occurs when only one parent organism or part of that organism produces a new organism. The new organism is genetically identical to the parent.

Mark the Text

Ask Questions As you read, write questions you have next to each paragraph. Read the lesson a second time and try to answer the questions. When you are done, ask your teacher any questions you have not been able to answer.

Sexual Reproduction

Sexual reproduction involves male and female sex cells. Each sex cell contributes genetic material to the offspring. Like animals, plants produce sperm (male sex cells) and eggs (female sex cells). Fertilization occurs when a sperm and an egg join, combining their genetic material. Sexual reproduction produces individuals that have a different genetic makeup than the parent organism or organisms. Seedless plants and seed plants can reproduce sexually.

Key Concept Check

1. Compare and Contrast How do asexual and sexual reproduction in plants compare and contrast?

Plant Life Cycles

All plants have two stages in their life cycle. They are called the gametophyte (guh MEE tuh fite) stage and the sporophyte (SPO ruh fite) stage.

Gametophyte Stage This stage begins with a spore, or haploid cell. Through mitosis and cell division, the spore produces a plant structure or an entire plant called a gametophyte. The gametophyte produces male and female sex cells through meiosis.

Visual Check

2. Identify the gametophyte and the sporophyte stages in this diagram.

____

____

____

____

____

Sporophyte Stage

During sexual reproduction, a male sex cell and a female sex cell combine. If fertilization occurs, a diploid cell forms, as shown in the figure at right. That diploid cell is the beginning of the sporophyte stage. This cell divides through mitosis and cell division and forms the sporophyte. In some plants, the sporophyte is a small structure. In other plants, such as an apple tree, the sporophyte is the tree.

Plant Life Cycle

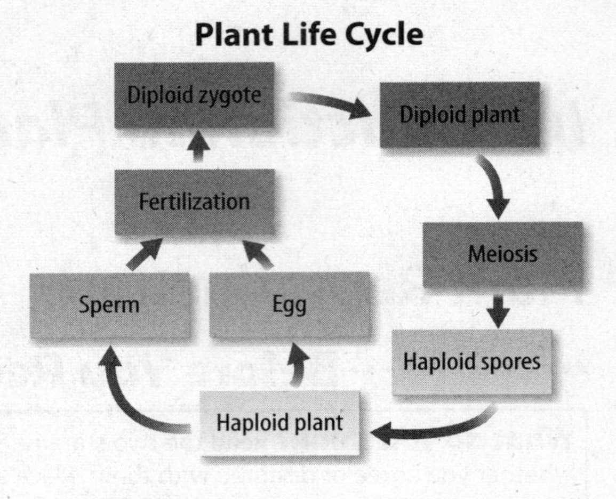

Seedless Plants

Plants that do not produce seeds are called seedless plants. They can reproduce by asexual reproduction or by producing spores. The sporophyte produces the spores. Recall that the sporophyte results from sexual reproduction. Mosses, liverworts, and ferns are seedless plants. The life cycle of a fern is shown in the figure below.

Visual Check

3. Draw a circle around the sex cells in the figure.

Life Cycle of a Fern

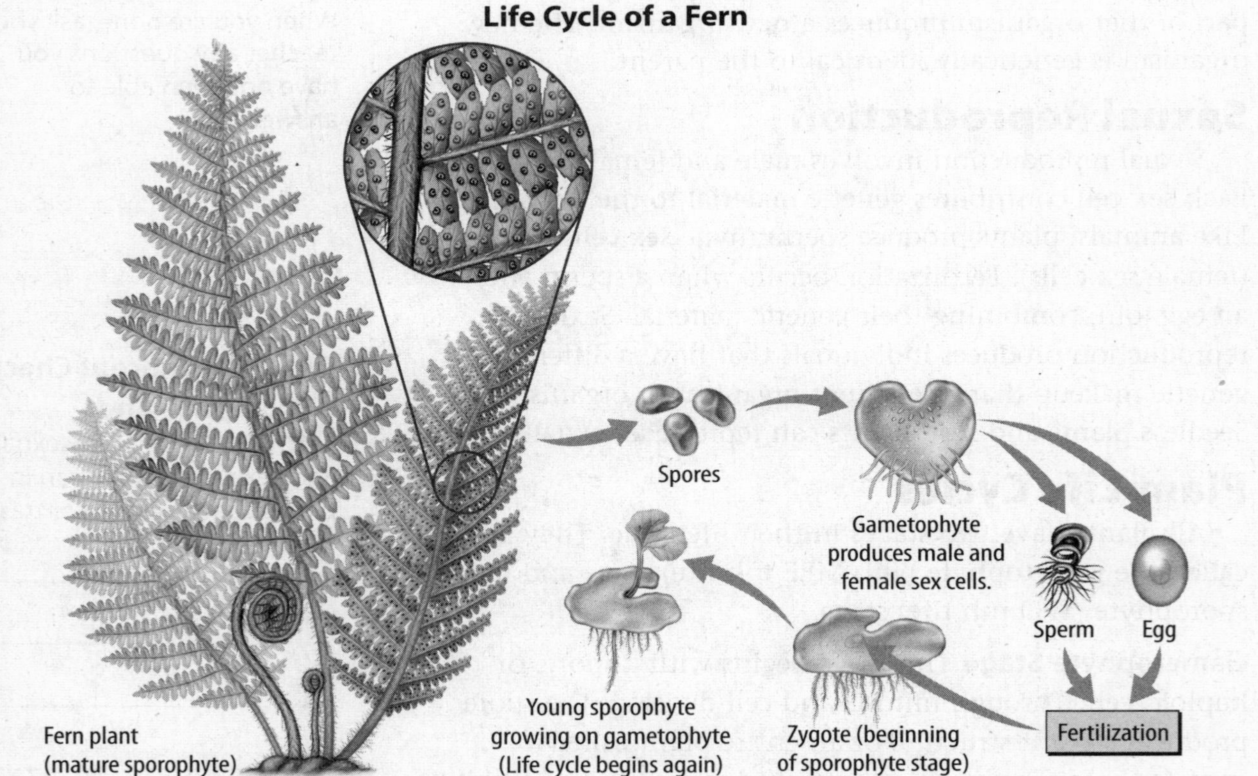

Seed Plants

Most plants produce seeds that result from sexual reproduction. The plants produce pollen grains, which contain sperm. Female structures in the plants produce one or more eggs. *The process that occurs when pollen grains land on a female plant structure of a plant in the same species is* **pollination**. If a sperm from a pollen grain joins with an egg, this is called fertilization. After fertilization occurs, the diploid cell undergoes many cell divisions, forming an embryo. The embryo is the beginning of the sporophyte stage of seed plants. The embryo and its food supply are enclosed within a protective coat. This is the seed, as shown below.

In most seed plants, the seed will go through **dormancy,** *which is a period of no growth.* Dormancy might last days, weeks, or even years. When environmental conditions are favorable, the seed will become active again. The process of a seed beginning to grow is called germination.

Gymnosperm Reproduction The life cycle shown in the figure below is typical of a gymnosperm. Notice that pollen is produced by the male cone; the eggs, and eventually the seeds, are contained within the female cone.

FOLDABLES®

Make a vertical three-tab book to compare and contrast the life cycles of seed plants and seedless plants.

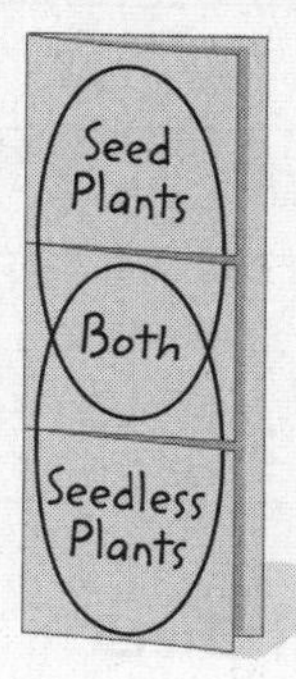

Visual Check

4. Examine Pollen is produced by ____. (Circle the correct answer.)

a. the female cone

b. the male cone

c. both the male and female cones

Gymnosperm Life Cycle

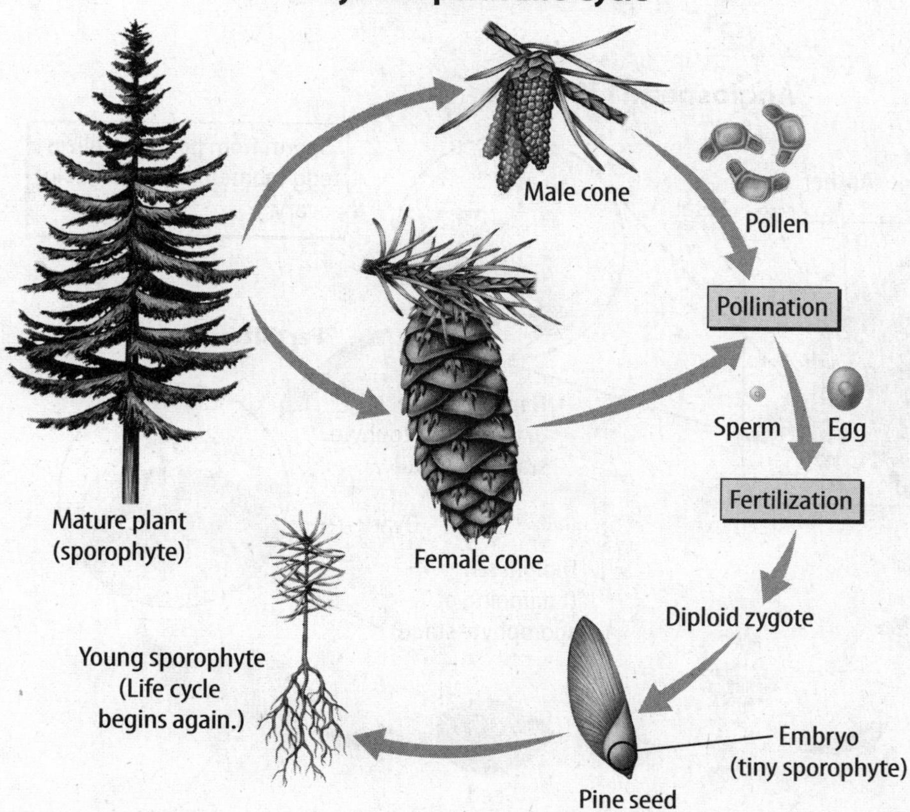

Key Concept Check
5. State What are the differences between life cycles of seedless and seed plants?

Visual Check
6. Describe what happens following fertilization.

Angiosperm Reproduction You probably have seen flowering plants in parks and around houses. Other plants, such as grasses and maple trees, also produce flowers. Some aquatic flowering plants are less than 1 mm in length.

Most flowers have four main structures. The petals, which attract insect or animal pollinators, might be brightly colored. The sepals usually are located beneath the petals and help protect the flower when it is a bud. *The female reproductive organ of a flower is the* **pistil**. It contains the ovary, where the seed develops. *The* **stamen** *is the male reproductive organ of a flower.* The anthers of the stamen produce pollen.

The figure below shows the parts of a flower. Some plants have flowers that have only the male or only the female structures. These are called male flowers or female flowers.

As shown in the figure below, the life cycle of a flowering plant includes both gametophyte and sporophyte stages. The gametophyte stage lasts a short time. It includes the production of eggs and sperm by a flower. When a sperm fertilizes an egg, the resulting diploid cell is the beginning of the sporophyte stage. In flowering plants, the sporophyte stage lasts much longer than the gametophyte stage.

Angiosperm Life Cycle

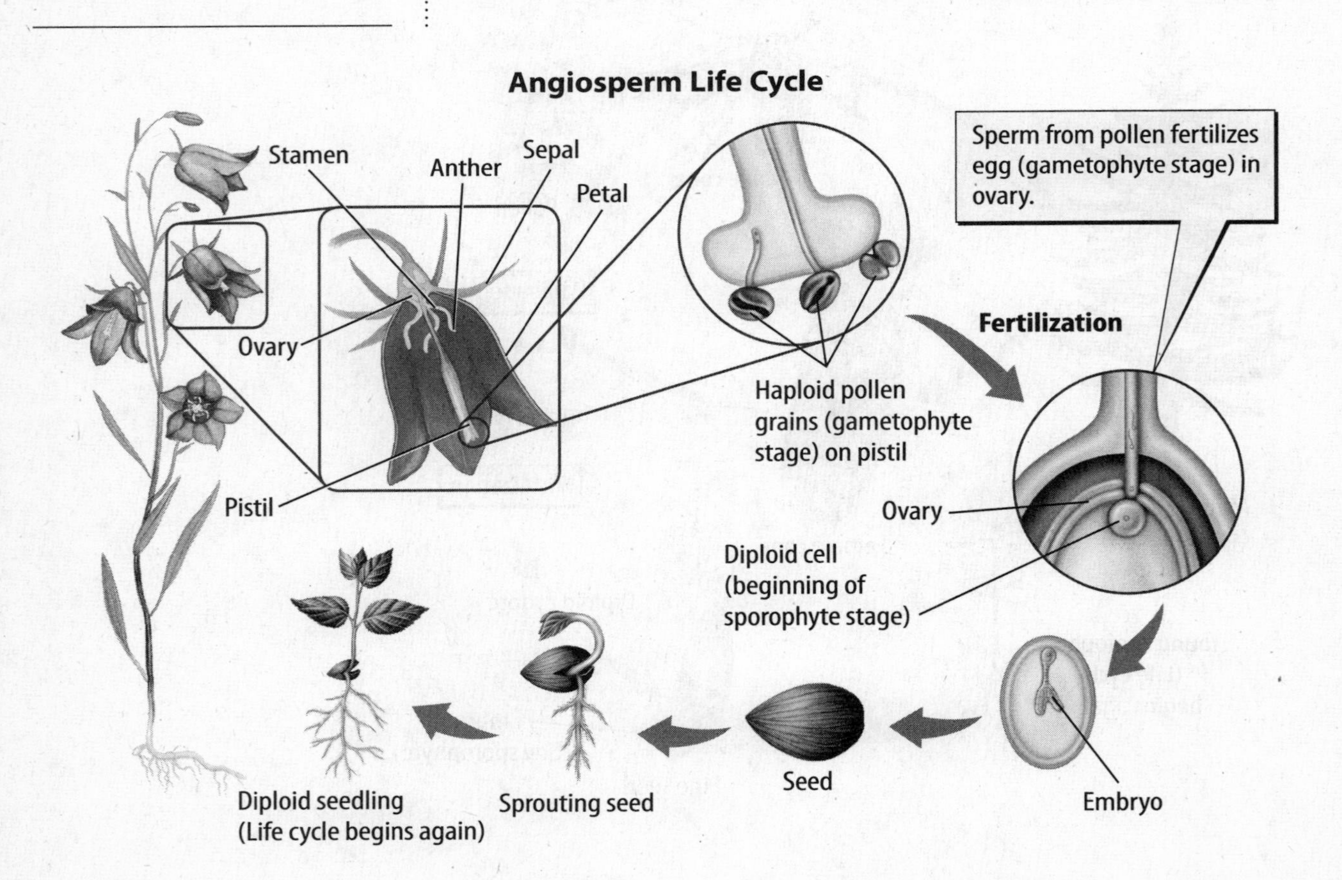

Growth Cycles Plants that grow from a seed and produce flowers in one growing season are called annuals. The seeds of annuals must be planted every year.

Some plants require two growing seasons to produce flowers. These are known as biennials and include carrots and beets. Biennials go through a period of dormancy between growing seasons. Many biennials have large roots that store food between growing seasons. ✓

Perennials are plants that grow and bud for many years. Some perennials, such as trees, can grow for hundreds of years. In cold climates, some perennials lose their leaves and become dormant for several months. When warmer temperatures return, the plant produces new leaves. They begin capturing sunlight for photosynthesis.

✓ **Reading Check**

7. Compare biennials with annuals.

After You Read

Mini Glossary

dormancy: a period of no growth

pistil: the female reproductive organ of a flower

pollination: the process that occurs when pollen grains land on a female plant structure of a plant in the same species

stamen: the male reproductive organ of a flower

1. Review the terms and their definitions in the Mini Glossary. Write a sentence describing the stamen and its purpose.

___

___

2. Use the diagram to compare sexual and asexual reproduction in plants. Write the letters of the phrases in the box below the type of reproduction it describes.

a. offspring have genetic makeup different than parents

b. requires only one parent organism

c. both male and female sex cells contribute genetic material

d. offspring are genetically identical to parent

e. requires sperm and egg

f. new plant grows from a plant part

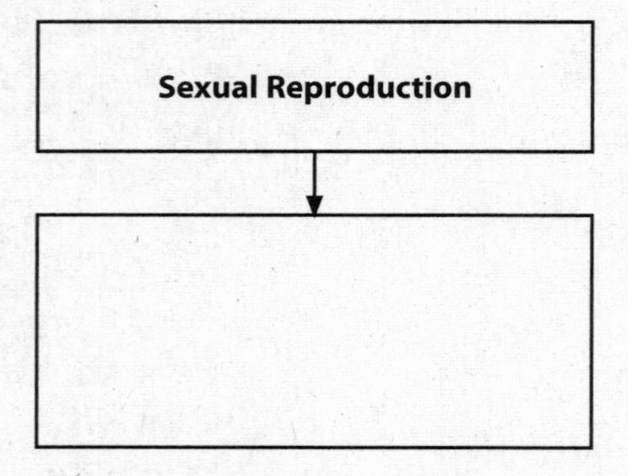

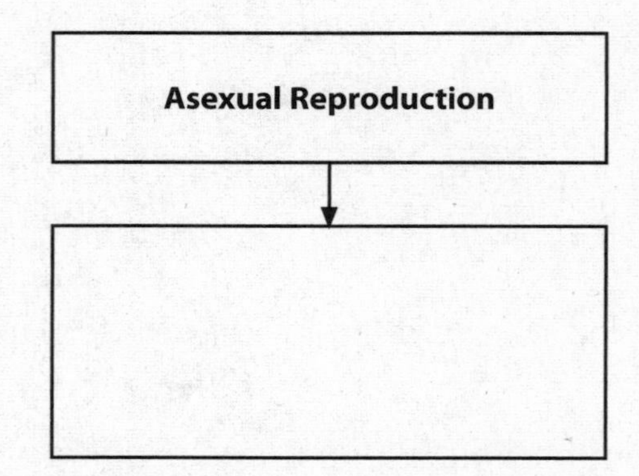

3. Give an example of dormancy in perennials.

___

___

What do you think NOW?

Reread the statements at the beginning of the lesson. Fill in the After column with an A if you agree with the statement or a D if you disagree. Did you change your mind?

Log on to ConnectED.mcgraw-hill.com and access your textbook to find this lesson's resources.

Introduction to Plants

Plant Processes

Before You Read

What do you think? Read the two statements below and decide whether you agree or disagree with them. Place an A in the Before column if you agree with the statement or a D if you disagree. After you've read this lesson, reread the statements to see if you have changed your mind.

Before	Statement	After
	5. Plants respond to their environments.	
	6. Because plants make their own food, they do not carry on cellular respiration.	

Key Concepts 

- What is the relationship between photosynthesis and cellular respiration?
- How do water and minerals move in vascular and nonvascular plants?
- How do plants respond to environmental changes?

Read to Learn

Photosynthesis and Cellular Respiration

All animal life depends upon plants. Even organisms that don't eat plants eat other organisms that do eat plants. Plants absorb light energy from the Sun and convert it into chemical energy through photosynthesis. Photosynthesis produces sugar that the plant uses as food. Organisms need energy for growth, movement, and other life processes. This energy comes from cellular respiration, which is the process of releasing energy by breaking down food.

Mark the Text

Building Vocabulary As you read, underline the words and phrases that you do not understand. When you finish reading, discuss these words and phrases with another student or your teacher.

Key Concept Check

1. Explain What is the relationship between photosynthesis and cellular respiration?

____

____

____

____

____

Making Sugars By Using Light Energy

For most plants, photosynthesis occurs in the leaves. Some leaf cells contain chloroplasts. Photosynthesis occurs inside these organelles. Chloroplasts contain chlorophyll, a green pigment that absorbs light energy. That energy splits apart water molecules into hydrogen atoms and oxygen atoms. Some of the oxygen leaves the plant through the stomata. Carbon dioxide in the leaf combines with the hydrogen atoms and forms glucose, a type of sugar. Photosynthesis can be shown by the following equation:

$$6CO_2 + 6H_2O \xrightarrow[\text{Chlorophyll}]{\text{Light energy}} C_6H_{12}O_6 + 6O_2$$

Breaking Down Sugars

Cellular respiration breaks down the glucose produced during photosynthesis and releases the sugar's energy. This process occurs in the cytoplasm and the mitochondria. As shown below, oxygen also is used during cellular respiration. The equation for cellular respiration is as follows:

$$C_6H_{12}O_6 + 6O_2 \longrightarrow 6CO_2 + 6H_2O + \text{ATP (Energy)}$$

During cellular respiration, glucose molecules release more energy than cells can use at one time. That excess energy is stored in a molecule called adenosine triphosphate (uh DEN uh seen • tri FAHS fayt), or ATP. It is used later for other cell processes. ✓

Reading Check

2. State What enables cells to use the energy from glucose molecules?

____

____

The Importance of Photosynthesis and Cellular Respiration

Do you know why plants are so important to life on Earth? One answer can be found in the two equations you just learned. Organisms, such as humans, need oxygen. Each time you inhale, your lungs fill with air that contains oxygen. Your body uses that oxygen for cellular respiration.

In your body's cells, cellular respiration breaks down food and stores the energy from food in ATP. Cellular processes such as growth, repair, and reproduction all use ATP. During cellular respiration, carbon dioxide and water are given off as waste products. Plants use these two compounds for photosynthesis.

Most organisms, including humans, use the products of photosynthesis—sugars and oxygen—during cellular respiration. Plants and some other organisms can use the waste products of cellular respiration—carbon dioxide and water—during photosynthesis. It is important to remember that plants also carry on cellular respiration, so they will use some of the oxygen released during photosynthesis.

Math Skills

A proportion is an equation with two ratios that are equivalent. The cross products of the ratios are equal. Proportions can be used to solve problems such as the following: In a cell, when one molecule of glucose breaks down completely to carbon dioxide and water, 36 ATP molecules are produced. How many ATP molecules are produced when 30 glucose molecules break down?

Set up the proportion.

$$\frac{\text{1 molecule glucose}}{\text{36 molecules ATP}} = \frac{\text{30 molecules glucose}}{x\text{ molecule ATP}}$$

Cross multiply.

$x = 30 \times 36$

$x = 1{,}080$ molecules ATP

3. Use Proportions During photosynthesis, 18 ATP molecules are required to produce 1 glucose molecule. How many ATP molecules would be required to produce 2,500 glucose molecules?

____

Movement of Nutrients and Water

In order for plants to carry on processes such as photosynthesis and cellular respiration, water and nutrients must move inside the plants. In nonvascular plants, this movement or transport of materials occurs through diffusion and osmosis. In vascular plants, water and nutrients move inside specialized vascular tissues. Osmosis and diffusion also move materials once they are outside the vascular tissues.

Absorption

Roots and rhizoids of plants absorb water and nutrients from the soil. Once inside a plant, water and nutrients move to cells, where they are used in cellular processes.

Water is used for photosynthesis. It also is part of many other chemical reactions inside cells. Nutrients from the soil, such as minerals, are used for making many of the compounds needed for cell growth and maintenance.

Transpiration

Water is a waste product of cellular respiration. Plants release excess water as water vapor in a process called transpiration. **Transpiration** *is the release of water vapor from stomata in leaves.*

Transpiration helps move water from the roots, up through the vascular tissue, and to the leaves. This movement provides water for photosynthesis and helps cool a plant on hot days.

Examine the figure below. Follow the path of water from the soil, up through the plant, and out of the leaves.

Key Concept Check

4. Explain How do water and nutrients move in a nonvascular plant? In a vascular plant?

____

____

____

____

____

____

Visual Check

5. Trace the path of water through the plant.

Importance of Water and Nutrients

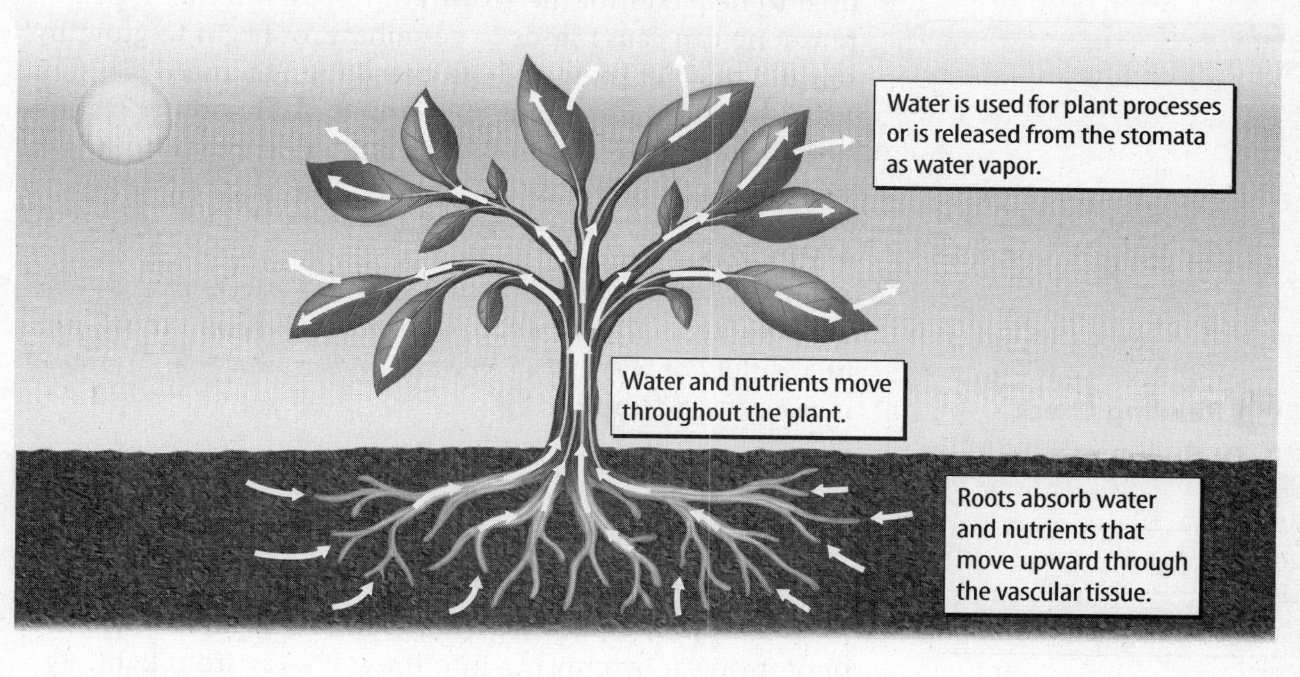

FOLDABLES®

Make a two-tab book to organize information about vascular and nonvascular plants.

Plant Responses

Can you remember the last time a loud noise startled you? You might have jumped and turned around to see what made the noise. **Stimuli** (singular, stimulus) *are any changes in an organism's environment that cause a response.*

Although a plant might not jump or turn around like a person would, a plant can respond to stimuli in a number of ways. For example, a mimosa plant responds to the stimulus of being touched by collapsing its leaves.

Types of Stimuli

Plants respond to external and internal stimuli. External stimuli include light, touch, and gravity. Internal stimuli occur inside a plant. Internal stimuli are chemicals, called hormones, that the plant produces.

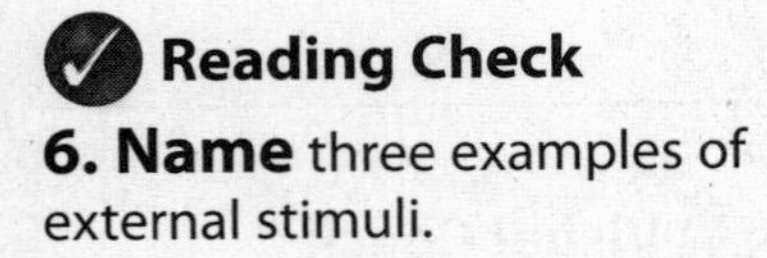

Reading Check

6. Name three examples of external stimuli.

Plants produce many different hormones. These hormones can affect growth, seed germination, or fruit ripening. The hormones that promote growth increase the rate of mitosis and cell divisions.

Certain hormones slow growth. These hormones can be used to help control the growth of weeds. Another type of hormone can cause seeds to germinate, or begin to grow, by starting the breakdown of the stored food in a seed. This releases energy needed for new growth. Still another plant hormone often is used to speed up the ripening of fruit to be sold in grocery stores.

Tropisms

Any external environmental stimulus affects plants. This includes light, gravity, and touch. Plants respond to stimuli in a variety of ways. *Plant growth toward or away from an external stimulus is called* **tropism**.

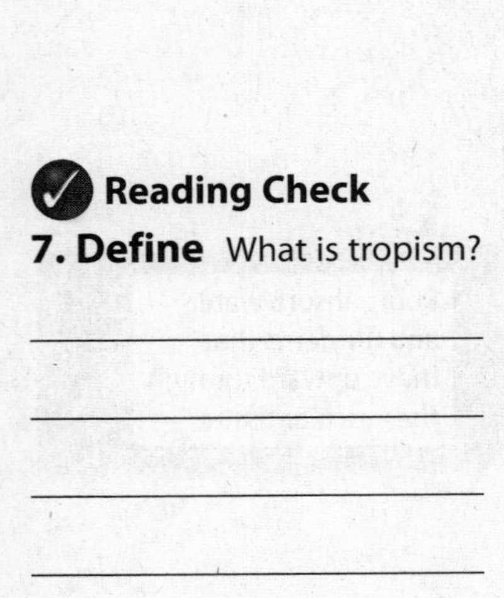

Reading Check

7. Define What is tropism?

Phototropism When a plant grows toward a light source, it is called positive phototropism. Growing toward a light source enables leaves and stems to receive the maximum amount of light for photosynthesis.

The roots of a plant generally exhibit negative phototropism by growing into the soil away from light. By growing into the soil, the roots are able to anchor the plant.

Gravitropism A plant's response to gravity is called gravitropism. The first root produced by a germinating seed grows downward. This is positive gravitropism. Gravitropism enables the new plant to be anchored in soil, where it can absorb water.

By contrast, a plant's stems and leaves grow upward and away from gravity. This is negative gravitropism. This response enables leaves to be exposed to light, making photosynthesis possible.

Thigmotropism Did you know that plants have a sense of touch? A plant's response to touch is called thigmotropism. The coiling of a vine's tendrils around another plant or object is an example of positive thigmotropism. A plant's roots exhibit negative thigmotropisn when they grow around a rock in the soil.

Reading Check

8. Describe an effect of positive gravitropism.

Key Concept Check

9. Summarize How do plants respond to environmental changes?

After You Read

Mini Glossary

stimulus: any change in an organism's environment that causes a response

transpiration: the release of water vapor from stomata in leaves

tropism: plant growth toward or away from an external stimulus

1. Review the terms and their definitions in the Mini Glossary. Write a sentence describing the benefits of transpiration to a plant.

___

___

2. This diagram describes three different plant responses to external stimuli. Write the correct term in the box below its definition.

phototropism **gravitropism** **thigmotropism**

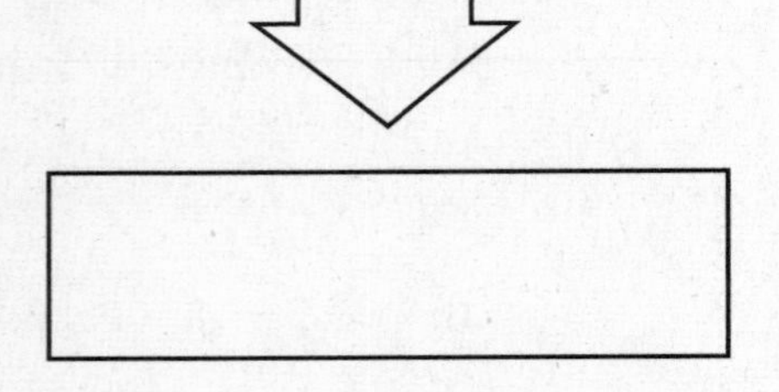

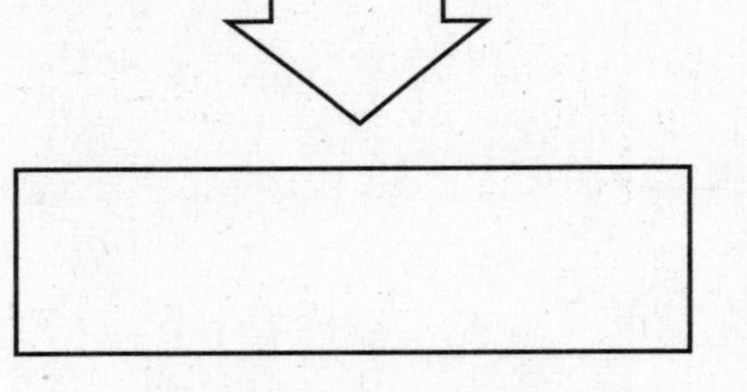

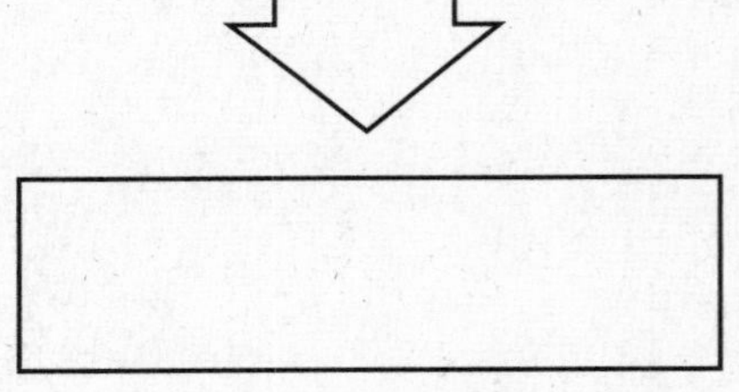

3. Suppose your friend believes that animals that eat only meat, such as lions, do not need plants to survive. Provide a reason why your friend is wrong.

___

___

What do you think NOW?

Reread the statements at the beginning of the lesson. Fill in the After column with an A if you agree with the statement or a D if you disagree. Did you change your mind?

Log on to ConnectED.mcgraw-hill.com and access your textbook to find this lesson's resources.

Introduction to Animals

What are animals?

Before You Read

What do you think? Read the two statements below and decide whether you agree or disagree with them. Place an A in the Before column if you agree with the statement or a D if you disagree. After you've read this lesson, reread the statements to see if you have changed your mind.

Before	Statement	After
	1. Animals must eat plants or other animals to live.	
	2. All animals have a left side and a right side that are similar.	

Key Concepts 

- What characteristics are common to all animals?
- How do scientists group animals?
- How are animal species adapted to their environments?

Read to Learn

Animal Characteristics

Zoos are interesting to people of all ages. Why do people keep coming back to the zoo year after year? To see the animals of course! In fact, *zoo* comes from the Greek word *zoion*, which means "living being" or "animal."

Like plants, all animals are multicellular. Also like plants, each animal cell has a nucleus at some point during its life. While cell walls support plant cells, a protein called collagen holds animal cells together. Animals are the only organisms that have nerve cells. Nerve cells conduct nerve impulses. Most animals also have muscle cells that help them move.

Animals cannot transform light energy into food energy as most plants can. All animals get energy from the food they take into their bodies. In most animals, food passes through their stomachs, and then their intestines absorb nutrients from the food.

All animals begin as a fertilized egg cell called a zygote. Recall that fertilization is the joining of an egg cell with a sperm cell. The zygote divides into more cells and forms an embryo. After many more cell divisions, the body of the animal is recognizable.

Mark the Text

Building Vocabulary As you read, circle all the words you do not understand. Highlight the part of the text that helps you understand these words. Review the marked words and their definitions after you finish reading the lesson.

Key Concept Check

1. Summarize What characteristics are common to all animals?

FOLDABLES®

Make a vertical two-column chart to organize your notes on animal characteristics and identification.

How do scientists group animals?

When you first learned to talk, you probably grouped things by what they looked like. For example, you might have called any round object "ball." As you developed, however, you came to know the differences between things, such as trees and cats. You knew that dogs were not birds. This was your first experience with classification. Scientists classify animals in many different ways.

Symmetry

One way to group animals is by looking at their symmetry, or how body parts are arranged. Animal bodies can have bilateral symmetry, radial symmetry, or no symmetry at all. The figure below illustrates the three types of symmetry.

Some animals have **bilateral symmetry,** *a body plan in which an organism can be divided into two parts that are nearly mirror images of each other.* Humans, frogs, and the gecko on the left in the figure below are examples of organisms with bilateral symmetry.

An animal with **radial symmetry** *has a body plan which can be divided anywhere through its central axis into two parts that are nearly mirror images of each other.* The sand dollar in the center of the figure below has radial symmetry.

Some animals do not have bilateral symmetry or radial symmetry. They have **asymmetry,** *meaning they have body plans which cannot be divided into any two parts that are nearly mirror images.* The sponge on the right in the figure below has asymmetry. ✓

Reading Check

2. Name the three types of symmetry.

Visual Check

3. Examine Can you find a way to divide the sponge into mirror image parts? Explain.

Symmetry

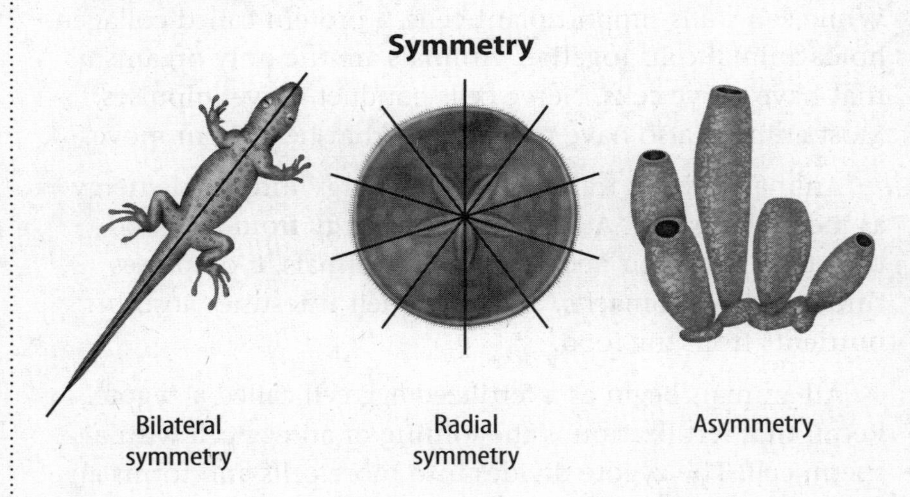

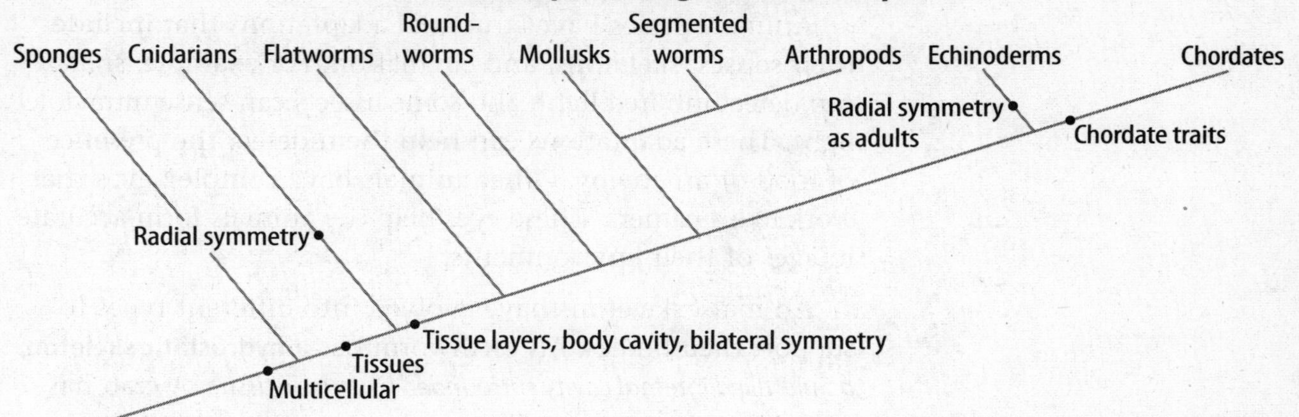

Groups of Animals

Scientists use a system called taxonomy to group organisms. This system groups organisms into levels called taxons. Taxons are groups of organisms that have certain traits in common.

Taxonomy The biggest groups in taxonomy are called domains. Because each animal cell has a nucleus at some point in its life, animals are in the Domain Eukarya. Kingdoms are the next level of taxons. Scientists use the traits you read about earlier to decide whether an organism belongs in the Kingdom Animalia. Animals then are classified into phyla (singular, phylum), genera (singular, genus), and species. Lessons 2 and 3 cover the nine most common animal phyla, shown along the top of the figure above.

Family Tree A family tree shows the relationships among and within generations of a family. Animal phyla also are organized by how they are related through time. A tree of phyla shows the relationships among animals. As each new feature evolved over time, the feature was placed on a new branch of the tree. For example, all animals above sponges have tissues, as shown in the diagram at the top of this page.

Animal Adaptations

Have you ever seen a film about gorillas in the wild? A gorilla eats when it is hungry. It has hands that can grasp objects. A gorilla's response to hunger and the structure of its hands are adaptations. *An* **adaptation** *is an inherited trait that increases an organism's chances of surviving and reproducing in its environment*. Animal adaptations can be structural, behavioral, or functional.

Visual Check

4. Identify Which animals are multicellular?

Key Concept Check

5. Classify How do scientists group animals?

Structural Adaptations

Animal species have structural adaptations that include their senses, skeletons, and circulation. For example, snakes can detect infrared light, and some insects can sense ultraviolet light. These adaptations can help them detect the presence of food or an enemy. Other animals have complex eyes that work like a camera. These eyes help the animals form accurate images of their environments.

Animals' skeletons have evolved into different types to support their bodies. An earthworm has a **hydrostatic skeleton,** *a fluid-filled internal cavity surrounded by muscle tissue.* A crab has soft internal structures. They are protected by a *thick, hard outer covering called an* **exoskeleton.** You probably are most familiar with *the internal rigid framework called an* **endoskeleton** *that supports you and other animals.* Your endoskeleton is made of bone. Your muscles attach to your bones and help you move. ✓

Reading Check

6. State What are the three types of animal skeletons?

Animal species also have structural adaptations for circulating blood. For example, ants have an open circulatory system. An ant's heart pumps blood into open spaces around its organs. However, an earthworm has a closed circulatory system. Its many hearts pump blood through a system of vessels. Other animals with closed circulation have only one heart.

Many animal species have evolved structural adaptations. For example, a chameleon has a long, sticky tongue to catch insects. A beaver's teeth have evolved a sharp edge that can cut through tree trunks.

Think it Over

7. Apply Give another example of a structural adaptation in a species.

Behavioral Adaptations

Animals are born with behaviors called instincts. These behaviors have evolved over time and help species survive in their environments. A male fly instinctively waves its wings to attract a female's attention. This action makes it more likely that the flies will breed and have offspring. Many tropical birds instinctively migrate when the number of daylight hours changes. These bird species have adapted to fly thousands of miles for food and breeding habitats. ✓

Reading Check

8. Define What are instincts?

The ability to learn behaviors also is an important animal adaptation. For example, young songbirds learn how to sing their songs by listening to their parents. Baby geese learn to follow their mothers soon after birth. This form of learned behavior is called imprinting. These behavioral adaptations increase a species' ability to survive and produce offspring.

Functional Adaptations

Animal species also have functional adaptations, which help them survive in their environment or maintain homeostasis. Some of these adaptations enable animals to reproduce successfully either in water or on land.

Most animals that live in water release large numbers of eggs or sperm into the water. If fertilization occurs in the water, the process is called external fertilization. If fertilization occurs inside a female, it is called internal fertilization. Most animals that live in the water produce many eggs and sperm. A water environment doesn't provide much protection for developing young. Many of the young do not survive. Fertilizing many eggs ensures that a few will survive.

Most animal species that live on land use internal fertilization. Because the fertilized eggs are inside a female, the female's body protects them. Therefore, these species need to fertilize only a few eggs to ensure that some young will survive.

Key Concept Check

9. Explain How are species adapted to their environments?

____

____

____

After You Read

Mini Glossary

adaptation: an inherited trait that increases an organism's chances of surviving and reproducing in its environment

asymmetry: a body plan in which an organism cannot be divided into any two parts that are nearly mirror images

bilateral symmetry: a body plan in which an organism can be divided into two parts that are nearly mirror images of each other

endoskeleton: an internal rigid framework that supports humans and other animals

exoskeleton: a thick, hard outer covering

hydrostatic skeleton: a fluid-filled internal cavity surrounded by muscle tissue

radial symmetry: a body plan in which an organism can be divided anywhere through its central axis into two parts that are nearly mirror images of each other

1. Review the terms and their definitions in the Mini Glossary. Write a sentence that shows that you know the difference between an endoskeleton and an exoskeleton.

2. Complete the diagram by identifying the taxonomy scientists use for organisms, from the largest group to the smallest. Write the classifications in the boxes.

genus **species** **domain** **phylum**

kingdom

3. Define one of the words that you circled as you read this lesson.

What do you think NOW?

Reread the statements at the beginning of the lesson. Fill in the After column with an A if you agree with the statement or a D if you disagree. Did you change your mind?

Log on to ConnectED.mcgraw-hill.com and access your textbook to find this lesson's resources.

Introduction to Animals

Invertebrates

Before You Read

What do you think? Read the two statements below and decide whether you agree or disagree with them. Place an A in the Before column if you agree with the statement or a D if you disagree. After you've read this lesson, reread the statements to see if you have changed your mind.

Before	Statement	After
	3. A sponge is not an animal because it cannot move.	
	4. There are more arthropods on Earth than all other kinds of animals combined.	

Key Concepts 

- What characteristics do invertebrates have in common?
- How do the groups of invertebrates differ?

Read to Learn

What is an invertebrate?

Have you ever seen sea anemones? Perhaps you saw them at an aquarium. Many people think sea anemones look like colorful flowers. But sea anemones are not flowers. They are animals. They trap food in their fingerlike tentacles. Anemones do not have backbones.

Recall how animals support their bodies. Most animals with an endoskeleton have a backbone for support. These animals are called vertebrates. Animals without backbones are called invertebrates. The bodies of most invertebrates are supported by either a hydrostatic skeleton—a fluid-filled internal cavity—or an exoskeleton—a hard outer covering. Some invertebrates have endoskeletons.

Invertebrates make up about 95 percent of all known animal species. In this lesson, you will read about eight of the most common invertebrate phyla. Recall that phyla are one of the levels of taxons.

Invertebrates have many adaptations for survival. Some invertebrates are parasites. **Parasites** *are animals that survive by living inside or on another organism, get food from the organism, and do not help in the organism's survival.* Other invertebrates hunt their food. Some invertebrates can change the color of their skin to match their environments.

Study Coach

Make an Outline As you read, summarize the information in the lesson by making an outline. Use the main headings in the lesson as the main headings in your outline. Use your outline to review the lesson.

Key Concept Check

1. Examine What characteristics do invertebrates have in common?

Academic Vocabulary
attach
(verb) to fasten

Visual Check
2. Explain How do sponges obtain food?

____

____

____

Visual Check
3. Identify Circle the part of the nematocyst that injects the poison.

Sponges

The oldest branch of the animal family tree, phylum Porifera (puh RIH fuh ruh), includes the sponges. Sponges are often called simple animals because they have only a few types of cells and no true tissues. Sponges live in water. Adult sponges cannot move. They attach to rocks and other underwater structures. Sponges take in food when water passes through their bodies, as shown in the figure to the right. Special cells inside the sponge filter out food particles in the water.

Sponge

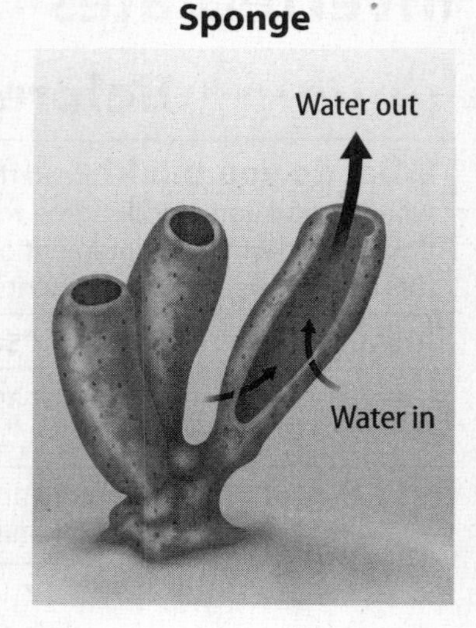

Sponges have tiny, stiff fibers that support their bodies. Scientists group sponges by the kinds of materials that make up these fibers. The most common group of sponges has fibers made of either silica or the protein spongin, or both. These sponges are harvested and sold as natural sponges. The stiff fibers make the sponges useful for scrubbing, but the fibers can scratch shiny surfaces.

Cnidarians

Corals, anemones, jellyfish, hydras, and Portuguese man-of-wars are all members of phylum Cnidaria (nih DAYR ee uh). The name *Cnidaria* comes from special cells these animals use to catch their prey. These cells—nematocysts (NE mah toh sihsts)—are shown in the figure to the right. Nematocysts can inject poison into animals that come in contact with them. Cnidarians have radial symmetry.

Nematocyst

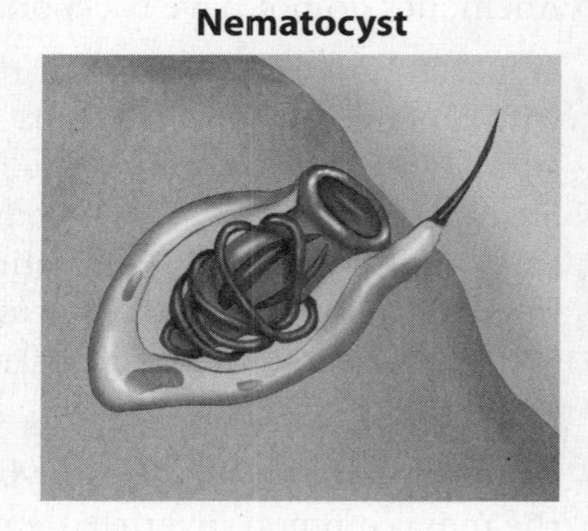

Cnidarians are more complex than sponges because cnidarians have true tissues. Some cnidarians, such as corals and anemones, remain attached to underwater surfaces for all of their adult lives. Others, such as jellyfish, can swim.

Flatworms

The common name for an animal in the phylum Platyhelminthes (pla tih hel MIHN theez) is flatworm. This name describes it accurately because its body shape is flat. Flatworms have bilateral symmetry; each worm has a left side and a right side that are similar.

Most flatworms live in water, either freshwater or salt water. Some flatworms are free-living. The planarian swims freely in water and ingests food through a tube on the underside of its body. Other flatworms are parasites. The liver fluke is a parasite and can infect humans.

Segmented Worms

Have you ever held an earthworm? Did you notice that its body was like a tube of tiny rings? The name for the phylum that includes earthworms, *Annelida* (ah NEL ud uh), means "little rings." These rings, as shown in the figure below, are called segments. Each segment is a fluid-filled compartment. Therefore, a segmented worm has a hydrostatic skeleton. Segmented worms have bilateral symmetry. 

Reading Check

4. Explain Why are Annelida called segmented worms?

Did you also notice that the sides of the earthworm's body felt prickly? The prickles are tiny, stiff hairs called setae (SEE tee). Setae help earthworms grip surfaces. As earthworms tunnel through soil, they take the soil into their bodies and absorb nutrients from it. Their tunnels help break up soil. Segmented worms also can be parasites. Leeches attach their mouths to other animals and suck blood.

5. Draw In the cross-section of the earthworm, draw a line to show the earthworm's bilateral symmetry.

Earthworm

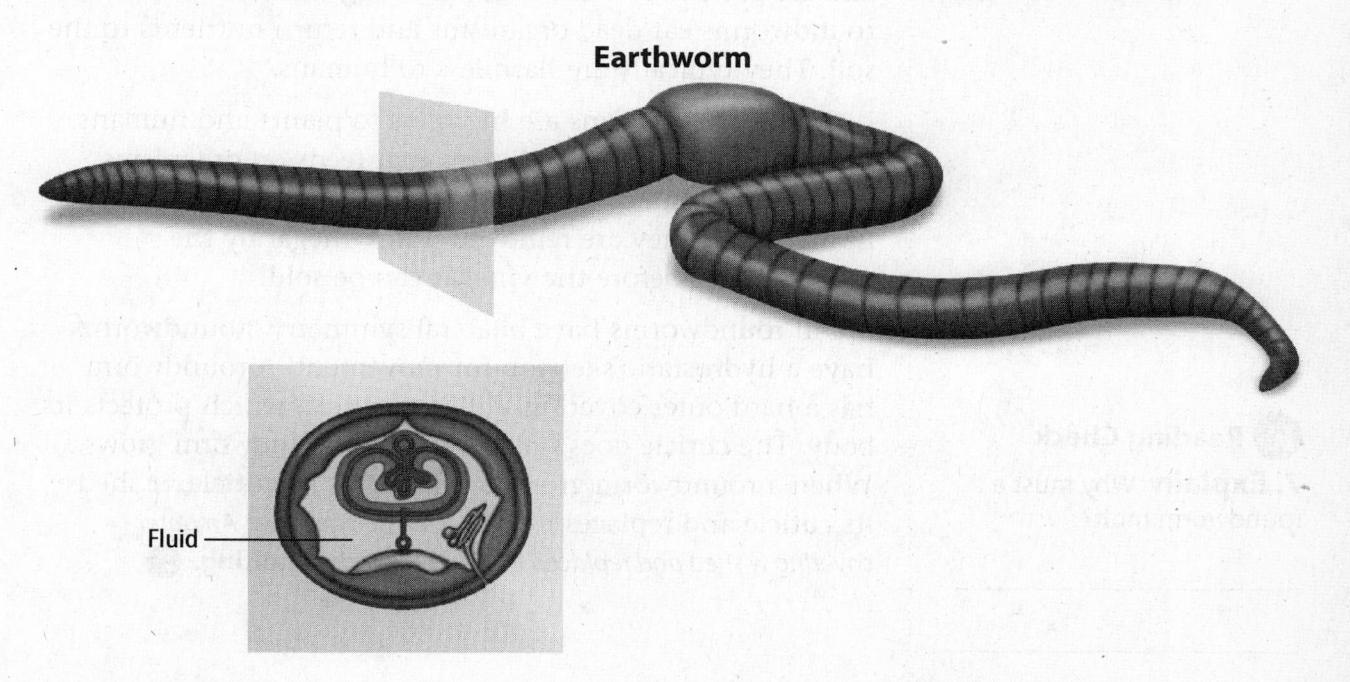

FOLDABLES

Make a horizontal two-tab book to identify similarities and differences in invertebrates.

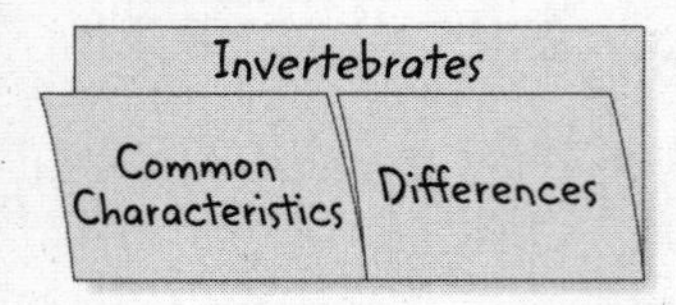

Mollusks

On summer mornings you might notice thin, slimy trails across a sidewalk. These trails likely were made during the night by snails or slugs searching for food. Snails and slugs are mollusks in phylum Mollusca (mah LUS kuh).

Most mollusks have a footlike muscle that the animal generally uses for movement. A mollusk also has a mass of tissue called a mantle. *A* ***mantle*** *is a thin layer of tissue that covers a mollusk's internal organs.* The mantle also is involved in making the shell of most mollusks.

A shell supports and protects a mollusk's soft body. Most mollusks, such as snails, have external shells. Some mollusks, such as slugs, do not have shells. Other mollusks, such as squids and octopuses, have internal shells. Mollusks have bilateral symmetry. ✓

✓ Reading Check

6. Identify Name one type of mollusk that does not have a shell.

____

Mollusks obtain food in different ways. Some mollusks, such as clams, oysters, and scallops, filter food particles from the water in which they live. Other mollusks, such as octopuses and squids, are predators and catch their prey in long, strong tentacles.

Roundworms

Animals in phylum Nematoda (ne muh TOH duh) are called nematodes or roundworms. Some roundworms are parasites. Some parasitic roundworms infect plant roots, while others infect humans. Most roundworms live in soil and are too small to see without a magnifying lens. These roundworms eat dead organisms and return nutrients to the soil. They typically are harmless to humans.

Other roundworms are harmless to plants and humans. The vinegar eel is a roundworm that feeds on organisms used in making vinegar. Vinegar eels are harmless to humans, but they are removed from vinegar by the manufacturer before the vinegar can be sold.

All roundworms have bilateral symmetry. Roundworms have a hydrostatic skeleton for movement. A roundworm has a hard outer covering, called a cuticle, which protects its body. The cuticle does not grow as the roundworm grows. When a roundworm grows too large for its cuticle, it sheds its cuticle and replaces it with a larger cuticle. *An outer covering is shed and replaced in a process called* ***molting.*** ✓

✓ Reading Check

7. Explain Why must a roundworm molt?

____

____

____

Arthropods

Can you imagine a billion billion of something? Scientists estimate that is how many individual arthropods exist on Earth. There are more animals in phylum Arthropoda (ar THRAH puh duh) than in all other animal phyla combined. Arthropods have bilateral symmetry.

Like a roundworm, an arthropod has a hard outer covering, so it must molt in order to grow. An arthropod has an exoskeleton for both movement and protection. Its muscles attach to the exoskeleton. An arthropod uses its muscles when moving its jointed appendages. An appendage is a structure, such as a leg or an arm, that extends from the central part of the body.

The body of an arthropod has three parts: a head, a thorax, and an abdomen. The head contains sense organs that see, feel, and taste the environment. The thorax is the part of the body where legs attach. The abdomen contains intestines and reproductive organs. ✓

Arthropods have open circulation. This means their blood is not in vessels. Instead, it washes over internal organs.

Insects Most arthropods are insects. Scientists call them hexapods because they have six legs. Insects are the only arthropods that have the ability to fly.

Another trait of insects is metamorphosis. In **metamorphosis,** *the body form of an animal changes as it grows from an egg to an adult.* The stages in the metamorphosis of a butterfly are egg, larva (caterpillar), pupa, and adult butterfly. ✓

Other Arthropod Groups There are three other major groups of arthropods. Spiders and scorpions are one group. They have eight legs used for walking and grasping.

Crabs and lobsters make up another group. Members of this group mostly live in water. They have chewing mouthparts and three or more pairs of legs. Some lobsters have as many as 19 pairs of appendages.

Centipedes and millipedes are in another group. They have the most appendages. Generally, a centipede has one pair of legs per segment and a millipede has two pairs per segment. Millipedes eat dead plants, but centipedes are predators. ✓

✓ **Reading Check**

8. Name What are the three parts of an arthropod's body?

____

____

____

✓ **Reading Check**

9. Paraphrase In your own words, describe metamorphosis.

____

____

____

✓ **Reading Check**

10. Distinguish Which group of arthropods has the most appendages?

____

____

Echinoderms

Have you ever seen a sea star, a sea cucumber, or a sea urchin? At first glance, they may appear fuzzy and soft. But if you touch an echinoderm (ih KI nuh durm), from the phylum Echinodermata (ih kin uh DUR muh tuh), you will discover that it feels the opposite of soft. *Echinoderm* means "spiny skin." An echinoderm feels spiny due to the hard endoskeleton just beneath its thin outer skin.

All echinoderms live in salt water. They move slowly with tiny suction-cuplike feet, called tube feet. Their tube feet are connected to larger tubes called canals. These canals connect to a central ring that controls water movement within the animal. Water moves back and forth through the canals and tube feet. This movement enables echinoderms to grab onto or let go of any surface they are moving across. Echinoderms have bilateral symmetry when they are young and radial symmetry as adults.

Echinoderms develop differently than the other invertebrate phyla. Their early development is similar to that of vertebrates. Because of this, echinoderm embryos often are used to study human development and early growth patterns.

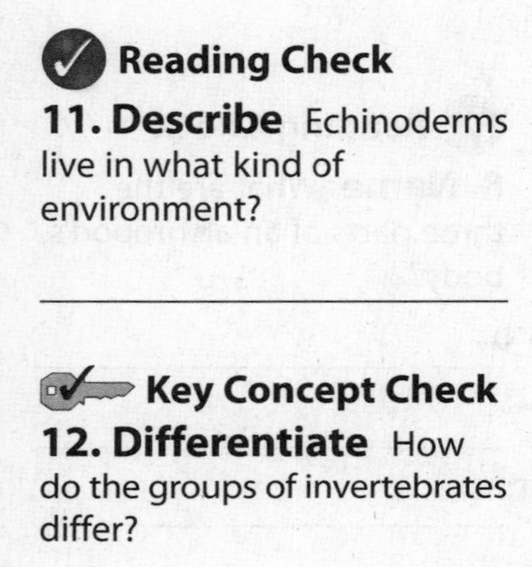

Reading Check

11. Describe Echinoderms live in what kind of environment?

Key Concept Check

12. Differentiate How do the groups of invertebrates differ?

After You Read

Mini Glossary

mantle: a thin layer of tissue that covers a mollusk's internal organs

metamorphosis: the process of change in an animal's body form as it grows from an egg to an adult

molting: a process by which an outer covering is shed and replaced

parasite: an animal that survives by living inside or on another organism, gets food from the organism, and does not help in the organism's survival

1. Review the terms and their definitions in the Mini Glossary. Write a sentence that explains the process of molting.

___

___

2. Identify the phylum and type of symmetry for each invertebrate described in the table. Some of them have been done for you.

Phylum	Type of Symmetry	Features	Examples
a.	asymmetry	few cell types; no true tissues; cannot move as adults; attach to underwater structures; stiff fibers support body	sponges
b.		use nematocysts to inject poison into their prey; have true tissues	corals, anemones, jellyfish
c. Platyhelminthes		flat body shape; most live in water; can be parasites or free-living	planarian, liver fluke
d.		divided into segments; have hydrostatic skeletons; some are parasites	earthworms, leeches
e.	bilateral	move with footlike muscle; have a mantle; most have shells	snails, slugs, clams, octopuses
f.		most live in soil and are too small to see without magnification; most harmless; some can infect humans or plants; molt exoskeleton	vinegar eel, other roundworms
g. Arthropoda		phylum contains more animals than all other animal phyla combined; hard exoskeleton; have head, thorax, and abdomen; open circulation	insects, spiders, crabs, centipedes
h.		hard, spiny endoskeleton just beneath skin; live in salt water; move with suction-cuplike feet; invertebrate	sea star, sea cucumber, sea urchin

What do you think NOW?

Reread the statements at the beginning of the lesson. Fill in the After column with an A if you agree with the statement or a D if you disagree. Did you change your mind?

Log on to ConnectED.mcgraw-hill.com and access your textbook to find this lesson's resources.

CHAPTER 9
LESSON 3

Introduction to Animals

Chordates

Key Concepts
- What characteristics do chordates have in common?
- What is the difference between vertebrate and invertebrate chordates?
- How do the groups of vertebrate chordates differ?

Before You Read

What do you think? Read the two statements below and decide whether you agree or disagree with them. Place an A in the Before column if you agree with the statement or a D if you disagree. After you've read this lesson, reread the statements to see if you have changed your mind.

Before	Statement	After
	5. All young mammals take in milk from their mothers.	
	6. Birds are the only animals that lay shelled eggs.	

Mark the Text

Sticky Notes As you read, use sticky notes to mark information that you do not understand. Read the text carefully a second time. If you still need help, write a list of questions to ask your teacher.

Key Concept Check

1. Name characteristics that all chordates share.

Read to Learn

What is a chordate?

Think of the zoo you read about at the beginning of this chapter. It's likely that most of the animals at the zoo were chordates. It also is likely that most of the animals at the zoo were mammals, like you. Chordates are animals that are grouped in the phylum Chordata. Mammals are chordates.

There are two types of chordates—vertebrate chordates and invertebrate chordates. A vertebrate is an animal with a backbone; an invertebrate chordate has no backbone. All chordates have four traits in common: a notochord, a tail, a nerve cord, and pharyngeal pouches. Each of these four traits exists at some time during the life of a chordate.

You are a chordate. When you were developing in the womb, you had a notochord. *A* **notochord** *is a flexible rod-shaped structure that supports the body of a developing chordate.* Your backbone replaced the notochord. You also had a tail. What is left of your tail is your tailbone. Before you had a brain and a spinal cord, you had a nerve cord. You also had pharyngeal (fuh run JEE uhl) pouches. **Pharyngeal pouches** *are grooves along the side of a developing chordate.* Your pharyngeal pouches developed into parts of your ears, head, and neck. Fish have pharyngeal slits that provide support for gills. These characteristics are evidence that you and other chordates have ancestors in common.

Invertebrate Chordates

As you have read, some chordates never develop backbones. These animals are called invertebrate chordates. What did the ancestor of chordates look like?

Lancelets

The earliest chordates probably looked similar to the lancelet in the figure at right. Lancelets are small animals that burrow in the sand just off ocean shores. Lancelets grow to only 5 cm in length. Lancelets can swim, but they often sit in the sand and catch food particles floating by. Lancelets have all four chordate traits, as shown in the figure.

Lancelet

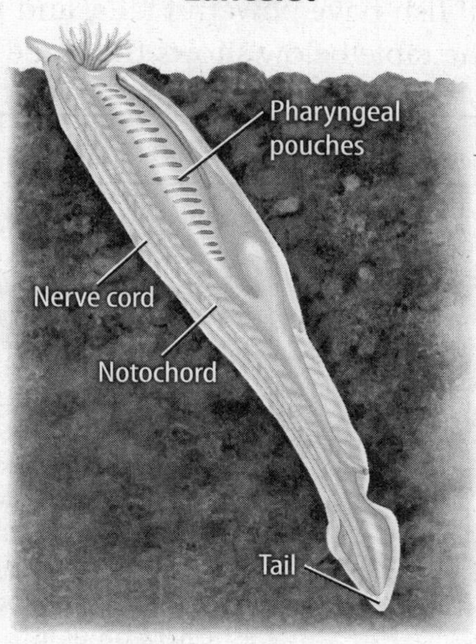

Tunicates

Adult tunicates look like sponges. Like sponges, adult tunicates live in the sea, attached to rocks or other fixed objects. However, adult tunicates have organized tissues and internal structures such as organs. They also have all the characteristics of chordates at some time in their lives. Before a tunicate becomes an adult, it looks and acts like a tadpole. Young tunicates can swim, and they have all four chordate traits.

Lancelets look like fish more than tunicates do. Therefore, scientists once thought lancelets were more closely related to vertebrates than tunicates were. But when scientists studied the DNA of all three groups, they discovered the opposite to be true. Vertebrate DNA shares more similarities with tunicate DNA than it does with lancelet DNA.

Vertebrate Chordates

Most of the animals you are familiar with probably are vertebrate chordates. This group includes cats, dogs, fish, snakes, frogs, and birds. Recall that vertebrates are animals with backbones. Most vertebrates also have jaws. As the bodies and skeletons of vertebrates continued to adapt, vertebrates became better at catching food and avoiding being eaten.

FOLDABLES®

Make a horizontal three-tab Venn book to compare and contrast the different types of chordates.

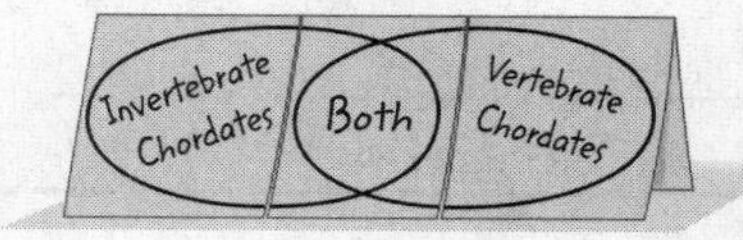

Visual Check

2. Circle the names of the traits that indicate the lancelet is a chordate.

Key Concept Check

3. Differentiate What is the difference between invertebrate and vertebrate chordates?

____

____

____

Fish

Interpreting Tables

4. Identify Which group of fish does not have paired fins?

____

You might think of a goldfish and a shark as fish, but what about a sea horse? All are fish and have traits that make them fish. All fish live in water and use gills for breathing. **Gills** *are organs that exchange carbon dioxide in the water for oxygen.* All fish have powerful tails, and most fish have paired fins. The table below shows the three major groups of fish.

Groups of Fish	
Jawless Fish Lampreys are jawless fish. The skeleton of jawless fish is made of cartilage. The tip of your nose and the flaps of your ears are made of cartilage. Some jawless fish get their nutrition from other fish. They have a circle of teeth that attach to the sides of other fish and make a wound. They then slowly suck out blood and other body fluid from the fish.	
	**Sharks and Rays** Most of the skeleton of sharks and rays is made of cartilage. However, shark skulls are made of bone. Sharks have paired fins. They are fast swimmers and also have powerful jaws. Their jaws make them dangerous predators of other animals, especially other fish.
Bony Fish All other fish have a bony skeleton, as well as paired fins and jaws. Bony fish, such as goldfish, also have a special sac called a swim bladder that the fish can fill with gas. This helps the fish move up and down in the water. Sea horses are unique bony fish because the males carry the young in their bodies as they develop.	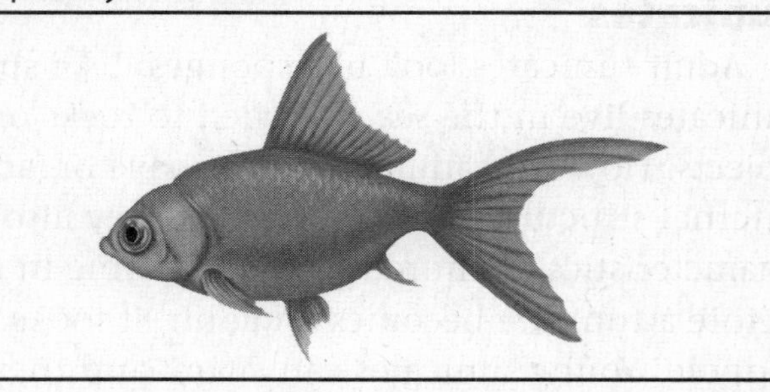

Amphibians

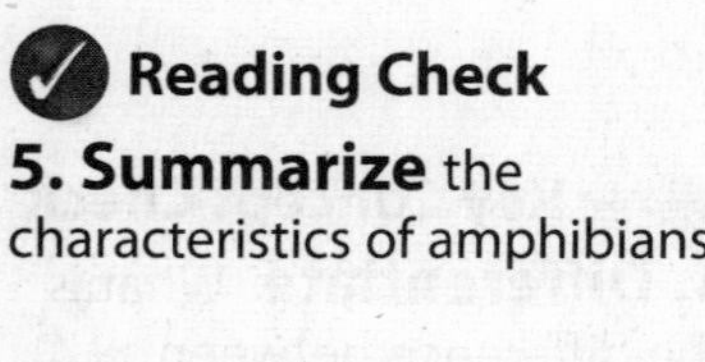

Reading Check

5. Summarize the characteristics of amphibians.

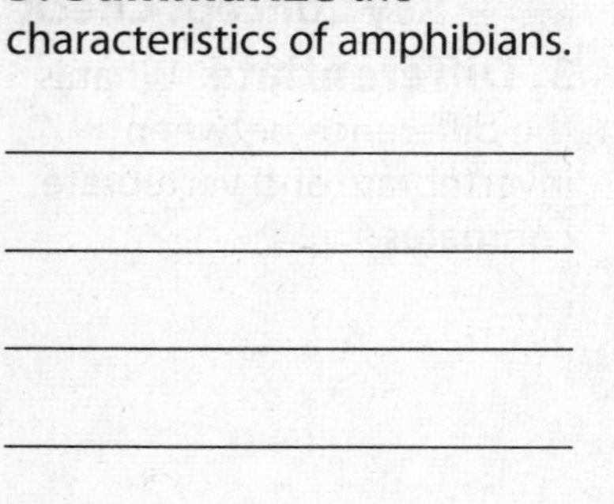

In Canada in 2004, scientists discovered the fossil of an animal that lived in shallow water long ago. The animal had both gills and lungs, a flexible neck, and fins with arm and hand bones. The fins could help the animal move in the water and on land. These scientists may have discovered one of the first tetrapods. A tetrapod is a vertebrate animal with four limbs.

Amphibians are a group of tetrapods that live on land but depend on water to survive and reproduce. *Amphibian* means "both ways of life." Most amphibians lay their eggs in water. Young amphibians, such as tadpoles, have gills and must spend most of their time in water. Most adults have lungs for breathing on land. However, amphibian skin is thin and moist. On land, amphibians must live in moist habitats to keep their bodies from drying out.

Types of Amphibians There are three types of amphibians. Salamanders and newts have tails and move by bending their bodies side-to-side. Frogs and toads do not have tails as adults. They have long legs that enable them to jump. Caecilians (sih SIHL yuhnz) are a group of amphibians that do not have legs. They look similar to earthworms and move by twisting their bodies back and forth like a snake.

Survival Scientists are concerned about the survival of amphibians. Many amphibian populations have become smaller since 1980. Some types of amphibians have not been seen for years. Scientists hypothesize that the size of the amphibian population is decreasing because of disease, climate change, herbicides, and habitat destruction.

Reptiles

Lizards and snakes, turtles, and alligators and crocodiles are the three most common groups of reptiles. Most reptiles live on land, and all have lungs for breathing. Scales on their skin prevent reptiles from drying out. Reptiles do not lay their eggs in water. Most reptiles lay shelled eggs that don't dry out. Inside the egg is **amnion,** *a protective membrane that surrounds the embryo.* An egg with an amnion is called an amniotic egg, as shown in the figure below. Reptiles, birds, and mammals have amniotic eggs.

Amniotic Egg

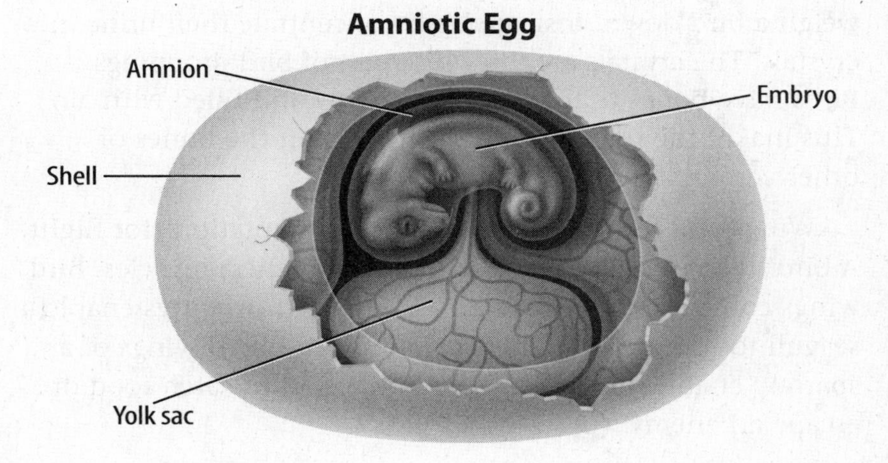

Lizards and Snakes One reptile group includes both lizards and snakes. Most lizards are small and would fit in your hand. However, one lizard, the Komodo dragon, can grow to 3 m in length. Snakes are legless reptiles. Many snakes are small. Some are several meters long. All snakes eat other animals. When snakes catch their prey, they can crush them or bite and poison them. Either way, most snakes swallow their prey whole!

Math Skills

The size of two organisms might be the same, but one floats while the other sinks. This is because the organisms have different densities. The formula for density is

$$\text{density} = \frac{\text{mass}}{\text{volume}}$$

For example, what is the density of a chicken's leg bone that has a mass of 5.5 g and a volume of 5.0 cm^3?

Replace the terms in the formula with the given values.

$$\text{density} = \frac{5.5\text{ g}}{5.0\text{ cm}^3}$$

Solve the problem.

$$\frac{5.5\text{ g}}{5.0\text{ cm}^3} = \frac{1.1\text{ g}}{\text{cm}^3}$$

6. Use a Formula A piece of a cow's leg bone with a volume of 10 cm^3 has a mass of 18 g. What is the density of the bone?

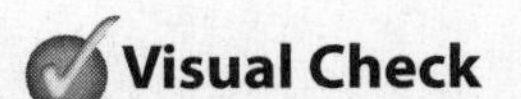

Visual Check

7. Identify What surrounds the developing embryo?

Turtles These animals are best known for their protective shells. Some turtles can live in the desert. Others, such as snapping turtles and sea turtles, live mostly in water.

Alligators and Crocodiles The third group of reptiles, alligators and crocodiles, lives in warm parts of the world. They live in or near water but lay their eggs in nests on the shore. They are fierce hunters and can move quickly for short distances. ✓

Reading Check

8. Name three characteristics of alligators and crocodiles.

How Reptiles Stay Warm Reptiles are **ectotherms,** *animals that heat their bodies from heat in their environments.* Warming, or basking, in sunlight is a behavioral adaptation of ectotherms. A reptile with a warm body can move faster and catch prey more easily. Reptiles move to cool, dark places to conserve energy when food is scarce.

Birds

Can you name a unique trait of birds? Did you think of flight? Many insects fly, and so do certain mammals. Some birds, such as penguins and emus, don't fly. Maybe you thought of wings. But most insects and some mammals also have wings. The one trait that makes birds different from all other animals is their feathers. ✓

Reading Check

9. Identify What trait is unique to birds?

Birds have many adaptations that enable them to fly. A bird does not have a urinary bladder. A full bladder would weigh a bird down. Instead, birds concentrate their urine into crystals. The crystals are the white part of bird droppings. Birds have bones that are nearly hollow and filled with air. This makes the bones of birds lighter than the bones of other vertebrates.

Wings and feathers are birds' major adaptations for flight. A bird's wings are connected to powerful chest muscles. Bird wings come in different shapes. Long, narrow wings enable a seagull to soar on long flights. The short, broad wings of a sparrow enable quick changes of direction to catch food or escape an enemy. ✓

Reading Check

10. Describe the different shapes of bird wings, and explain how the shapes help the birds survive.

Feathers also keep birds warm. Unlike reptiles, birds are **endotherms,** *animals that generate their body heat from the inside.* This enables birds to live in cold habitats. However, birds require much energy to keep their body temperatures high. Like you, birds shiver when they get cold. Shivering muscles help produce more body heat.

Mammals

Maybe the main reason to go to the zoo is to see the mammals. Lemurs, lions, alpacas, and apes all are mammals. You are a mammal, too. All mammals have hair and **mammary glands,** *special tissues that produce milk for young mammals.* Like birds, mammals are endotherms. The hair of mammals is an adaptation that helps keep them warm. Milk production also is an adaptation. The milk helps the young grow and survive when they are too young to find their own food. There are three groups of mammals: monotremes, marsupials, and placental mammals.

Monotremes One type of mammal, the monotreme, lays eggs. After the eggs hatch, the mother's milk nourishes the young. These mammals include the platypus and the echidna.

Marsupials Mammals that raise their young in pouches are called marsupials. The young are not fully developed when born. After birth, they crawl through their mother's hair into a pouch. Here they can feed on their mother's milk and continue to grow. Most marsupials are native to, or live in, Australia. Many marsupials resemble mammals that live in North America. There are marsupial squirrels, marsupial mice, and marsupial moles. The only marsupial native to North America is the opossum.

Placental Mammals The third group of mammals is called placental mammals. A placental mammal has a structure called a placenta. The young are attached to the placenta as they grow inside the mother. Of the three types of mammals, you are probably most familiar with different kinds of placental mammals, such as dogs, cats, horses, cows, and humans.

Reading Check

11. Explain How does hair benefit a mammal?

Key Concept Check

12. Contrast How do the groups of vertebrate chordates differ?

After You Read

Mini Glossary

amnion: a protective membrane that surrounds the embryo

ectotherm: an animal that heats its body from heat in its environment

endotherm: an animal that generates its body heat from the inside

gill: an organ that exchanges carbon dioxide in the water for oxygen

mammary gland: special tissue that produces milk for young mammals

notochord: a flexible, rod-shaped structure that supports the body of a developing chordate

pharyngeal (fuh run JEE uhl) pouch: a groove along the side of a developing chordate

1. Review the terms and their definitions in the Mini Glossary. Write a sentence that explains how mammary glands help mammals survive.

2. In the graphic organizer, identify one unique trait shared by all fish. Then do the same for mammals and birds.

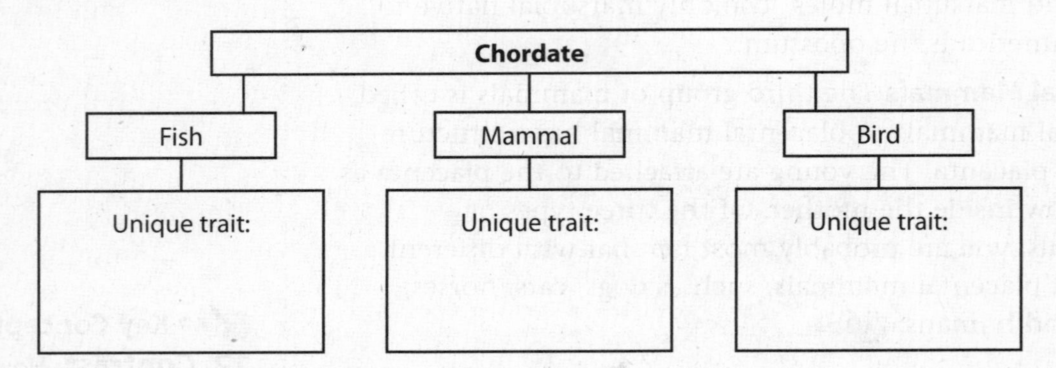

3. No lizards live in Antarctica. Based on what you learned about lizards in the lesson, give one reason for this.

What do you think NOW?

Reread the statements at the beginning of the lesson. Fill in the After column with an A if you agree with the statement or a D if you disagree. Did you change your mind?

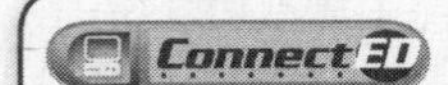

Log on to ConnectED.mcgraw-hill.com and access your textbook to find this lesson's resources.

Interactions of Life

Ecosystems

Before You Read

What do you think? Read the two statements below and decide whether you agree or disagree with them. Place an A in the Before column if you agree with the statement or a D if you disagree. After you've read this lesson, reread the statements to see if you have changed your mind.

Before	Statement	After
	1. An ecosystem is all the animals that live together in a given area.	
	2. A layer of decayed leaves that covers the soil in a forest is an example of a living factor.	

Key Concepts 

- How can you describe an ecosystem?
- What are the similarities and differences between the abiotic and biotic parts of an ecosystem?
- In what ways can populations change?

Read to Learn

What is an ecosystem?

Picture yourself at a park. You sit on the grass in the warm sunshine. A robin pulls an earthworm from the soil. A breeze blows dandelion seeds through the air. These interactions are just a few of the many interactions that can happen in an ecosystem. *An* **ecosystem** *is all the living things and nonliving things in a given area.*

There are many kinds of ecosystems on Earth, including forests, deserts, grasslands, rivers, beaches, and coral reefs. Ecosystems that have similar climates and contain similar types of plants are grouped together into biomes. For example, the tropical rain forest biome includes ecosystems full of lush plant growth located near the equator in places where rainfall averages 200 cm per year and the temperature averages 25°C.

Mark the Text

Ask Questions As you read, write questions you may have in the margin. Read the lesson a second time and try to answer the questions. When you are done, ask your teacher any questions you still have.

Key Concept Check

1. Describe How can you describe an ecosystem?

Abiotic Factors

The nonliving parts of an ecosystem are called **abiotic factors.** They include sunlight, temperature, air, water, and soil. Abiotic factors provide many of the resources organisms need for survival and reproduction. Abiotic factors in an ecosystem determine the kinds of organisms that can live there. For example, only organisms that can survive with little water can live in a desert.

Sunlight and Temperature Sunlight is essential for almost all life on Earth. It supplies the energy for photosynthesis—the chemical reactions that produce sugars. Photosynthesis occurs in most plants and some bacteria and protists.

Sunlight also provides warmth. An ecosystem's temperature depends in part on the amount of sunlight it receives. In some ecosystems, such as a desert, temperatures can be around 49°C during the day and fall below freezing at night.

REVIEW VOCABULARY

atmosphere
the whole mass of air surrounding Earth

Atmosphere The gases in Earth's atmosphere include nitrogen, oxygen, and carbon dioxide. Plants need nitrogen to grow. Some bacteria in soil take nitrogen from the air and change it to a form that plants can use. Most organisms need oxygen for cellular respiration—the process that releases energy in cells. The process of photosynthesis uses carbon dioxide from the air.

FOLDABLES®

Make a vertical shutterfold book to organize your notes about abiotic and biotic factors.

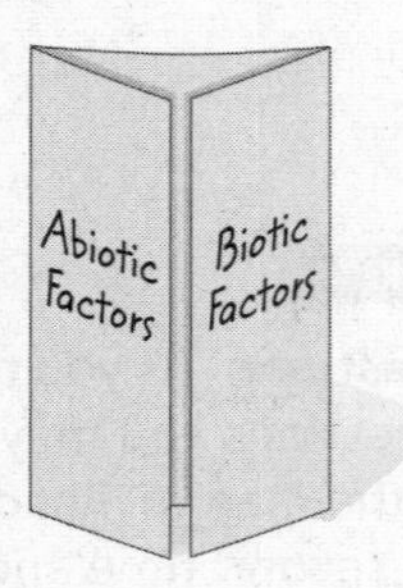

Water Without water, life would not be possible. All life processes that occur inside cells—such as cellular respiration, digestion, and photosynthesis—require water. Stream ecosystems can support many forms of life because water is plentiful. Areas with little water support fewer organisms.

Soil Why is soil important for healthy plants? Soil contains a mixture of living and nonliving things. The biotic part of soil is humus (HEW mus). Humus is the decayed remains of plants, animals, bacteria, and other organisms. Deserts have thin soil with little humus. Forest soils usually are thick and fertile, with a higher humus content. Abiotic factors in soil include minerals and particles of rock, sand, and clay.

Many animals, such as the gopher, insects, and earthworms shown in the figure, live in soil. Their tunnels loosen the soil. This helps move water and air through soil for plant roots and soil-dwelling organisms.

Soil-Dwelling Organisms

Visual Check

2. Explain How do soil-dwelling organisms benefit soil?

Biotic Factors

Living or once-living things in an ecosystem are called **biotic factors.** Biotic factors include all living organisms—from the smallest bacterium (plural, bacteria) to the largest redwood tree. Biotic factors also include the remains of dead organisms, such as fallen leaves or decayed plant matter in soil.

Species are adapted to the abiotic and biotic factors of the ecosystems in which they live. Algae, fungi, and mosses live in moist ecosystems such as forests, ponds, and oceans. Many cactus species can survive in a desert because they have thick stems that store water. Gophers live in burrows underground. Gophers have large front claws for digging through soil and strong teeth for loosening soil and chewing plant roots.

Key Concept Check

3. Compare and Contrast What are the similarities and differences between abiotic and biotic factors?

Habitats

Every organism in an ecosystem has its own place to live. *A* **habitat** *is the place within an ecosystem that provides food, water, shelter, and other biotic and abiotic factors an organism needs to survive and reproduce.*

Reading Check

4. Define What is a habitat?

Organisms have a variety of habitats. For example, house martins sometimes live in meadows or grasslands, but these birds also might find a habitat under the eaves of a building. Crickets live in damp, dark places with plenty of plant material and fungi to eat. Skunks live in areas where they can find food such as mice, insects, eggs, and fruit. During the day, skunks take shelter near their food supply—in hollow logs, under brush piles, and beneath buildings.

Plants have their own habitats, too. You have read that cacti live in desert habitats. The wood sorrel is a plant species that grows in deep shade under redwood trees.

When biotic or abiotic factors in an ecosystem change, habitats can change or disappear. A wildfire quickly can destroy the habitats of thousands of animals that live in forests or grasslands. Erosion or flooding can wash away soil, destroying plant habitats.

Think it Over

5. Name another way that a habitat can be destroyed.

Populations

Every ecosystem includes many individuals of many different species. *A* **population** *is all the organisms of the same species that live in the same area at the same time.*

Vacant-Lot Community

Visual Check

6. Identify What abiotic factors are included in this ecosystem?

____

____

____

____

____

Examples of Populations All the dandelions growing in the vacant lot in the figure above form a population. All the ants in the vacant lot make up another population. All the pigeons form a third population. The grasses are yet another population in this ecosystem.

Populations and Community *All the populations living in the same area at the same time form a* **community.** As shown in the figure, a vacant-lot community might include populations of grasses, dandelions, spiders, ants, and pigeons. A community combined with all the abiotic factors in the same area forms an ecosystem. The populations that make up the community interact in the ecosystem.

Population Density

Suppose your classroom has an aquarium. The aquarium contains guppies, water ferns, and a few algae-eating snails. Keeping your aquarium community healthy includes cleaning the tank and feeding the fish. However, it also means making sure the fish don't become overcrowded. Overcrowding can lead to stress and disease.

How can you determine if the aquarium contains too many fish? You could calculate the population density. **Population density** *is the size of a population compared to the amount of space available*. You can calculate population density by using the following formula:

$$\text{Population density} = \frac{\text{number of individuals}}{\text{unit area or volume of space}}$$

An aquarium expert has recommended that no more than 10 guppies live in a 20-gallon aquarium. Using the formula, you can calculate the recommended population density for your aquarium:

$$\text{Population density} = \frac{10 \text{ fish}}{20 \text{ gallons}} = 0.5 \text{ fish per gallon}$$

When population density is high, organisms live closer together and might not be able to obtain all the resources needed for life. Diseases also spread more easily when organisms are forced to live too close together.

Population Change

On a hike one summer, you notice a few wild sunflowers growing among grasses in an overgrown field. Two years later you return and find the field covered with sunflowers. What caused the population to increase? Each sunflower plant produces hundreds of seeds. Even if only a few of the seeds from each plant sprout and grow, the number of sunflowers will increase. But if a drought prevents seeds from sprouting or if a farmer plants the field with corn, soybeans, or another crop, the sunflower population will decrease.

Most populations change over time. Production of offspring increases the size of a population. The deaths of individuals reduce population size. If births outnumber deaths, the population grows.

Changes in the abiotic or biotic factors in an ecosystem can cause organisms to move away or die out. For example, if a forest begins to burn, birds, deer, and other fast-moving animals can escape to other areas. Many other animals, however, could be trapped and die.

Math Skills

A formula shows the relationship among several factors. The formula for population density determines the number of individuals in a unit area or a volume of space.

$$\text{Population density} = \frac{\text{number of individuals}}{\text{unit area or volume of space}}$$

For example, what is the population density of insects if 140 insects are found in a patch of ground measuring 3.0 m^2?

a. Replace the terms in the equation with the given values.

$$\text{Population density} = \frac{140 \text{ insects}}{3.0 \text{ m}^2}$$

b. Solve the problem.

$$\frac{140 \text{ insects}}{3.0 \text{ m}^2} = \frac{46.6 \text{ insects}}{\text{m}^2}$$

c. Round the answer to significant figures.

$$\frac{46.6 \text{ insects}}{\text{m}^2} = \frac{47 \text{ insects}}{\text{m}^2}$$

7. Use a Formula There are 20 small tropical fish swimming in a 55-gallon aquarium. What is the population density?

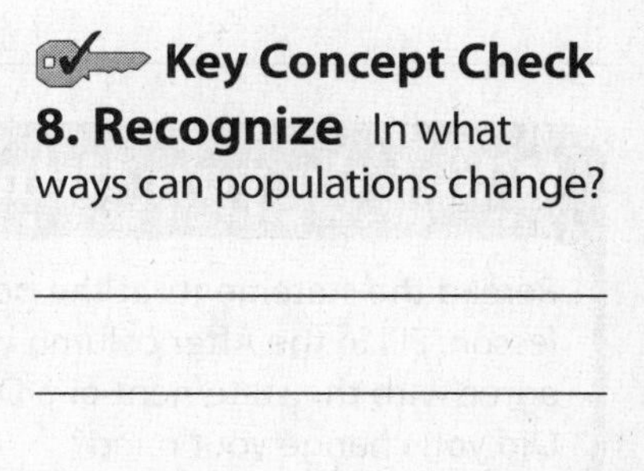

Key Concept Check

8. Recognize In what ways can populations change?

After You Read

Mini Glossary

abiotic factor: a nonliving part of an ecosystem

biotic factor: a living or once-living thing in an ecosystem

community: all the populations living in the same area at the same time

ecosystem: all the living things and nonliving things in a given area

habitat: the place within an ecosystem that provides food, water, shelter, and other biotic and abiotic factors an organism needs to survive and reproduce

population: all the organisms of the same species that live in the same area at the same time

population density: the size of a population compared to the amount of space available

1. Review the terms and their definitions in the Mini Glossary. Write a sentence explaining the relationship between community and population.

2. Identify five abiotic factors in ecosystems to complete the graphic organizer.

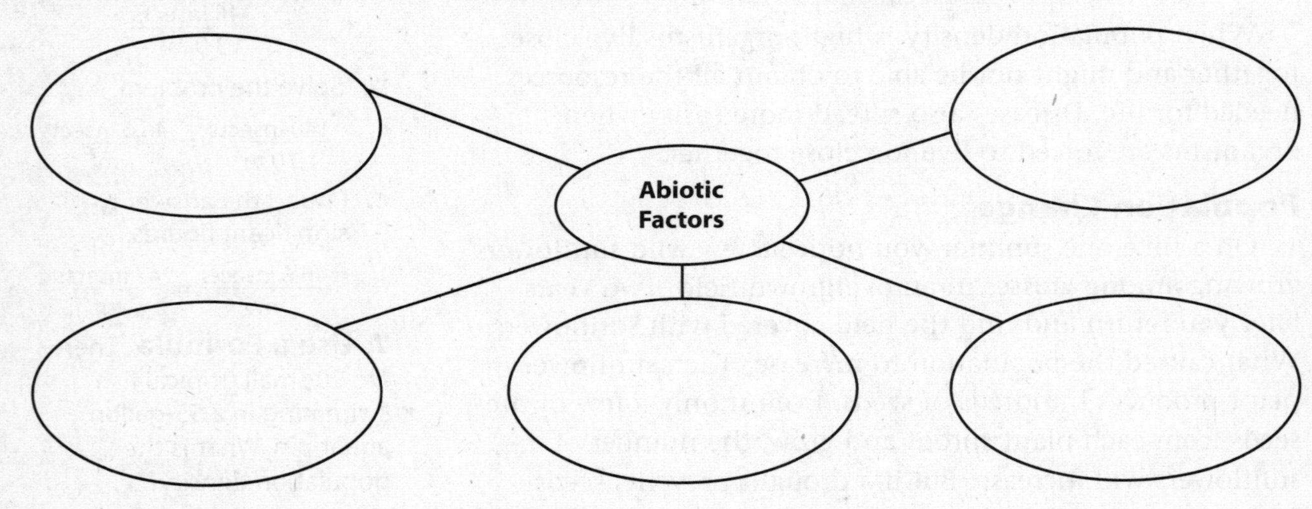

3. How does population density affect organisms?

What do you think NOW?

Reread the statements at the beginning of the lesson. Fill in the After column with an A if you agree with the statement or a D if you disagree. Did you change your mind?

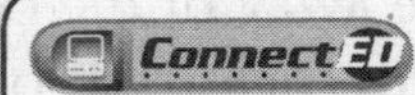

Log on to ConnectED.mcgraw-hill.com and access your textbook to find this lesson's resources.

Interactions of Life

Relationships Within Ecosystems

Before You Read

What do you think? Read the two statements below and decide whether you agree or disagree with them. Place an A in the Before column if you agree with the statement or a D if you disagree. After you've read this lesson, reread the statements to see if you have changed your mind.

Before	Statement	After
	3. A niche is the place where an animal lives.	
	4. Symbiosis is a close relationship between two species.	

Key Concepts 

- How does a niche differ from a habitat?
- In what ways can organisms interact in an ecosystem?

Read to Learn

Niches

Recall that a habitat is the area within an ecosystem that provides an organism with the resources it needs for life. Most organisms don't have a habitat all to themselves. Usually many species share a habitat.

For example, hundreds of species share a coral reef habitat. During the day, the spiny lobsters hide under the coral. They come out at night and feed on worms, shrimp, clams, and dead fish. Angelfish use their rough teeth to scrape sponges and sea squirts from the surface of the coral. Filefish eat the algae that they scrape from the coral.

Each species that shares a habitat has a separate niche. *A* **niche** *(NICH) is the way a species interacts with abiotic and biotic factors to obtain food, find shelter, and fulfill other needs.* Species share habitats, but no two species share the same niche. For example, two species of crabs on a reef might share a habitat, but one might eat algae and the other might eat snails.

Study Coach

Make Flash Cards For each head in this lesson, write a question on one side of a flash card and the answer on the other side. Quiz yourself until you know all of the answers.

Key Concept Check

1. Differentiate How does a niche differ from a habitat?

______

______

______

______

______

______

______

Competition

In springtime, robins find mates, build nests, and raise their young. A male robin chooses a safe nesting site with plenty of food and water nearby. He sings to attract a female and to keep other males away. If another male comes too close to his territory, he chases the competitor away.

What is competition? **Competition** *describes the demand for resources, such as food, water, and shelter, in short supply in a community.* Competition can take place among the members of a population or between two populations of different species. For example, the plants that grow in the same area compete for nutrients and living space.

Reading Check
2. Define What is competition?

Competition and Population Size Competition limits population size. If a community has more robins than nesting sites, competition for these sites increases. Some robins will leave the area. Others might not nest. The availability of nesting sites limits the size of the robin population.

Overpopulation

White-tailed deer live near the edges of forests and meadows. They eat leaves, twigs, acorns, and fruit. Deer populations in some areas have become so large that they harm forest habitats, destroy crops, and even invade home gardens. **Overpopulation** *occurs when a population becomes so large that it causes damage to the environment.*

When deer overpopulate an ecosystem, the deer eat plants at a rate faster than the plants can regrow. This reduces the available habitat for the deer and other species. The deer, as well as other organisms in the area, must compete for a limited amount of resources.

When deer overpopulate an ecosystem, they often move into areas where they are not normally found, such as home gardens. Overpopulation also means that the deer live closer together. Disease can spread more easily within populations when they are overcrowded.

Graph of Overpopulation

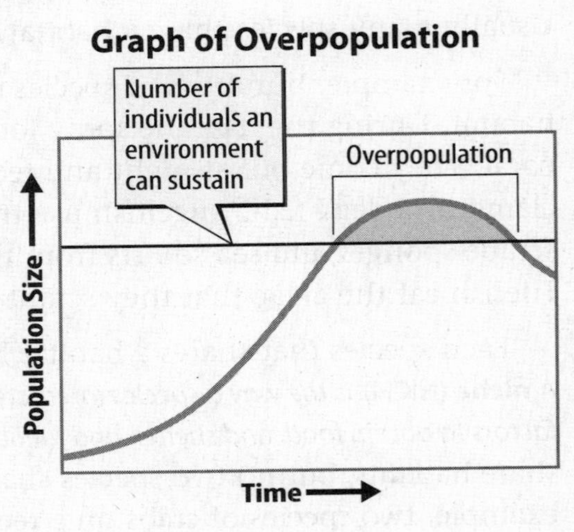

Visual Check
3. Draw an *X* on the graph line where the population size is highest.

Reading Check
4. Explain Why is overpopulation temporary?

Overpopulation is temporary. When food and other resources eventually run out, some animals will move elsewhere, starve, or die from disease. Then the population quickly shrinks, as shown in the graph above. This allows the resources in the environment to slowly return to normal.

Competing with Humans

Humans need food, living space, and water—some of the same biotic and abiotic factors that other organisms need. To meet these needs, people take certain actions. They plow grasslands to plant food crops. People clear forests and fill in wetlands for constructing roads and buildings. They divert water from lakes and streams to supply irrigation for crops and drinking water for cities and towns. Actions such as these put humans in competition with other species for the same resources. 

You might have heard news reports about raccoons raiding garbage cans, snakes living under houses, or squirrels moving into attics. When humans replace natural environments with homes and other structures, they disturb the natural habitats for these and other organisms. One example is roads that can make it dangerous for animals to move safely from one part of their habitat to another.

Sometimes humans compete with other organisms in less-obvious ways. The North American population of monarch butterflies spends the winter in small forested regions in Mexico. Logging by humans endangers the monarchs' winter habitat. With fewer trees for protection and to rest in, many monarchs do not survive for the return trip north in spring.

Predation

A predator is an organism that hunts and kills other organisms for food. Prey are the organisms that are hunted or eaten by a predator. **Predation** *is the act of one organism, a predator, feeding on another organism, its prey.* Predator and prey populations influence each other. Predators help control the size of prey populations. When prey populations decrease, the number of predators usually decreases because less food is available.

Symbiosis

Competition and predation are two types of interactions that take place between organisms in an ecosystem. Symbiosis (sim bee OH sus) is another type of interaction that occurs. **Symbiosis** *is a close, long-term relationship between two species that usually involves an exchange of food or energy.* Three types of symbiosis are mutualism, commensalism, and parasitism. Each is discussed on the next page.

Reading Check

5. Identify three biotic or abiotic factors that humans compete with other organisms to obtain.

____

____

____

FOLDABLES®

Make a horizontal three-tab concept map book to organize your notes on symbiotic relationships.

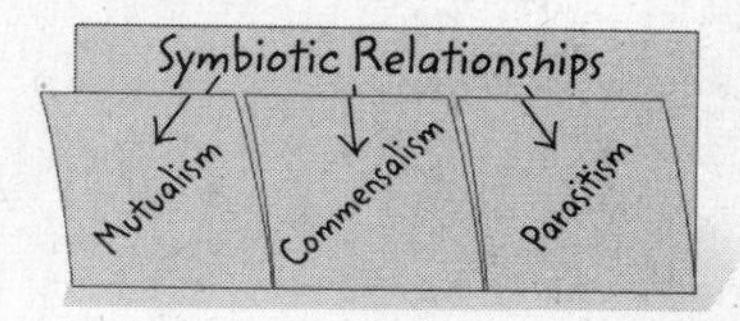

Think it Over

6. Name one organism that is a predator. Name one organism that is prey for this predator.

____

____

The Three Types of Symbiosis

Mutualism

Commensalism

Parasitism

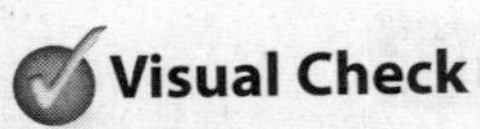

Visual Check

7. Circle the symbiotic relationship(s) in the figure in which one of the organisms is harmed.

Key Concept Check

8. Describe In what ways can organisms interact in an ecosystem?

Mutualism *A symbiotic relationship in which both organisms benefit is* **mutualism.** For example, on the left in the figure above, a cleaner shrimp is removing tiny organisms from the fish's body. The fish benefits by getting rid of unwanted organisms. The cleaner shrimp benefit by getting food.

Commensalism *A symbiotic relationship in which one organism benefits but the other neither benefits nor is harmed is* **commensalism.** Clumps of moss growing on the bark of the tree in the center of the above figure is an example of a commensal relationship. The moss benefits by having somewhere to grow. The presence of the moss neither benefits nor harms the tree.

Parasitism *A symbiotic relationship in which one organism benefits while the other is harmed is* **parasitism.** The organism that benefits is a parasite. One example is shown on the right in the figure above. A parasitic wasp lays its eggs in the caterpillar's body. When the eggs hatch, the larvae develop and eventually chew their way out of the caterpillar, killing it. The organism that is harmed is the host, in this case, the caterpillar.

After You Read

Mini Glossary

commensalism: a symbiotic relationship in which one organism benefits but the other neither benefits nor is harmed

competition: the demand for resources, such as food, water, and shelter, in short supply in a community

mutualism: a symbiotic relationship in which both organisms benefit

niche (NICH): the way a species interacts with abiotic and biotic factors to obtain food, find shelter, and fulfill other needs

overpopulation: occurs when a population becomes so large that it causes damage to the environment

parasitism: a symbiotic relationship in which one organism benefits while the other is harmed

predation: the act of one organism, a predator, feeding on another organism, its prey

symbiosis (sim bee OH sus): a close, long-term relationship between two species that usually involves an exchange of food or energy

1. Review the terms and their definitions in the Mini Glossary. Write a sentence that defines the term *niche* in your own words.

____

____

2. For each description in the diagram below, identify the type of symbiosis.

A honeybee eats pollen from a rose. The bee carries the pollen to other roses.

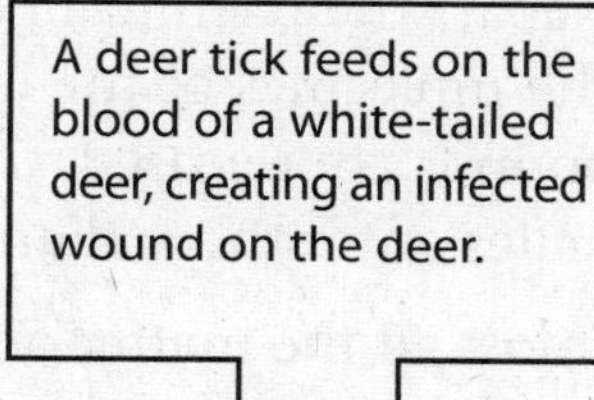

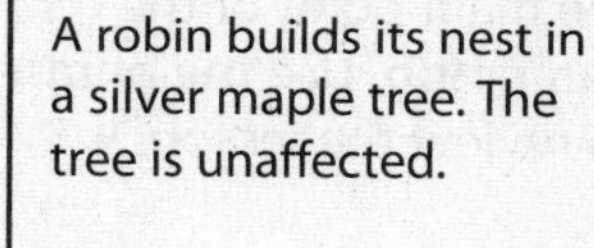

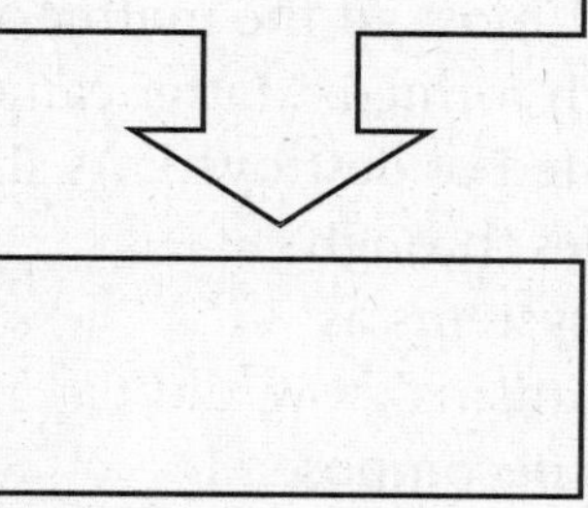

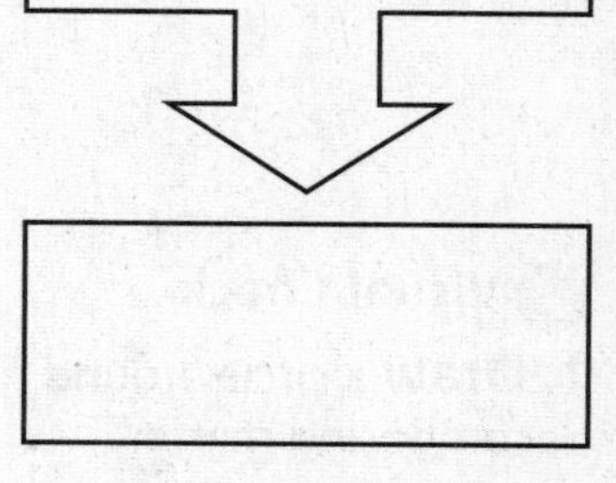

3. Select and define a word from one of the flash cards you created as you read the lesson.

____

What do you think NOW?

Reread the statements at the beginning of the lesson. Fill in the After column with an A if you agree with the statement or a D if you disagree. Did you change your mind?

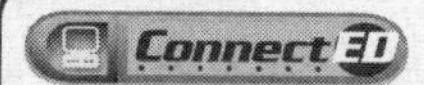

Log on to ConnectED.mcgraw-hill.com and access your textbook to find this lesson's resources.

Interactions of Life

Matter and Energy in Ecosystems

Key Concepts
- How do matter and energy move through ecosystems?
- How do organisms obtain energy?
- What are the differences between a food chain and a food web?

Before You Read

What do you think? Read the two statements below and decide whether you agree or disagree with them. Place an A in the Before column if you agree with the statement or a D if you disagree. After you've read this lesson, reread the statements to see if you have changed your mind.

Before	Statement	After
	5. Energy from sunlight is the basis for almost every food chain on Earth.	
	6. A plant creates matter when it grows.	

Mark the Text

Identify the Main Ideas Write a phrase beside each paragraph that summarizes the main point of the paragraph. Use the phrases to review the lesson.

Read to Learn

Matter and Energy

A leaf drops to the ground. Over time, bacteria and fungi break apart the chemical bonds that hold together the atoms and the molecules of the leaf. Energy, water vapor, and other compounds are released. Carbon compounds and water molecules enter the soil and are used by plants.

Almost all the matter on Earth today has been here since Earth formed. Matter can change form, but it cannot be created or destroyed. As shown in the figure, some matter cycles through ecosystems as organisms grow, die, and decompose.

Energy is not like matter. Energy cannot be recycled. However, energy can be converted. For example, the chemical energy in a log converts to thermal energy and light energy when the log burns.

Matter Cycling Through Ecosystem

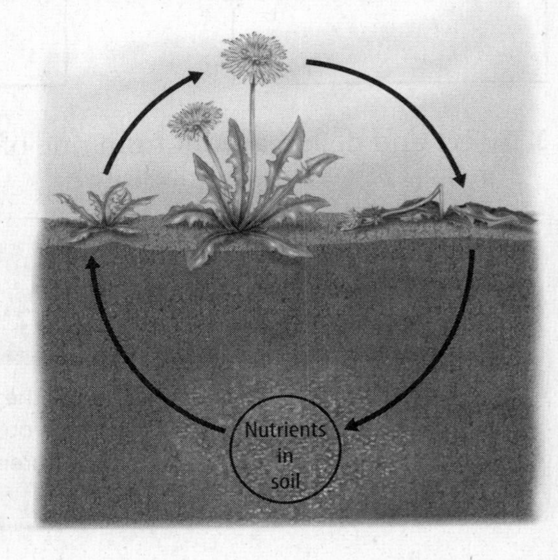

Visual Check

1. Draw a circle around the decomposing matter.

Key Concept Check

2. Explain How do matter and energy move through ecosystems?

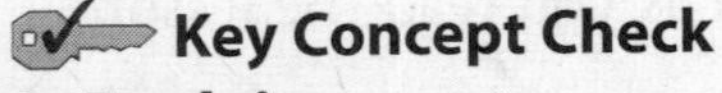

Obtaining Energy

When you eat a sandwich, your body gets atoms and molecules that it needs to make new cells and tissues. Your cells also get the energy they need to make proteins and carry out other life processes. All organisms need a constant supply of energy to maintain life. Where does that energy come from?

Producers

Most of the energy used by all organisms on Earth comes from the Sun. Photosynthesis is the process during which some organisms use carbon dioxide, water, and light energy, usually from the Sun, to make sugars. These sugars serve as food for living organisms.

Producers *are organisms that use an outside energy source, such as the Sun, and produce their own food.* The energy in food molecules is in the chemical bonds that hold the molecules together. During cellular respiration, these bonds break. This releases energy that fuels the producer's life processes. As shown in the figure below, producers also release waste products during cellular respiration.

3. Define What is a producer?

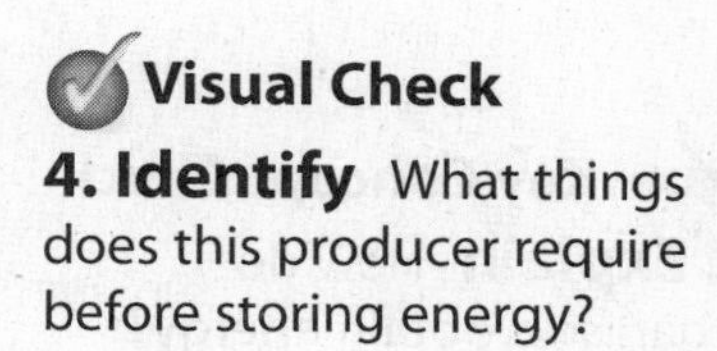

4. Identify What things does this producer require before storing energy?

Producers

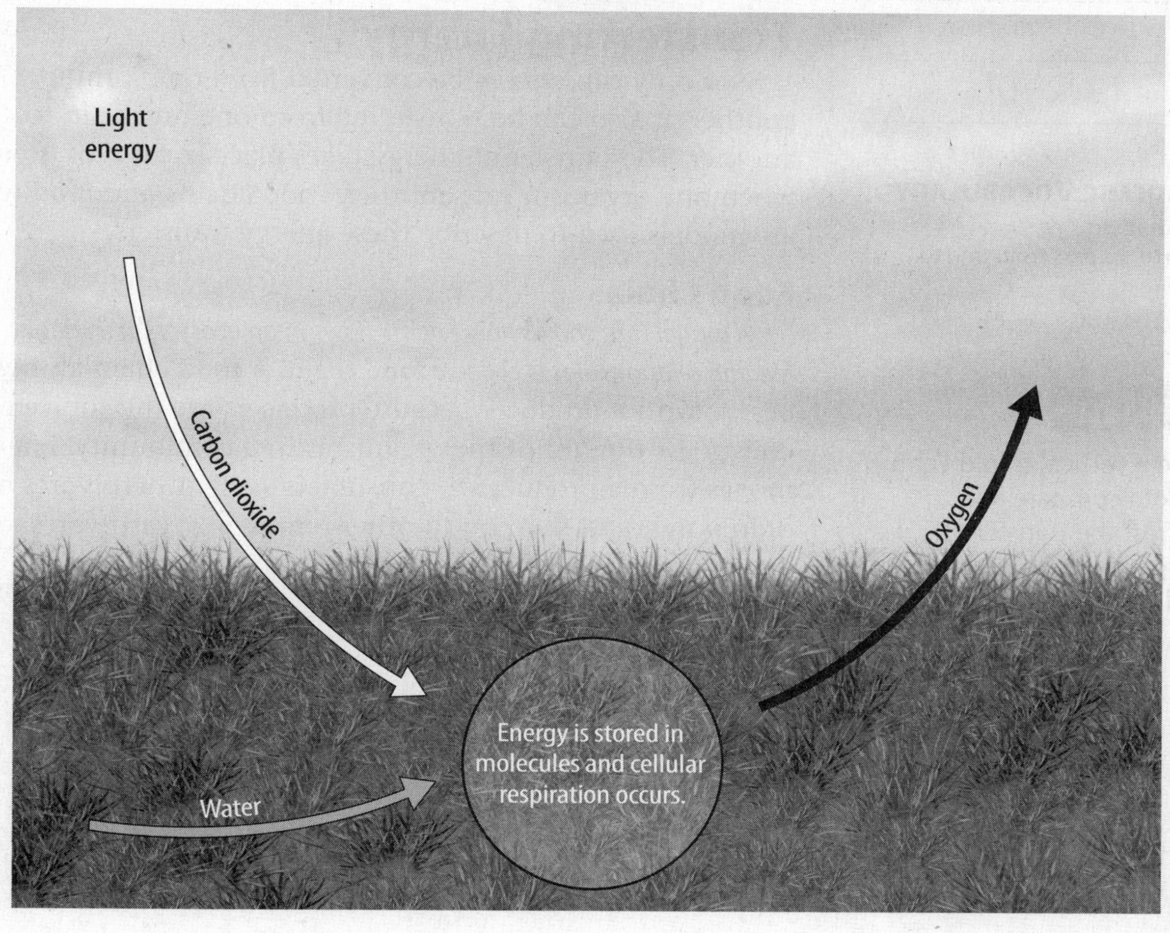

Consumers

The energy-rich molecules formed by producers provide food for other organisms. **Consumers** *are organisms that cannot make their own food. Consumers obtain food by eating producers or other consumers.* Ecosystems include several different kinds of consumers:

- Herbivores eat only plants and other producers.
- Carnivores eat herbivores and other consumers.
- Omnivores eat producers and consumers.
- Decomposers break down dead organisms.

Herbivores eat plants and other producers. Examples of herbivores include snails, rabbits, deer, and bees. Carnivores eat herbivores and other consumers. Cats, snakes, hawks, frogs, and spiders are carnivores. Omnivores eat producers and consumers. Omnivores include bears, robins, pigs, rats, and humans. Decomposers break down the bodies of dead organisms into compounds that living organisms can use. Without decomposers, matter could not be recycled. Decomposers include fungi, bacteria, wood lice, termites, and earthworms.

Key Concept Check

5. Explain How do organisms obtain energy?

____

____

____

____

Transferring Energy

Not only can energy be converted from one form to another, it also can be transferred from one organism to another. The transfer of energy takes place in an ecosystem when one organism eats another. Food chains and food webs are models used to describe these energy transfers.

ACADEMIC VOCABULARY

transfer
(verb) to pass from one to another

Food Chains

A model that shows how energy flows in an ecosystem through feeding relationships is called a **food chain.** A food chain always begins with a producer because producers are the source of energy for the rest of the organisms in a community. Energy moves from a producer to consumers such as herbivores or omnivores, and then on to other omnivores, carnivores, or decomposers.

A simple food chain from a community of organisms living in a vacant lot might look like this:

Grass → Mouse → Cat

The arrows show the directions of the energy transfer.

FOLDABLES

Make a vertical trifold Venn book to compare and contrast the transfer of energy in a food chain and a food web.

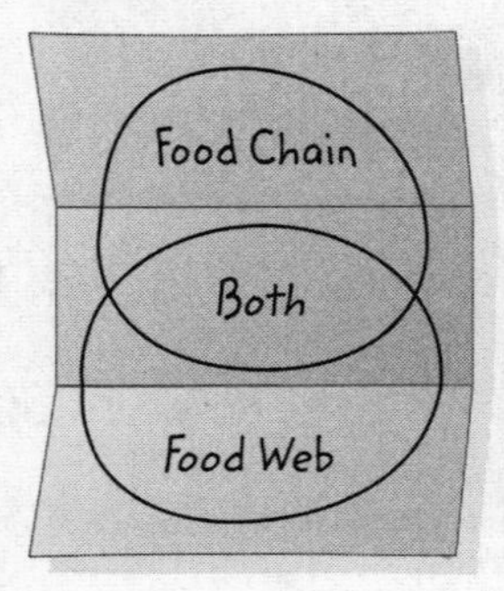

Food Web in a Vacant Lot

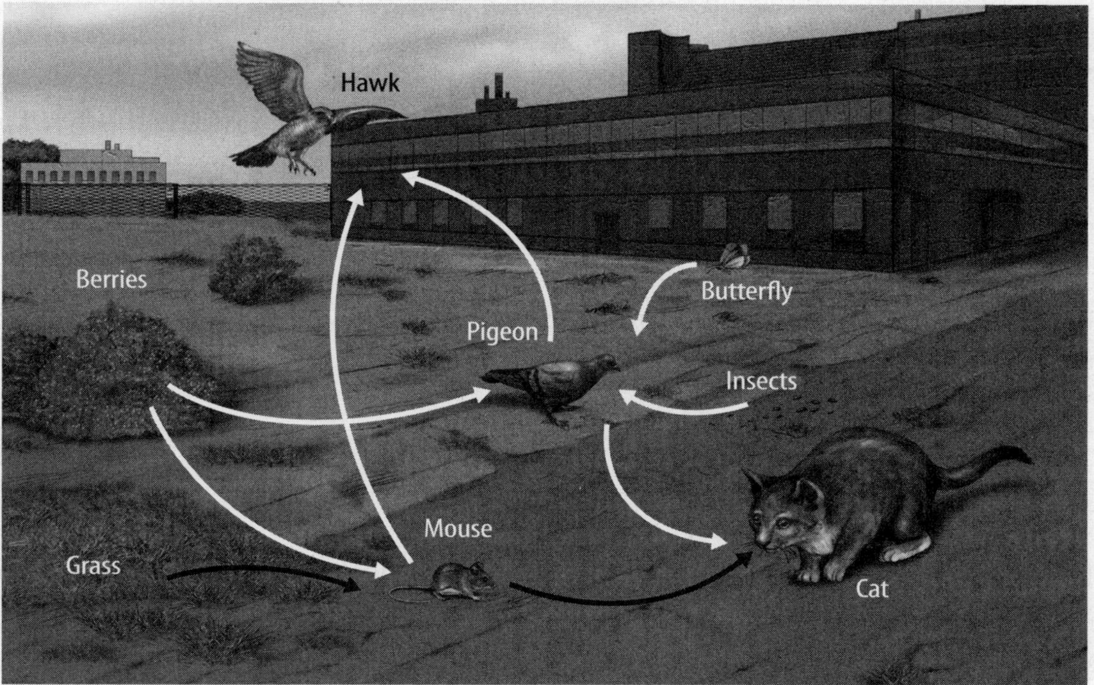

Food Webs

Most ecosystems contain many food chains. *A* **food web** *is a model of energy transfer that can show how the food chains in a community are interconnected.* A food web contains many food chains. For example, the figure above shows a food web in a vacant lot. The pigeons eat berries and insects. They are prey for hawks and cats.

Visual Check

6. Identify two species that are prey for the hawk.

____

____

Key Concept Check

7. Contrast What are the differences between a food chain and a food web?

____

____

____

____

After You Read

Mini Glossary

consumer: an organism that cannot make its own food

food chain: a model that shows how energy flows in an ecosystem through feeding relationships

food web: a model of energy transfer that can show how the food chains in a community are interconnected

producer: an organism that uses an outside energy source, such as the Sun, and produces its own food

1. Review the terms and their definitions in the Mini Glossary. Write a sentence defining a consumer in your own words.

____

____

2. Write the following words in the boxes to show the order of energy transfer that takes place in the food chain.

giraffe **lion** **leaves and twigs**

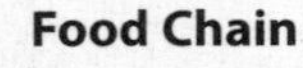

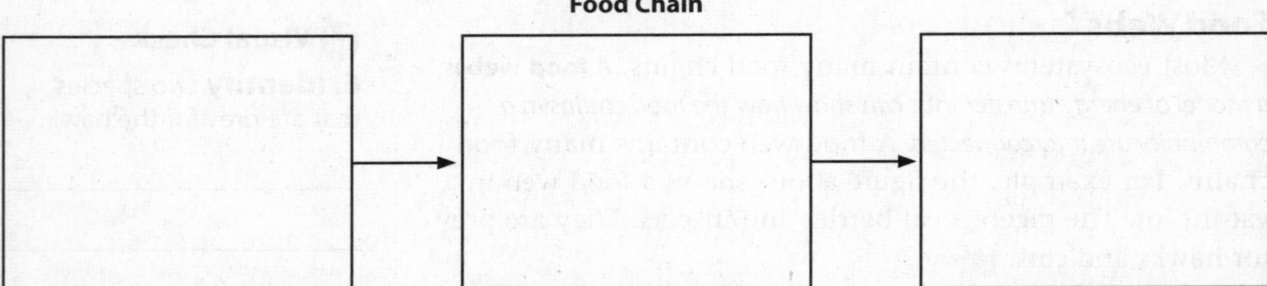

3. In an ecosystem, mice and rabbits eat grasses and berries, weasels eat mice and rabbits, and foxes eat mice and berries. Identify the producers, herbivores, omnivores, and carnivores in this ecosystem.

____

____

____

What do you think NOW?

Reread the statements at the beginning of the lesson. Fill in the After column with an A if you agree with the statement or a D if you disagree. Did you change your mind?

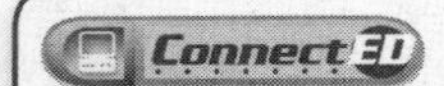

Log on to ConnectED.mcgraw-hill.com and access your textbook to find this lesson's resources.

Matter and Atoms

Substances and Mixtures

Before You Read

What do you think? Read the three statements below and decide whether you agree or disagree with them. Place an A in the Before column if you agree with the statement or a D if you disagree. After you've read this lesson, reread the statements to see if you have changed your mind.

Before	Statement	After
	1. Things that have no mass are not matter.	
	2. The arrangement of particles is the same throughout a mixture.	
	3. An atom that makes up gold is exactly the same as an atom that makes up aluminum.	

Key Concepts 

- What is the relationship among atoms, elements, and compounds?
- How are some mixtures different from solutions?
- How do mixtures and compounds differ?

Read to Learn

What is matter?

Have you ever gone windsurfing? You lean back to balance the board as the force of the wind pushes the sail. You can feel the spray of water against your face. Everything around you, whether you are windsurfing or sitting at your desk, is made of matter. **Matter** *is anything that has mass and takes up space.* Matter is everything you can see, such as water and trees. It is also some things you cannot see, such as air. You know that air is matter because you can feel its mass when it blows against your skin. You can see that it takes up space when it inflates a sail or a balloon.

Anything that does not have mass or volume is not matter. Types of energy, such as heat, sound, and electricity, are not matter. Forces, such as gravity, are not forms of matter.

Mark the Text

Identify Main Ideas
Highlight the main idea of each paragraph. Highlight two details that support each main idea with a different color. Use your highlighted copy to review what you studied in this lesson.

Reading Check

1. Identify two characteristics that a thing must have to be called matter.

____

____

What is matter made of?

All solids, liquids, and gases are made of atoms. *An* **atom** *is a small particle that is the building block of matter.* In this chapter, you will read that an atom is made of even smaller particles. There are many types of atoms. Each type of atom has a different number of smaller particles. Atoms can combine with each other in many ways. It is the many kinds of atoms and the ways they combine that form the different types of matter.

Classifying Matter

All the different types of matter around you are made of atoms, so they must have characteristics in common. But why do all types of matter look and feel different? How is the matter that makes up a pure gold ring similar to the matter that makes up your favorite soda or your body? How are these types of matter different?

Scientists place matter into one of two groups—substances or mixtures. Pure gold is in one group. Soda and your body are in the other. What determines whether a type of matter is a substance or a mixture? The difference is in the composition.

Reading Check
2. Name the two groups into which scientists place matter.

____

____

What is a substance?

What is the difference between a gold ring and a can of soda? What is the difference between table salt and trail mix? Pure gold is always made up of the same type of atom, but soda is not. Similarly, table salt, or sodium chloride, is always made up of the same types of atoms, but trail mix is not. This is because sodium chloride and gold are substances.

A **substance** *is matter with a composition that is always the same.* A certain substance always contains the same kinds of atoms in the same combination. Soda and trail mix are another type of matter that you will read about later in this lesson.

Gold is a substance. So anything that is pure gold will have the same composition. Bars of gold are made of the same atoms as those in a pure gold ring. Sodium chloride is a substance. So the atoms that make up salt will be the same whether you are salting your food in Alaska or in Ohio. If the composition of a given substance changes, you will have a new substance.

Reading Check
3. Summarize Why is pure gold a substance?

____

____

____

Elements

Some substances, such as gold, are made of only one kind of atom. Others, such as sodium chloride, are made of more than one kind of atom. *An* **element** *is a substance made of only one kind of atom.*

All atoms of an element are alike, but atoms of one element are different from atoms of other elements. For example, the element gold is made of only gold atoms, and all gold atoms are alike. But gold atoms are different from silver atoms, oxygen atoms, and atoms of every other element.

Key Concept Check
4. Explain How are atoms and elements related?

____

____

____

____

What is the smallest part of an element? If you could break down an element into its smallest part, that part would be one atom. Most elements, such as carbon and silver, are made up of a large group of individual atoms. Some elements, such as hydrogen and bromine, are made of molecules.

A **molecule** (MAH lih kyewl) *is two or more atoms that are held together by chemical bonds and act as a unit.* Examples of elements made of individual atoms and molecules are shown in the figure to the right.

Atoms and Molecules

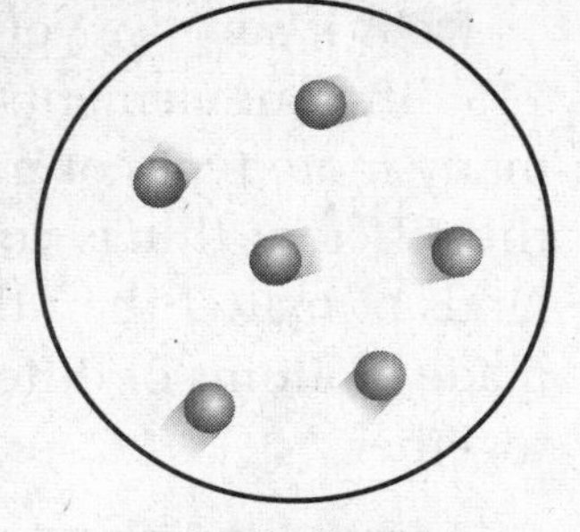

Individual atoms

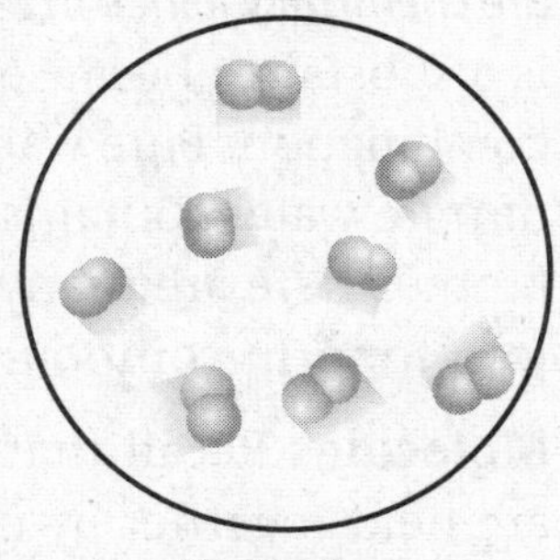

Molecules

Elements on the Periodic Table You probably can name many elements, such as carbon, gold, and oxygen. Did you know that there are about 115 elements? The figure below shows that each element has a symbol, such as C for carbon, Au for gold, and O for oxygen. Element symbols have either one or two letters. Temporary symbols have three letters. The periodic table printed in the back of this book gives other information about each element. You will learn more about elements in the next lesson.

Reading Check

5. Distinguish What is the smallest part of an element?

Visual Check

6. Analyze How many atoms does each molecule in the figure to the left have?

Visual Check

7. Draw Circle the blocks used for elements that have not yet been verified.

The Periodic Table

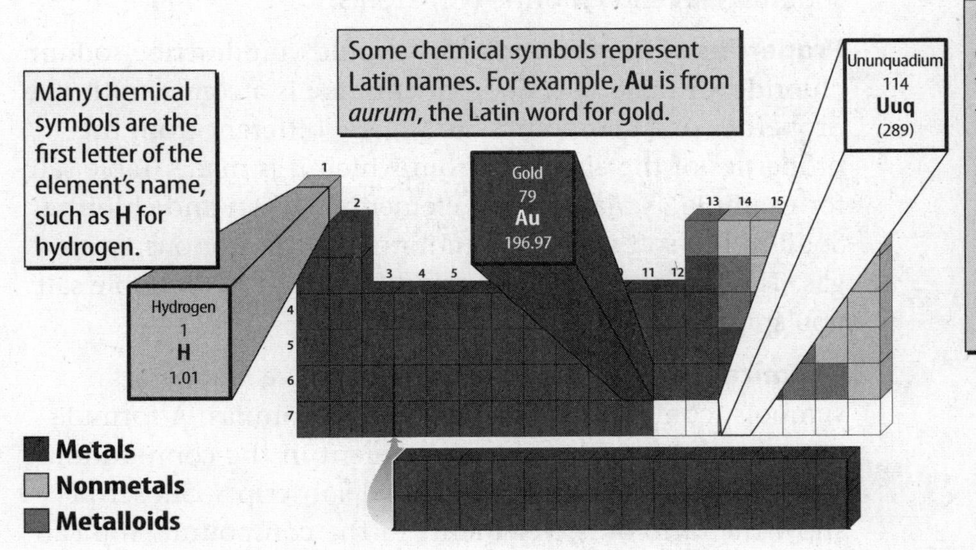

Compounds

Does it surprise you to learn that there are only about 115 different elements? After all, you probably can name many more types of matter than this if you think about all the different things you see each day. Why are there more kinds of matter than there are elements? Most matter is made of atoms of different types of elements bonded together.

*A **compound** is a substance made of two or more elements that are chemically joined in a specific combination.* A compound is a substance because it is made of atoms in a specific combination. Pure water (H_2O) is a compound. Every sample of pure water contains two hydrogen atoms to every oxygen atom. There are many types of matter because elements can join to form compounds.

Key Concept Check
8. Contrast How do elements and compounds differ?

______

______

______

______

Molecules Recall that a molecule is two or more atoms that are held together by chemical bonds and that act as a unit. Is a molecule the smallest part of a compound? A molecule is the smallest part of many compounds. Many compounds exist as molecules. One example is water. In water, two hydrogen atoms and one oxygen atom always exist together and act as a unit. Carbon dioxide (CO_2) and sugar ($C_6H_{12}O_6$) are also compounds that are made of molecules.

But some compounds are not made of molecules. In some compounds, such as table salt, or sodium chloride, no specific atoms travel together as a unit. However, table salt (NaCl), is still a substance because it always contains only sodium (Na) and chlorine (Cl) atoms.

FOLDABLES®
Make a vertical two-tab book to review properties of elements and compounds.

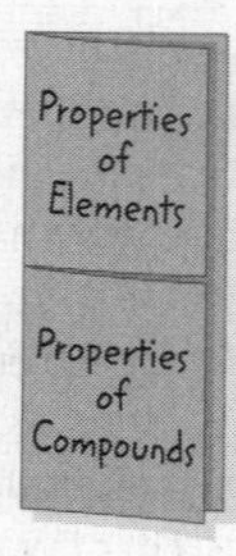

Properties of Compounds How would you describe sodium chloride, or table salt? Sodium chloride is a compound. The properties of a compound are usually different from the properties of the elements from which it is made. Table salt, for example, is made of the elements sodium and chlorine. Sodium is a soft metal, and chlorine is a poisonous green gas. These properties are much different from the table salt you sprinkle on food!

Reading Check
9. Summarize What information do the subscripts in a chemical formula provide?

______

______

______

Chemical Formulas Just as elements have chemical symbols, compounds have chemical formulas. A formula includes the symbols of each element in the compound. It also includes small numbers called subscripts. Subscripts show the ratio of the elements in the compound. You can see the formulas for some compounds in the table at the top of the next page.

Properties of Common Nitrogen Compounds	
Formula	**Properties/Uses**
N_2O Nitrous oxide	colorless gas used as an anesthetic
NO_2 Nitrogen dioxide	brown gas, toxic, air pollutant
N_2O_3 Dinitrogen trioxide	blue liquid

Different Combinations of Atoms Sometimes the same elements combine to form different compounds. For example, nitrogen and oxygen can form six different compounds. The chemical formulas are N_2O, NO, N_2O_3, NO_2, N_2O_4, and N_2O_5. Each compound contains the same elements, but because the combinations of atoms are different, each compound has different properties. Three examples are shown in the table above.

What is a mixture?

Can you tell whether the clear liquid in a glass is lemon-lime soda or water? Lemon-lime soda is almost clear. Someone might confuse it with water, which is a substance. Recall that a substance is matter with a composition that is always the same.

Sodas are a combination of substances such as water, carbon dioxide, sugar, and other compounds. In fact, most solids, liquids, and gases you see each day are mixtures. *A* **mixture** *is matter that can vary in composition.* A mixture is made of two or more substances that are blended but are not chemically bonded.

What would happen if you added more sugar to a glass of soda? You would still have soda, but it would be sweeter. Changing the amount of one substance in a mixture does not change the identity of the mixture or its individual substances.

Air and tap water are also mixtures. Air is a mixture of nitrogen, oxygen, and other substances. But the composition of air can vary. Air in a scuba tank usually contains more oxygen and less of the other substances. Tap water might look like pure water. However, tap water is a mixture of pure water (H_2O) and small amounts of other substances. The composition of tap water can vary because the substances that make up tap water are not bonded together. This is true for all mixtures.

Interpreting Tables

10. Analyze What is the ratio of nitrogen atoms to oxygen atoms in dinitrogen trioxide?

11. Apply Which pair of compounds has a ratio of one nitrogen atom to two oxygen atoms? (Circle the correct answer.)

a. NO_2 and N_2O_4
b. N_2O and NO_2
c. NO and N_2O_4

Reading Check

12. Restate Why does the composition of mixtures vary?

Types of Mixtures

How do trail mix, soda, and air differ? One difference is that trail mix is a solid, soda is a liquid, and air is a gas. This tells you that a mixture can be any state of matter. Another difference is that you can see the individual parts that make up trail mix. But you cannot see the parts that make up soda or air. This is because trail mix is a different type of mixture than soda and air.

ACADEMIC VOCABULARY
individual
(adjective) single; separate

Two types of mixtures can be made—one is heterogeneous (he tuh roh JEE nee us), and the other is homogeneous (hoh muh JEE nee us). The prefix *hetero-* means "different," and the prefix *homo-* means "the same." Heterogeneous and homogeneous mixtures differ in how evenly the substances that comprise them are mixed.

Heterogeneous Mixtures

Suppose you take a bag of trail mix and pour it into two identical bowls. What might you notice? At first glance, each bowl appears the same. But if you look closely, you might notice that one bowl has more nuts and another bowl has more raisins. The contents of the bowls differ because trail mix is a heterogeneous mixture.

A **heterogeneous mixture** *is a mixture in which the substances are not evenly mixed.* So if you take two samples from the same mixture, such as trail mix, the samples might have different amounts of the individual substances. Trail mix, granite, and smoke are examples of heterogeneous mixtures.

13. Explain why vegetable soup is classified as a heterogeneous mixture.

Homogeneous Mixtures

If you poured soda from a bottle into two glasses, the amounts of water, carbon dioxide, sugar, and other substances in the mixture would be the same in both glasses. Soda is an example of a **homogeneous mixture**—*a mixture in which two or more substances are evenly mixed, but not bonded together*.

Evenly Mixed Parts In a homogeneous mixture, the substances are so small and evenly mixed that you cannot see the boundaries between substances in the mixture. Brass is a mixture of copper and zinc. It is a homogeneous mixture because the copper atoms and the zinc atoms are evenly mixed. You cannot tell which atoms are which even under most microscopes. Lemonade and air are also examples of homogeneous mixtures for the same reason.

Reading Check
14. Classify Is yellow mustard a heterogeneous mixture or a homogeneous mixture?

Solution Another name for a homogeneous mixture is a solution. A solution is made of two parts—a solvent and one or more solutes. The solvent is present in the largest amount. The solutes dissolve, or break apart, and mix evenly in the solvent. In a mixture of water, salt, and pepper, salt is soluble in water. So water is the solvent and salt is the solute. But pepper does not dissolve in water. No solution forms between pepper and water, so it is insoluble in water.

Brass is a solution of solid copper and solid zinc. Brass is a metal often used to make musical instruments, such as trumpets and tubas. The natural gas used in a gas stove is a solution of methane, ethane, and other gases. Ammonia, often used as a cleaner, is a solution of water and ammonia gas. Note that a solvent or a solute in a solution can be any of the three states of matter—solid, liquid, and gas.

Compounds v. Mixtures

Think again about the trail mix in two bowls. If you add peanuts to one bowl, you still have trail mix in both bowls. The substances that make up a mixture are not bonded. So adding more of one substance does not change the identity or the properties of the mixture. It also does not change the identity or the properties of each individual substance. Adding more peanuts to a mixture of peanuts, pretzels, and raisins will not change the properties of the individual parts. Peanuts and raisins don't bond and become something new.

In a solution of soda or air, the substances do not bond together and form something new. Carbon dioxide, water, sugar, and other substances in soda are mixed together. Nitrogen, oxygen, and other substances in air also keep their properties because air is a mixture. If air were a compound, the parts would be bonded and would not keep their separate properties.

Compounds and Solutions Differ

Compounds and solutions are alike in that they both look like pure substances. Think about lemon-lime soda. The soda is a solution. A solution might look like a substance because the elements and the compounds that make up a solution are evenly mixed. But compounds and solutions differ in one important way. The atoms that make up a given compound are bonded together. So the composition of a given compound is always the same. Changing the composition results in a new compound.

Key Concept Check
15. Apply How are some mixtures different from solutions?

Key Concept Check
16. Contrast How do mixtures and compounds differ?

Differences Between Solutions and Compounds		
	Solutions	**Compounds**
Composition	Made up of substances (elements and compounds) evenly mixed together; the composition can vary in a given mixture.	Made up of atoms bonded together; the combination of atoms is always the same in a given compound.
Changing the Composition	The solution is still the same with similar properties. However, the relative amounts of substances might be different.	The compound has changed into a new compound with new properties.
Properties of Parts	The substances keep their own properties when they are mixed.	The properties of the compound are different from the properties of the atoms it is made from.

Interpreting Tables

17. Identify Which substance always has the same combination of atoms?

However, the substances that make up a solution are not bonded together. So adding more of one substance in a solution will change the composition of the solution. It will just change the ratio of the substances in the solution. These differences are described in the table above.

Separating Mixtures

Have you ever picked something you did not like off a slice of pizza? If you have, you have separated a mixture. Because the parts of a mixture are not combined chemically, you can use a physical process to separate the mixture. One way to do this is to remove the parts by hand. The identity of the parts does not change. Separating a compound back into the parts from which it was made is more difficult. The elements that make up a compound are combined chemically. Only a chemical change can separate them.

REVIEW VOCABULARY

chemical change
a change in matter in which the substances that make up the matter change into other substances with different chemical and physical properties

Separating Heterogeneous Mixtures Separating the parts of a pizza is easy because the pizza has large, solid parts. Heterogeneous mixtures can also be separated in other ways. For example, a strainer, or sieve, can be used to filter larger rocks from a mixture of rocks and dirt. Oil and vinegar is also a heterogeneous mixture because the oil floats on the vinegar. You can separate this mixture by carefully removing the floating oil.

Reading Check

18. Name three methods of separating heterogeneous mixtures.

Other properties also might be useful for separating the parts. For example, if one of the parts is magnetic, you could use a magnet to remove it. In a mixture of solid powders, you might dissolve one part in water and then pour it out, leaving the other part behind. In each case, to separate a heterogeneous mixture, you use differences in the physical properties of the parts.

Separating Homogeneous Mixtures Imagine trying to separate soda into water, carbon dioxide, sugar, and the other substances it is made from. The parts are so small and evenly mixed that separating a homogeneous mixture such as soda can be difficult. But you can separate some homogeneous mixtures by boiling or evaporation. If you place a bowl of sugar water outside on a hot day, the water will evaporate slowly, leaving the sugar behind. Rock candy is made by evaporating water from a sugar solution.

Visualizing Classification of Matter

Think about all the types of matter you have read about in this lesson. As shown in the figure below, matter can be classified as either a substance or a mixture. Substances are either elements or compounds. The two kinds of mixtures are homogeneous mixtures and heterogeneous mixtures. Notice that all substances and mixtures are made of atoms. Matter is classified according to the types of atoms and the arrangement of atoms in matter. In the next lesson, you will study the structure of atoms.

19. Infer Sea salt is harvested by evaporating seawater. Is seawater a homogeneous mixture or a heterogeneous mixture?

____

____

20. Identify Circle the name of the mixture made of substances that might be seen without a microscope.

Classifying Matter

Matter
- Anything that has mass and takes up space
- Most matter on Earth is made up of atoms.
- Two classifications of matter: substances and mixtures

Substances
- Matter with a composition that is always the same
- Two types of substances: elements and compounds

Element
- Consists of just one type of atom
- Organized on the periodic table
- Each element has a chemical symbol.

Compound
- Two or more types of atoms bonded together
- Properties are different from the properties of the elements that make it up
- Each compound has a chemical formula.

Substances physically combine to form mixtures.

Mixtures can be separated into substances by physical methods.

Mixtures
- Matter that can vary in composition
- Substances are not bonded together.
- Two types of mixtures: heterogeneous and homogeneous

Heterogeneous Mixture
- Two or more substances unevenly mixed
- Different substances are visible by an unaided eye or a microscope.

Homogeneous Mixture—Solution
- Two or more substances evenly mixed
- Different substances cannot be seen even by a microscope.

After You Read

Mini Glossary

atom: a small particle that is the building block of matter

compound: a substance made of two or more elements that are chemically joined in a specific combination

element: a substance made of only one kind of atom

heterogeneous (he tuh roh JEE nee us) mixture: a mixture in which the substances are not evenly mixed

homogeneous (hoh muh JEE nee us) mixture: a mixture in which two or more substances are evenly mixed, but not bonded together

matter: anything that has mass and takes up space

mixture: matter that can vary in composition

molecule (MAH lih kyewl): two or more atoms that are held together by chemical bonds and act as a unit

substance: matter with a composition that is always the same

1. Review the terms and their definitions in the Mini Glossary. Write a sentence to compare and contrast a mixture and a compound.

____

____

2. Use what you have learned about matter to complete the graphic organizer.

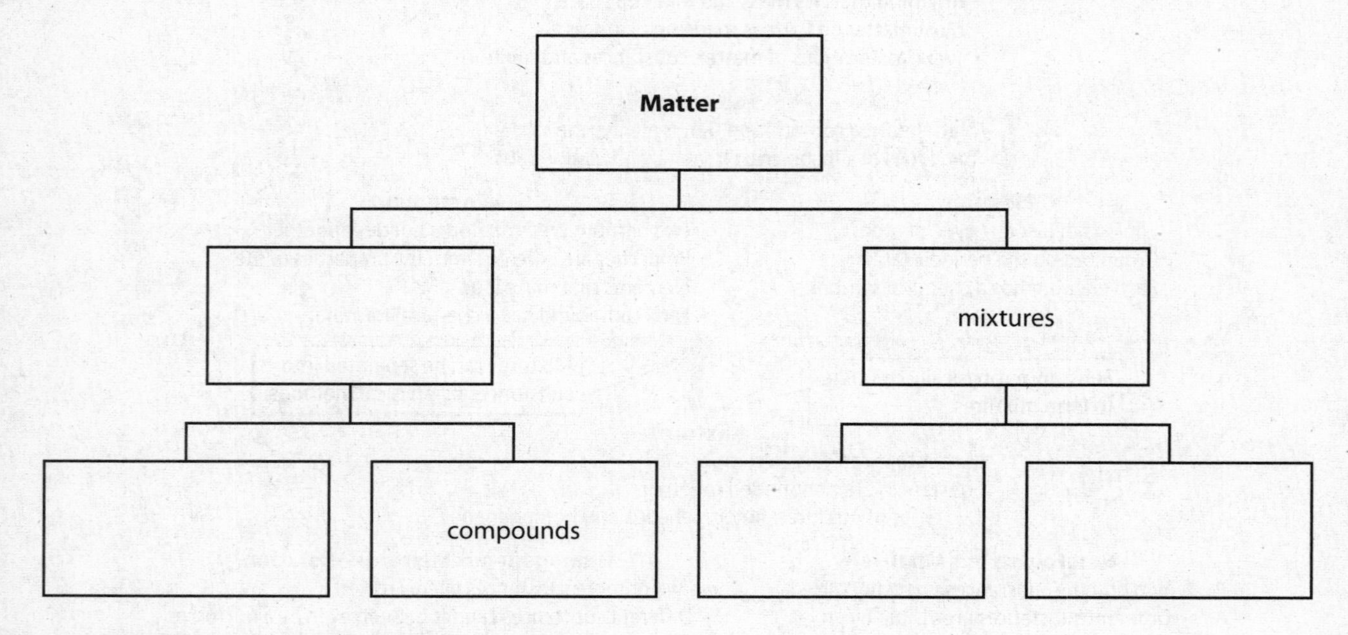

What do you think NOW?

Reread the statements at the beginning of the lesson. Fill in the After column with an A if you agree with the statement or a D if you disagree. Did you change your mind?

Log on to ConnectED.mcgraw-hill.com and access your textbook to find this lesson's resources.

Matter and Atoms

The Structure of Atoms

Before You Read

What do you think? Read the three statements below and decide whether you agree or disagree with them. Place an A in the Before column if you agree with the statement or a D if you disagree. After you've read this lesson, reread the statements to see if you have changed your mind.

Before	Statement	After
	4. An atom is mostly empty space.	
	5. If an atom gains electrons, the atom will have a positive charge.	
	6. Each electron is a cloud of charge that surrounds the center of an atom.	

Key Concepts 

- Where are protons, neutrons, and electrons located in an atom?
- How is the atomic number related to the number of protons in an atom?
- What effect does changing the number of particles in an atom have on the atom's identity?

Read to Learn

The Parts of an Atom

Now that you have read about ways to classify matter, you can probably recognize the different types of matter you see each day. You might see pure elements, such as copper and iron. You also might see many compounds, such as table salt.

Table salt is a compound because it contains the atoms of two different elements in a specific combination. These elements are sodium and chlorine. You also probably see many mixtures. The silver often used in jewelry is a homogeneous mixture of metals. The metals are evenly mixed, but they are not bonded together.

As you read in Lesson 1, the many types of matter are possible because there are about 115 different elements. Each element is made up of a different type of atom. Atoms can combine in many different ways. They are the basic parts of matter.

What makes the atoms of each element different? Atoms are made of several types of tiny particles. The number of each of these particles in an atom is what makes atoms different from each other. It is what makes so many types of matter possible.

Study Coach

Identify the Main Ideas As you read, write one sentence to summarize the main idea in each paragraph. Write the main ideas on a sheet of paper or in your notebook to study later.

Reading Check

1. Contrast What makes the atoms of different elements different from each other?

FOLDABLES®

Make a two-column chart book to organize information about the particles in an atom.

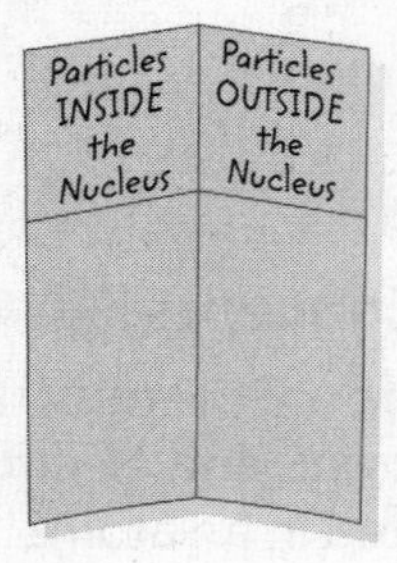

The Nucleus—Protons and Neutrons

The basic structure of all atoms is the same. The basic structure of an atom is shown in the figure below.

An atom has a center region with a positive charge. One or more negatively charged particles move around this center region. *The* **nucleus** *is the region at the center of an atom that contains most of the mass of the atom.*

Two kinds of particles make up the nucleus. *A* **proton** *is a positively charged particle in the nucleus of an atom. A* **neutron** *is an uncharged particle in the nucleus of an atom.*

Parts of an Atom

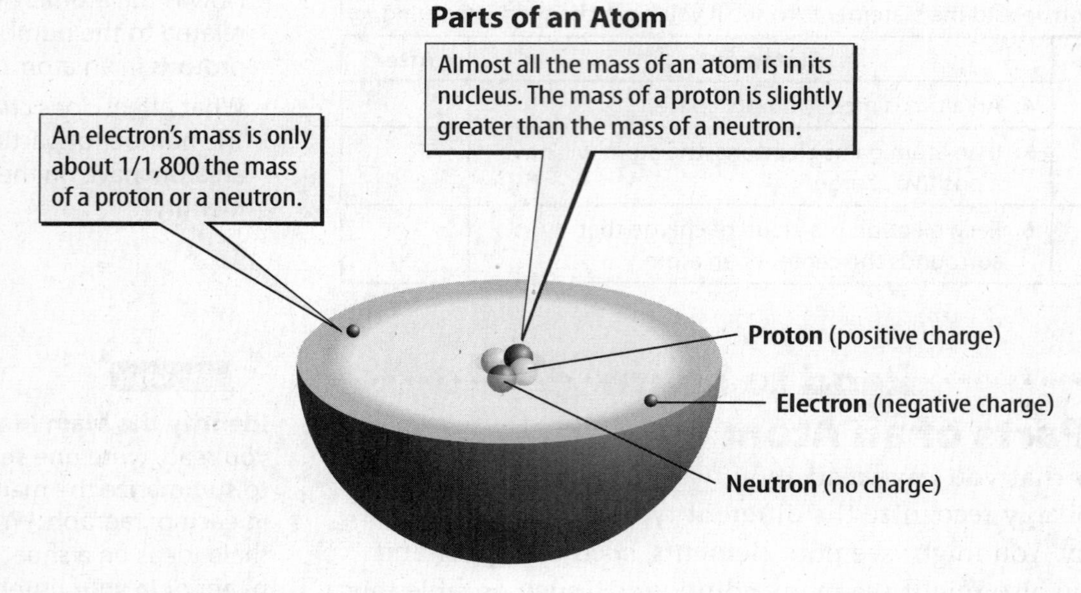

Visual Check

2. Identify How many protons and how many electrons does this atom have?

Electrons

Atoms have no electric charge unless they change in some way. So there must be a negative charge that balances the positive charge of the nucleus. *An* **electron** *is a negatively charged particle that occupies the space in an atom outside the nucleus.*

Electrons are small and move quickly. Because of this, scientists are unable to tell exactly where a given electron is located at any specific time. So scientists describe the positions of electrons around the nucleus as a cloud rather than specific points.

All atoms have a positively charged nucleus surrounded by one or more electrons. An electron is shown in the model of an atom in the figure above.

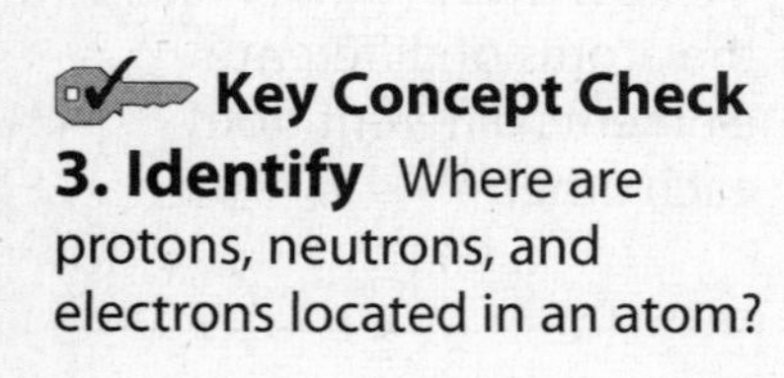

Key Concept Check

3. Identify Where are protons, neutrons, and electrons located in an atom?

An Electron Cloud Drawings of an atom often show electrons circling the nucleus like planets orbiting the Sun. Scientists have conducted experiments that show the movement of electrons is more complex than this. The modern idea of an atom is called the electron-cloud model. *An* ***electron cloud*** *is the region surrounding an atom's nucleus where one or more electrons are most likely to be found.* It is important to understand that an electron is not a cloud of charge. An electron is one tiny particle. An electron cloud is mostly empty space. At any moment in time, electrons are located at specific points within that area.

Electron Energy You have read that electrons are constantly moving around the nucleus in a region called the electron cloud. But some electrons are closer to the nucleus than others. Electrons occupy certain areas around the nucleus according to their energy. Electrons close to the nucleus are strongly attracted to it and have less energy than electrons farther from the nucleus. Electrons farther from the nucleus are less attracted to the nucleus and have more energy than electrons closer to the nucleus.

The Size of Atoms

It might be difficult to visualize an atom. But every solid, liquid, and gas is made of millions and millions of atoms. Your body, your desk, and the air you breathe are all made of tiny atoms. Suppose you could increase the size of everything around you. If you could make everything larger by multiplying an object's width by 100 million, or 1×10^8, an atom would be the size of an orange, and an orange would be the size of Earth!

Differences in Atoms

In some ways, atoms are alike. Each has a positively charged nucleus surrounded by a negatively charged electron cloud. But atoms can differ from each other in several ways. Atoms can have different numbers of protons, neutrons, or electrons.

Protons and Atomic Number

Look at the periodic table in the back of this book. The number under the element name in each block shows how many protons each atom of the element has. For example, each oxygen atom has eight protons.

The ***atomic number*** *is the number of protons in the nucleus of an atom of an element.* If there are 12 protons in the nucleus of an atom, that element's atomic number is 12.

Math Skills

Scientists write very large and very small numbers using scientific notation. A gram of carbon has about 50,000,000,000,000,000,000 atoms. Express this in scientific notation.

a. Move the decimal until one nonzero digit remains on the left:

5.0000000000000000000

b. Count how the places you moved. In this case, 19 places left.

c. Show that number as a power of 10. The exponent is negative if the decimal moves right and positive if it moves left.

Answer: 5×10^{19}

d. Reverse the process to change scientific notation back to a whole number.

4. Use Scientific Notation The diameter of a carbon atom is 2.2×10^{-8} cm. Write this as a whole number.

____

Key Concept Check

5. Recognize How is the atomic number related to the number of protons in an atom?

____

____

____

Protons and the Atomic Number

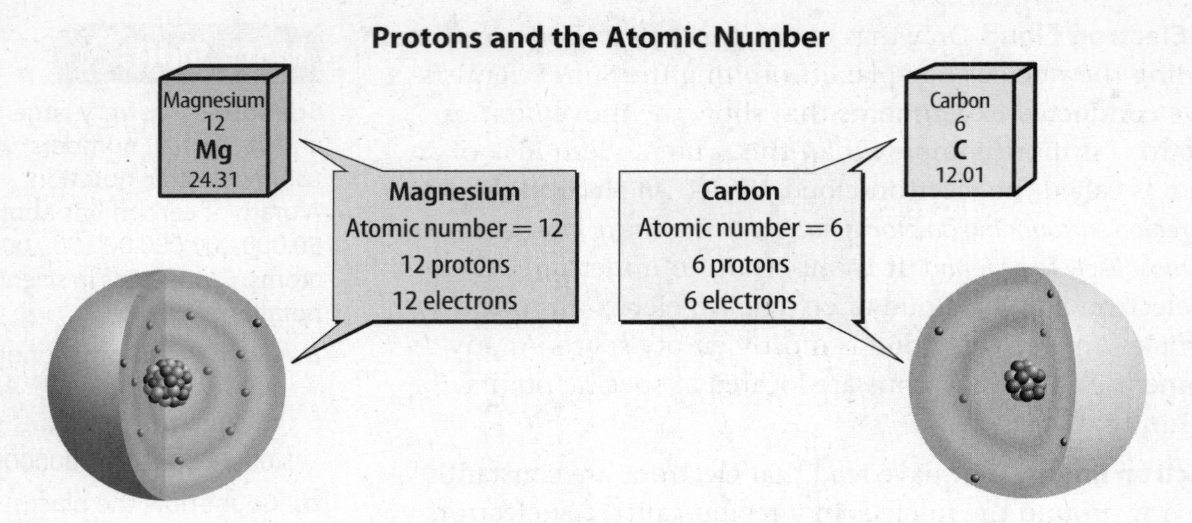

Visual Check

6. Locate In the figure above, circle the atomic number in the cube that represents the carbon atom.

Examine the figure above. Notice that the atomic number of magnesium is the whole number above its symbol. The atomic number of carbon is 6. This means that each carbon atom has 6 protons. Every element in the periodic table has a different atomic number.

You can identify an element if you know either its atomic number or the number of protons its atoms have. If an atom has a different number of protons, it is a different element.

Isotopes

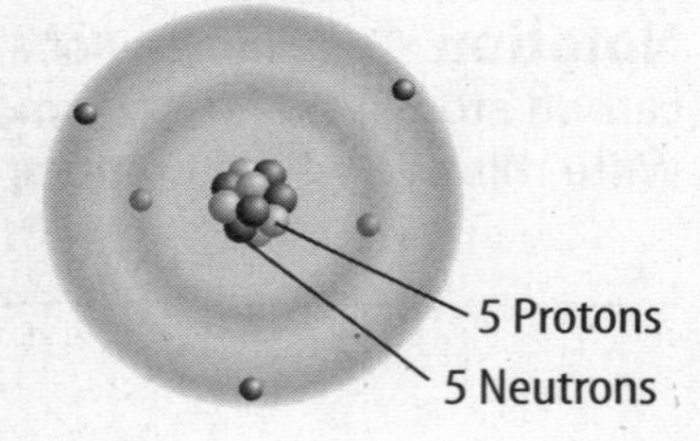

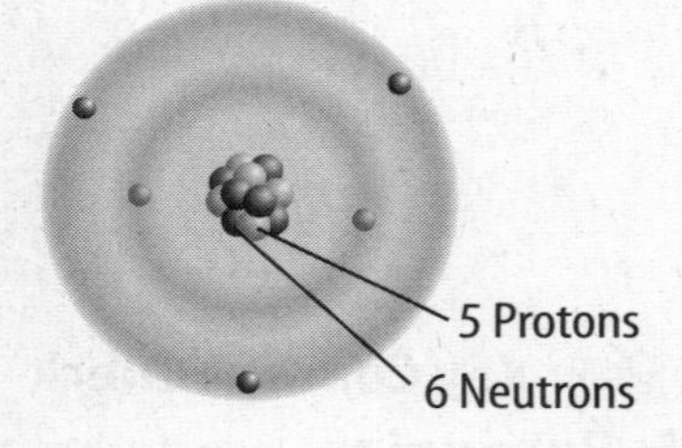

Visual Check

7. Describe What makes boron-10 and boron-11 isotopes?

____

____

Neutrons and Isotopes

Each atom of an element contains the same number of protons, but the number of neutrons can vary. *An* **isotope** (I suh tohp) *is one of two or more atoms of an element having the same number of protons, but a different number of neutrons.* Boron-10 and boron-11 are isotopes of boron, as shown in the figure to the left. Notice that boron-10 has ten particles in its nucleus. Boron-11 has 11 particles in its nucleus.

Electrons and Ions

You read that atoms can differ by the number of protons or neutrons they have. The figure at the top of the next page shows a third way atoms can differ—by the number of electrons.

A neutral, or uncharged, atom has the same number of positively charged protons and negatively charged electrons. As atoms bond, their numbers of electrons can change. Because electrons are negatively charged, a neutral atom that has lost an electron has a positive charge. A neutral atom that has gained an electron has a negative charge. *An* **ion** (I ahn) *is an atom that has a charge because it has gained or lost electrons.* An ion is the same element it was before it gained or lost electrons because the number of protons is unchanged.

Ions

Neutral atom

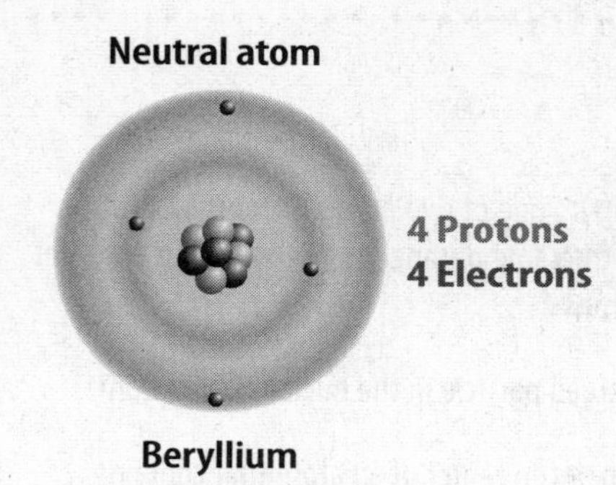

4 Protons
4 Electrons

Beryllium

A neutral atom has the same number of electrons and protons. The atom has no charge.

Positive ion

3 Protons
2 Electrons

Lithium

If an atom loses an electron during chemical bonding, it has more protons than electrons. It is now positively charged.

Negative ion

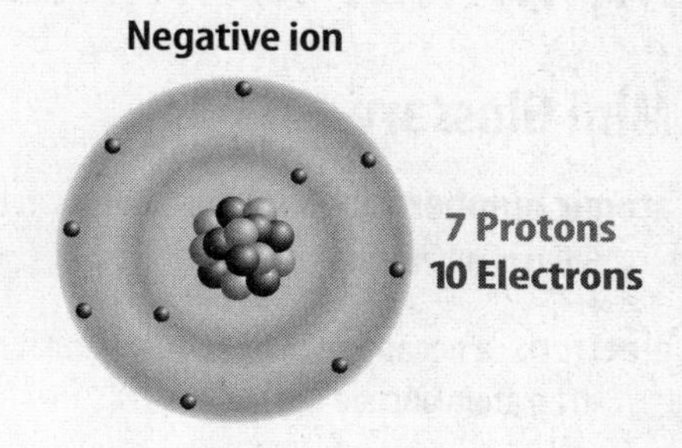

7 Protons
10 Electrons

Nitrogen

If an atom gains an electron during chemical bonding, it has more electrons than protons. It is now negatively charged.

In the previous lesson, you read that each particle of a compound is two or more atoms of different elements bonded together. One of the ways a compound forms is when one or more electrons move from an atom of an element to an atom of a different element. This results in a positive ion for one element and a negative ion for the other element.

Visual Check

8. Explain How can a neutral atom become a positive ion or a negative ion?

Atoms and Matter

You have read that a substance has a composition that is always the same, but the composition of a mixture can vary. All types of matter are made of atoms. The atoms of a certain element always have the same number of protons, but the number of neutrons can vary. When elements combine to form compounds, the number of electrons in the atoms can change. The different ways in which atoms can change are summarized in the table below. The ways in which the atoms combine result in the many different kinds of matter around you.

Key Concept Check

9. Relate What effect does changing the number of particles in an atom have on the atom's identity?

Possible Changes in Atoms

Neutral Atom	Change	Results
Carbon • 6 protons • 6 neutrons • 6 electrons	**Protons** add one proton	**New element—nitrogen** • 7 protons • 7 neutrons • 7 electrons
	Neutrons add one neutron	**Isotope** • 6 protons • 7 neutrons • 6 electrons
	Electrons add one electron	**Ion** • 6 protons • 6 neutrons • 7 electrons

Interpreting Tables

10. Interpret Adding one proton to carbon creates what new element?

After You Read

Mini Glossary

atomic number: the number of protons in the nucleus of an atom of an element

electron: a negatively charged particle that occupies the space in an atom outside the nucleus

electron cloud: the region surrounding an atom's nucleus where one or more electrons are most likely to be found

ion (I ahn): an atom that has a charge because it has gained or lost electrons

isotope (I suh tohp): one of two or more atoms of an element having the same number of protons, but a different number of neutrons

neutron: an uncharged particle in the nucleus of an atom

nucleus: the region at the center of an atom that contains most of the mass of the atom

proton: a positively charged particle in the nucleus of an atom

1. Review the terms and their definitions in the Mini Glossary. Write a sentence to explain how two atoms can form a compound by forming ions.

2. Use what you have learned about the particles of an atom to complete the table.

Particle	Charge	Location
electron		
		inside nucleus
	no charge	

3. How did studying your list of main ideas help you understand this lesson?

What do you think NOW?

Reread the statements at the beginning of the lesson. Fill in the After column with an A if you agree with the statement or a D if you disagree. Did you change your mind?

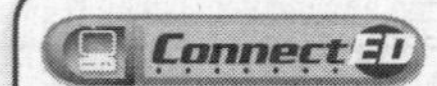

Log on to ConnectED.mcgraw-hill.com and access your textbook to find this lesson's resources.

Matter: Properties and Changes

Matter and Its Properties

Before You Read

What do you think? Read the three statements below and decide whether you agree or disagree with them. Place an A in the Before column if you agree with the statement or a D if you disagree. After you've read this lesson, reread the statements to see if you have changed your mind.

Before	Statement	After
	1. The particles in a solid object do not move.	
	2. Your weight depends on your location.	
	3. The particles in ice are the same as the particles in water.	

Key Concepts
- How do particles move in solids, liquids, and gases?
- How are physical properties different from chemical properties?
- How are properties used to identify a substance?

Read to Learn

What is matter?

Look around you. All the objects that you see are made of matter. Matter can be in different forms and can have different properties. As you read, you will learn about matter, its properties, and its uses.

Matter is anything that has mass and takes up space. You, your book, your desk, and the water you drink are matter because they have mass and take up space. The air you breathe is matter, even though you can't see it. Air has mass and takes up space. Light from the Sun is not matter because it does not have mass and does not take up space. Sounds, forces, and energy are not matter because they do not have mass and do not take up space.

Matter has many different properties. For example, a helmet you wear while biking is hard and shiny. The water in the stream is cool and clear. You will learn about some of the physical properties and chemical properties of matter in this chapter. Learning about these properties will help you to identify many types of matter and their uses.

Study Coach

Use an Outline As you read, make an outline to summarize the information in the lesson. Use the main headings in the lesson as the main headings in the outline. Complete the outline with the information under each heading.

REVIEW VOCABULARY
matter
anything that has mass and takes up space

States of Matter

One property of a substance is its state of matter. You can tell the state of a material by answering the following questions:

- Can its shape change?
- Can its volume change?

Volume *is the amount of space a sample of matter occupies*. Three states of matter are solids, liquids, and gases. The table below shows whether shape and volume change for a solid, a liquid, and a gas when moved from one container to another.

	Solid	**Liquid**	**Gas**
When moved from one container to another	does not change shape	does change shape	does change shape
	does not change volume	does not change volume	does change volume

Solids, Liquids, and Gases

The table above shows that *a* **solid** *is a state of matter with a definite shape and volume. A* **liquid** *is a state of matter with a definite volume but not a definite shape. A* **gas** *is a state of matter without a definite shape or a definite volume.*

Moving Particles

All matter is made of tiny particles. The particles of matter are always moving. Particles in solids move quickly or vibrate back and forth in all directions. They can't move from place to place. In liquids, particles are farther apart. They can slide past each other. In a gas, particles move freely rather than staying close together.

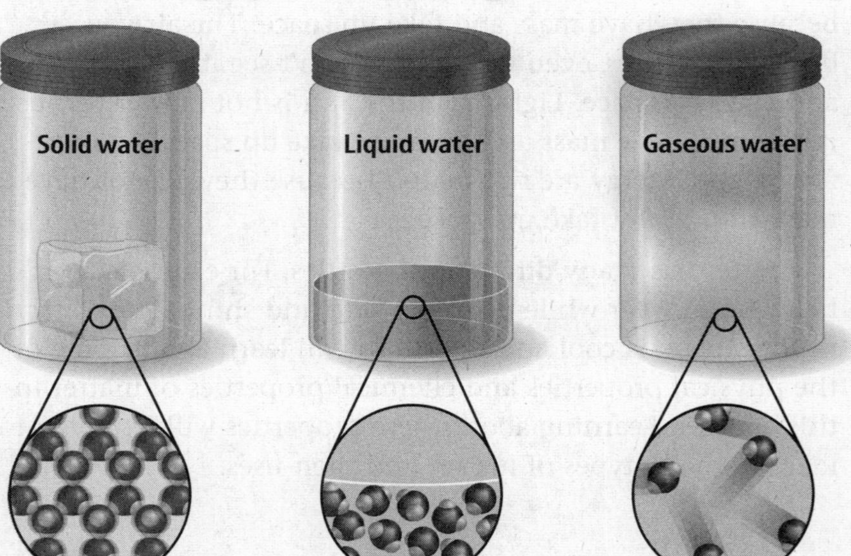

Reading Check

1. Identify Which state of matter does not change shape or volume?

Key Concept Check

2. Explain How do particles move in solids, liquids, and gases?

Visual Check

3. Identify Circle the matter that moves freely.

Attraction Between Particles

Particles of matter that are close to each other attract, or pull on, each other. The stronger the attraction on each other, the closer together the particles are. Because particles of a solid are close together, they attract each other strongly. Particles of a liquid can flow because the forces between the particles are weaker. Particles of a gas are so far apart that they are not held together by attractive forces.

What are physical properties?

Matter has physical properties. *A* **physical property** *is any characteristic of matter that you can observe without changing the identity of the substances that make it up*. Examples of physical properties are state of matter, shape, mass, volume, density, solubility, and temperature.

Mass and Weight

Some physical properties, such as mass and weight, depend on the size of the sample. **Mass** *is the amount of matter in an object*. Weight is the gravitational pull on an object.

Weight depends on the location of an object. Mass does not. The mass of an object is the same on Earth as it is on the Moon. An object's weight, however, is greater on Earth than it is on the Moon because Earth's gravity is stronger than the Moon's gravity. 

Volume

Like mass and weight, the volume of an object is a physical property. Volume depends on the size or amount of the sample. You can measure the volume of a liquid by pouring it into a measuring cup or a graduated cylinder. You can measure the volume of a solid in two ways. If a solid has a regular geometric shape, multiply its length, width, and height together. You can find the volume of a solid with an irregular shape by using the displacement method that is shown below.

Volume of an Irregular-Shaped Solid

The volume of an irregular-shaped object can be measured by displacement. The volume of the object is the difference between the water level before and after placing the object in the water. The common unit for liquid volume is the milliliter (mL).

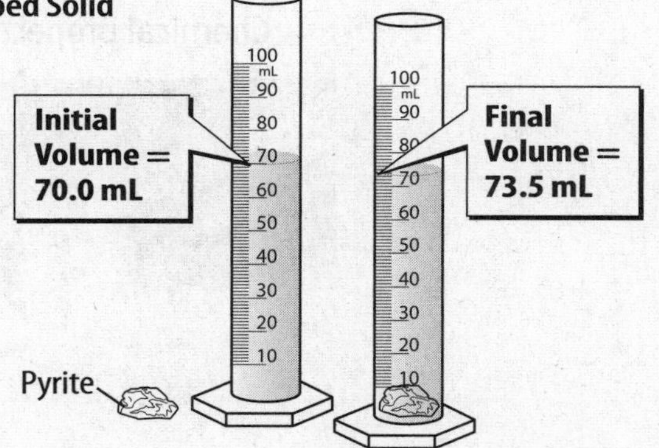

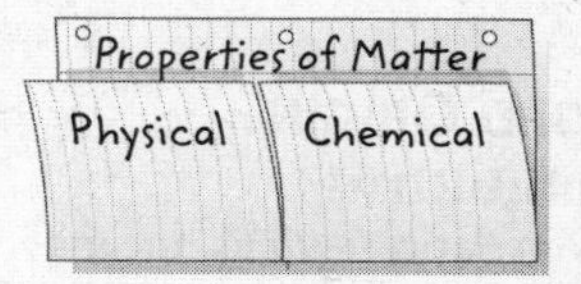

FOLDABLES

Make the following two-tab book to organize your notes about the properties of matter.

Reading Check

4. Describe How do mass and weight differ?

Visual Check

5. Highlight the part of the water in the second cylinder that is equal in volume to the rock in the cylinder.

Density

Another physical property of matter is density. ***Density** is the mass of a substance divided by the volume of the substance.* Density does not depend on the size or amount of the sample. The density of a substance never changes.

Solubility

WORD ORIGIN
solubility
from Latin *solubilis,* means "capable of being dissolved"

You can observe another physical property of matter when a solid, such as sugar, dissolves in water. *To dissolve* means "to mix evenly." **Solubility** (sahl yuh BIH luh tee) *is the ability of one material to dissolve in another material.*

Melting and Boiling Point

Each material has a melting point and a boiling point. Melting point and boiling point do not depend on the size or the amount of the material. The melting point is the temperature at which a solid changes to a liquid. The boiling point is the temperature at which a liquid changes to a gas. Melting point and boiling point are physical properties. ✓

✓ Reading Check
6. Explain How does a material change at its melting point and at its boiling point?

____

____

____

____

Additional Physical Properties

There are many physical properties that make materials useful. Some materials conduct electricity, some are magnetic, and some are malleable. Malleable materials can be bent or pulled into different shapes.

What are chemical properties?

Substances undergo chemical reactions when they change into other substances. *A* **chemical property** *is the ability or inability of a substance to combine with or change into one or more new substances.* When substances react, their particles combine to form different substances.

Flammability

Flammability is a chemical property. Flammability is the ability of a type of matter to burn easily, as shown below. Some substances, such as wood and paper, are flammable. Rocks and sand are not flammable.

Chemical property

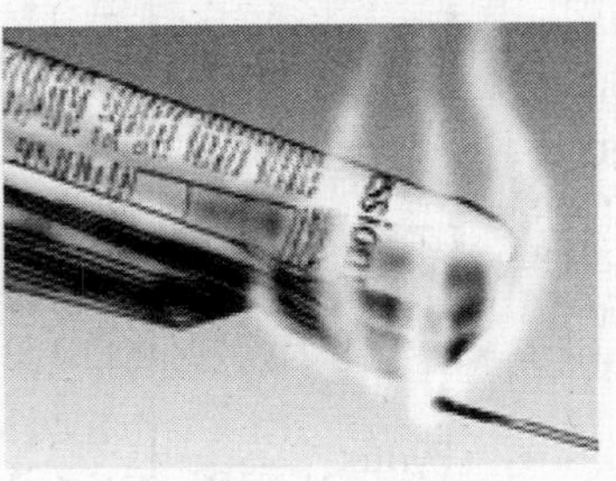

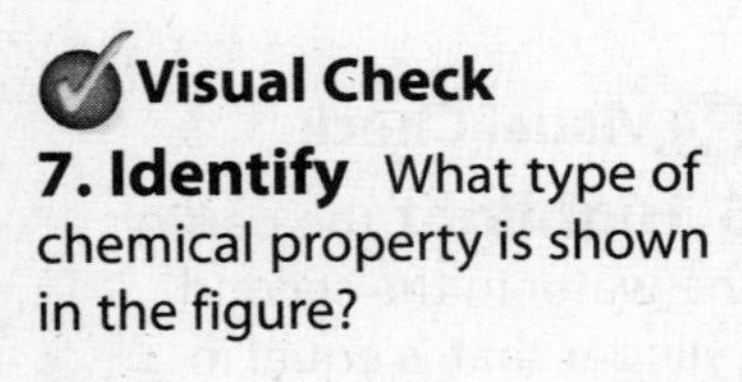

Visual Check
7. Identify What type of chemical property is shown in the figure?

____

Ability to Rust

You have probably seen objects, such as old cars, that have begun to rust. Rust is a substance that forms when iron reacts with oxygen and water in the air. The ability to rust is a chemical property of iron or metals that contain iron.

Key Concept Check

8. Contrast What is the difference between chemical properties and physical properties?

Identifying Matter Using Physical Properties

Physical properties can be used to identify unknown substances. Look at the table below of substances and their physical properties. You can identify the unknown substance by comparing its physical properties to the physical properties of the known substances.

Substance	Color	Mass g	Melting Point °C	Density g/cm^3
Table salt	white	14.5	801	2.17
Sugar	white	11.5	148	1.53
Baking soda	white	16.0	50	2.16
Unknown	white	16.0	801	2.17

All the substances are white. So, you cannot identify the unknown substance by its color. You also cannot identify it by its mass. Mass and volume are properties of matter that change with the amount of the sample. However, melting point and density are physical properties that do not depend on the size or amount of the sample. The unknown substance has the same melting point and density as table salt, so it must be table salt.

Sorting Materials Using Properties

Both physical properties and chemical properties are used for sorting materials. You probably often sort materials by their properties without realizing it. Objects are usually sorted based on the physical and chemical properties they have in common.

Separating Mixtures Using Physical Properties

Physical properties can be used to separate different types of matter that are mixed. Size, for example, can be used to separate a mixture of grains by sifting the mixture. Boiling point can be used to separate salt from water. The liquid water changes to gas, leaving the salt behind.

Math Skills

A statement that two expressions are equal is an equation. For example, look at the density equation:

$$D = \frac{m}{V}$$

This equation shows that density, *D*, is equal to mass, *m*, divided by volume, *V*. To solve a one-step equation, place the variables you know into the equation. Then solve for the unknown variable. For example, if

$$m = 52\text{ g}$$
$$V = 4\text{ cm}^3$$

the density would be:

$$D = \frac{52\text{ g}}{4\text{ cm}^3} = 13\text{ g/cm}^3$$

9. Solve Equation A cube of metal measures 3 cm on each side. It has a mass of 216 g. What is the density of the metal?

After You Read

Mini Glossary

chemical property (KEM ih kul • PRAH pur tee): the ability or inability of a substance to combine with or change into one or more new substances

density (DEN sih tee): the mass per unit volume of a substance

gas: a state of matter without a definite shape or a definite volume

liquid (LIH kwud): a state of matter with a definite volume but not a definite shape

mass: the amount of matter in an object

physical property (FIH zih kul • PRAH pur tee): any characteristic of matter that you can observe without changing the identity of the substances that make it up

solid (SAH lud): a state of matter with a definite shape and volume

solubility (sahl yuh BIH luh tee): the ability of one substance to dissolve in another

volume (VAHL yum): the amount of space a sample of matter occupies

1. Review the terms and their definitions in the Mini Glossary. Write a sentence that describes how some of the physical properties of a substance might be measured.

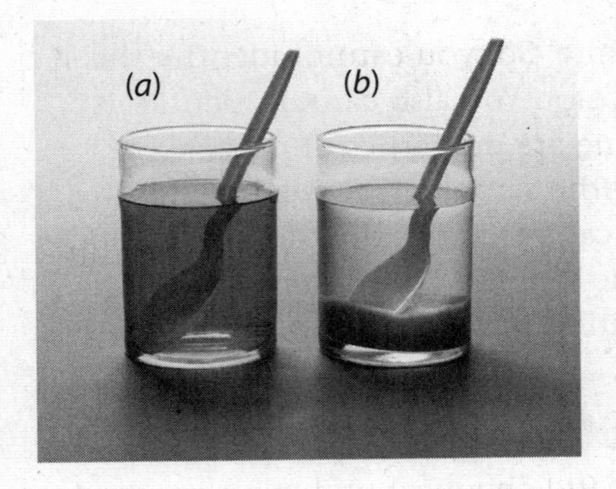

2. Look at the picture above. Which of the two glasses, *a* or *b*, contains a dissolved mixture? Describe the difference between the particles in the dissolved mixture and the contents of the other glass.

What do you think NOW?

Reread the statements at the beginning of the lesson. Fill in the After column with an A if you agree with the statement or a D if you disagree. Did you change your mind?

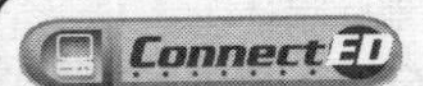

Log on to ConnectED.mcgraw-hill.com and access your textbook to find this lesson's resources.

Matter: Properties and Changes

Matter and Its Changes

Before You Read

What do you think? Read the three statements below and decide whether you agree or disagree with them. Place an A in the Before column if you agree with the statement or a D if you disagree. After you've read this lesson, reread the statements to see if you have changed your mind.

Before	Statement	After
	4. Mixing powdered drink mix with water causes a new substance to form.	
	5. If you combine two substances, bubbling is a sign that a new type of substance might be forming.	
	6. If you stir salt into water, the total amount of matter decreases.	

Key Concepts 

- How are physical changes different from chemical changes?
- How do physical and chemical changes affect mass?

Read to Learn

Changes of Matter

Matter can change physically and chemically. The identity of a substance does not change in a physical change. A substance does change into a different substance in chemical changes. Chemical changes happen when substances react with one another. In any change, matter can only be rearranged. It cannot be created or destroyed.

Matter can change in many ways. You can notice many of these changes in the world around you. Temperature, state of matter, shape, and color are all changes in matter that you have seen and felt many times. These changes can be either physical or chemical.

Mark the Text

Underline Main Ideas As you read, underline the main ideas under each heading. After you finish reading, review the main ideas that you have underlined.

What are physical changes?

A change in the size, shape, form, or state of matter that does not change the matter's identity is a **physical change.** You can change the shape of a ball of clay, and it is still clay. When a physical change occurs, the chemical properties of the matter stay the same.

1. Identify Circle the image that shows the sugar completely dissolved in the water.

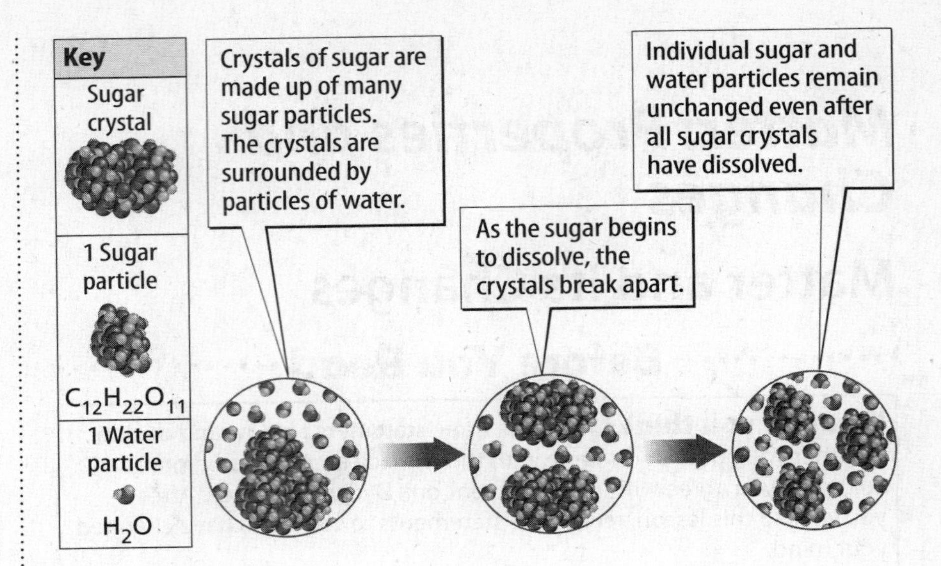

Dissolving

You read in Lesson 1 that solubility is a physical property. Solubility is the ability of one material to dissolve in another. *To dissolve* means "to mix evenly." The picture above shows what happens to sugar molecules when sugar crystals dissolve in water. Dissolving is a physical change because the identities of the substances do not change.

2. Explain why dissolving is classified as a physical change.

____

____

____

Changing State

In Lesson 1 you read about three states of matter—solid, liquid, and gas. When matter changes from one state to another, it goes through a physical change.

Melting and Boiling When water freezes, it changes into ice. When ice melts, it changes into water. When water boils, it changes into a gas. In each case, the substances that make up water are still the same.

Energy and Change in State The energy of particles and distances between the particles are different for a solid, a liquid, and a gas. Changes in energy cause changes in the state of matter. Energy must be added to a substance to change it from a solid to a liquid or from a liquid to a gas. Adding energy to a substance can increase its temperature. When the temperature reaches the substance's melting point, the solid changes to a liquid. At the boiling point, the liquid changes to a gas.

FOLDABLES®

Make a half-book to compare information about physical and chemical changes.

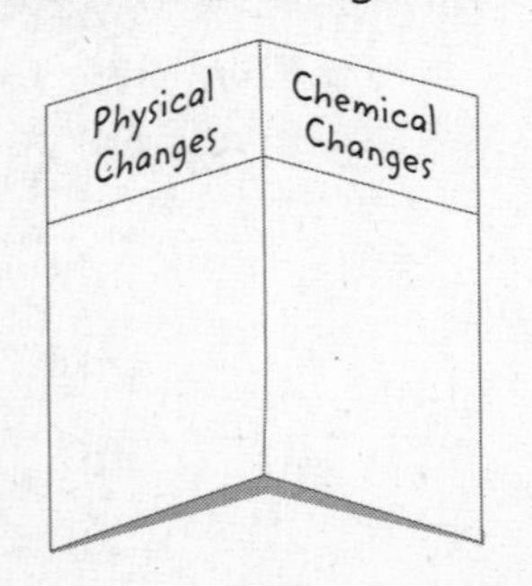

What are chemical changes?

Some changes in matter change the identity of the substance. *A* **chemical change** *is a change in matter in which the substances that make up that matter change into substances with different chemical and physical properties.* Another name for a chemical change is a chemical reaction. In a chemical reaction, the particles that make up two or more substances react, or combine, with each other and form a new substance.

Signs of a Chemical Change

A chemical change causes changes in physical properties. For many chemical reactions, changes in color, density, or state of matter are signs that a chemical change has taken place. The only sure sign of a chemical reaction is the formation of a new substance.

Formation of Gas When a liquid boils, the bubbles show that it is changing into a gas. This is a physical change. When you add a medicine tablet to water, gas bubbles form. The water does not boil, so the gas must be a new substance formed by a chemical reaction.

Formation of a Precipitate Some chemical reactions result in a solid forming when two liquids combine. This solid is called a precipitate (prih SIH puh tut). A precipitate is not the result of a change in state from a liquid to a solid. It is a chemical change. A precipitate forms when the particles that were dissolved in the two liquids react to form the particles of the solid, a new substance.

Color Change A change in color is another sign of a chemical change. If you painted the walls of your room a different color, this would not be a chemical change. But if a substance changes color, it is a sign that a chemical change has taken place. When you roast marshmallows or leave an apple slice on a plate, the change in color is a sign that there has been a chemical change.

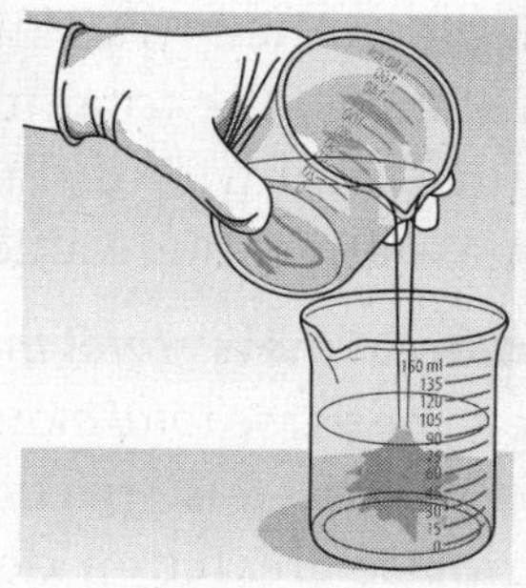

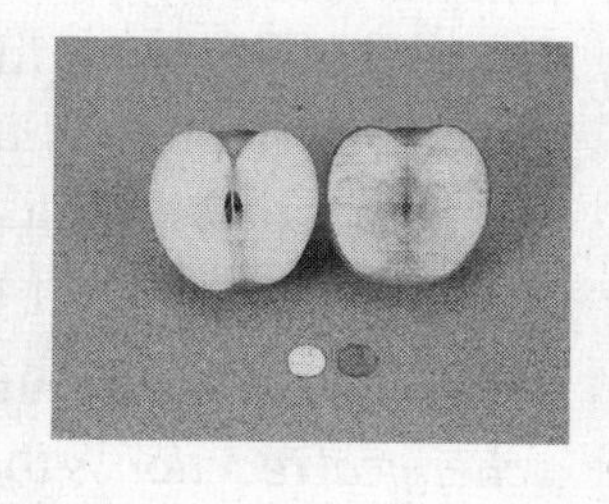

Key Concept Check
3. Describe how chemical changes are different from physical changes.

Reading Check
4. Describe how you can tell whether bubbles are a result of a chemical change or a physical change.

Visual Check
5. Explain In each of the images to the left, what new substance can you see that has formed?

Energy and Chemical Change

Chemical changes involve a change in energy. Think about a fireworks display. The release of thermal energy, light, and sound are signs of chemical changes. All chemical reactions involve energy changes.

Chemical reactions can also take in energy. Thermal energy is often needed for a chemical change to take place. To bake bread, for example, you have to put dough in a hot oven.

Some chemical reactions need energy in the form of light. Plants and some unicellular organisms use the Sun's energy for photosynthesis. Photosynthesis is the chemical reaction by which these organisms make sugar and oxygen. This process occurs only if the organisms are exposed to light.

ACADEMIC VOCABULARY
expose
(verb) to uncover; to make visible

Can changes be reversed?

Think again about a fireworks display. After the fireworks explode, they cannot go back to the chemicals they were before. Chemical changes cannot be reversed.

Some physical changes cannot be reversed, either. If you cut an apple, it cannot be put back together. Other physical changes can be reversed. For example, if you dissolve salt in a pan of water and then boil the mixture, the water will change to a gas and the salt will be left behind.

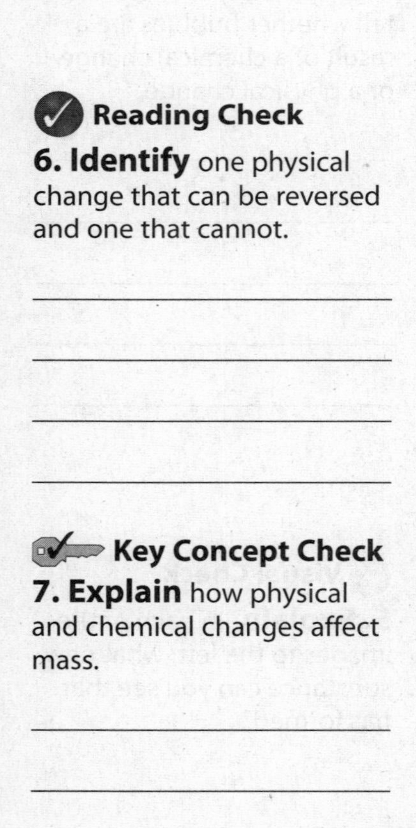

Reading Check
6. Identify one physical change that can be reversed and one that cannot.

Key Concept Check
7. Explain how physical and chemical changes affect mass.

Conservation of Mass

Physical changes do not affect the masses of substances. For example, when an ice cube melts, the mass of liquid water will be the same as the mass of the ice cube. If you cut a sheet of paper into pieces, the total mass of the pieces will be the same as the mass of the paper you started with. Mass is conserved, or unchanged, during a physical change.

Mass is also conserved during chemical changes. Antoine Lavoisier, a French chemist, discovered this in the 1700s. The masses of two substances that will chemically react can be measured and added together. After the two substances react to form new substances, the total mass after the reaction can be measured. You will find that the total mass before and the total mass after the reaction are the same.

The **law of conservation of mass** *states that the total mass before a chemical reaction is the same as the total mass after the chemical reaction*. This is always true because particles are only rearranged. They cannot be created or destroyed, so the total mass cannot increase or decrease.

Comparing Physical and Chemical Changes

The difference between a physical change and a chemical change is that the identity of matter changes during a chemical change but does not change during a physical change. You might not be able to tell just by looking at a substance whether its identity has changed. The particles that make up the matter can look the same.

You might have to look for more than one clue to tell whether a change is a physical change or a chemical change. The table below gives a summary of physical changes and chemical changes.

Comparing Physical and Chemical Changes

Type of Change	Examples	Characteristics
Physical change	• melting • boiling • changing shape • mixing • dissolving • changing temperature	• The substance is the same before and after the change. • Only physical properties change.
Chemical change	• changing color • burning • rusting • formation of gas • formation of precipitate • spoiling food • tarnishing silver • digesting food	• The substance changes into something else. • Both physical and chemical properties change.

Reading Check

8. Identify some clues you can use to determine whether a change is a physical change or a chemical change.

Visual Check

9. State What type of change occurs when ice melts? When a log burns?

After You Read

Mini Glossary

chemical change (KEM ih kul • CHANJ): a change in matter in which the substances that make up the matter change into other substances with different chemical and physical properties

law of conservation (kun SURV ay shun) of mass: states that the total mass before a chemical reaction is the same as the total mass after the chemical reaction

physical change (FIH zih kul • CHANJ): a change in the size, shape, form, or state of matter that does not change the matter's identity

1. Review the terms and their definitions in the Mini Glossary. Write a sentence that describes some ways in which physical changes and chemical changes are similar and some ways in which they are different.

2. A chemical change is shown in the picture below. Name three signs that you could use to tell that it is a chemical change.

3. How did underlining the main ideas help you learn the material in this lesson?

What do you think NOW?

Reread the statements at the beginning of the lesson. Fill in the After column with an A if you agree with the statement or a D if you disagree. Did you change your mind?

Log on to ConnectED.mcgraw-hill.com and access your textbook to find this lesson's resources.

Energy and Energy Transformations

Forms of Energy

Before You Read

What do you think? Read the three statements below and decide whether you agree or disagree with them. Place an A in the Before column if you agree with the statement or a D if you disagree. After you've read this lesson, reread the statements to see if you have changed your mind.

Before	Statement	After
	1. A fast-moving baseball has more kinetic energy than a slow-moving baseball.	
	2. A large truck and a small car moving at the same speed have the same kinetic energy.	
	3. A book sitting on a shelf has no energy.	

Key Concepts 
- What is energy?
- What are potential and kinetic energy?
- How is energy related to work?
- What are different forms of energy?

Read to Learn

What is energy?

Think of the last time you saw a fireworks display. When fireworks explode, you can see bursts of color in the night sky. Fireworks release energy when they explode. **Energy** *is the ability to cause change.* The energy in the fireworks causes the changes that you see as flashes of light and that you hear as loud booms.

Energy also causes other changes. Plants use energy from the Sun to make food for growth and other processes. Energy can cause changes in the motions or positions of objects. When a hammer hits a nail, energy from the hammer moves the nail. The explosion of fireworks, the growth of a flower, and the motion of a hammer involve energy.

Study Coach

Sticky Notes As you read, use sticky notes to mark information that you do not understand. Ask your teacher to explain.

WORD ORIGIN

energy
from Greek *energeia,* means "activity"

Key Concept Check

1. Define What is energy?

Kinetic Energy—Energy of Motion

Have you ever been to a bowling alley? When you rolled the ball and it hit the pins, a change occurred—the pins fell over. This change occurred because the ball had a form of energy called kinetic (kuh NEH tik) energy. **Kinetic energy** *is energy due to motion.* All moving objects have kinetic energy. The kinetic energy of a moving object depends on two factors: the object's speed and its mass.

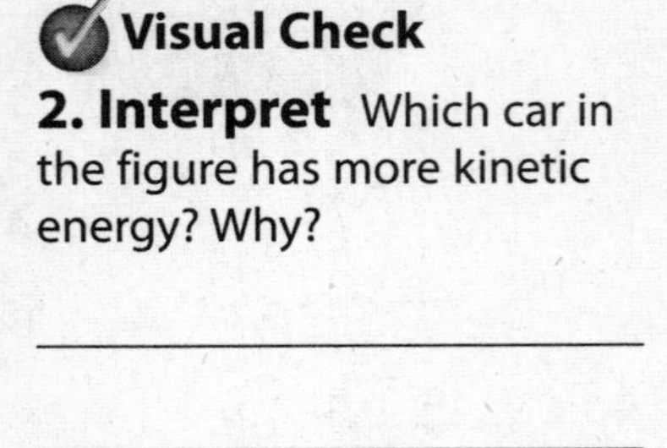

Visual Check

2. Interpret Which car in the figure has more kinetic energy? Why?

FOLDABLES

Make a two-pocket book. Organize information about the forms of energy on quarter sheets of paper and put them in the pockets.

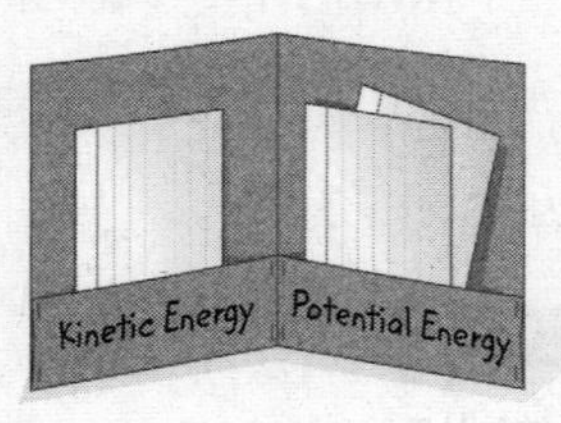

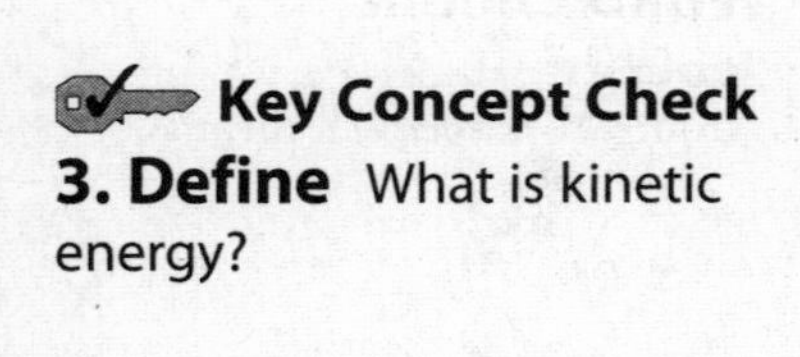

Key Concept Check

3. Define What is kinetic energy?

Speed, Mass, and Kinetic Energy

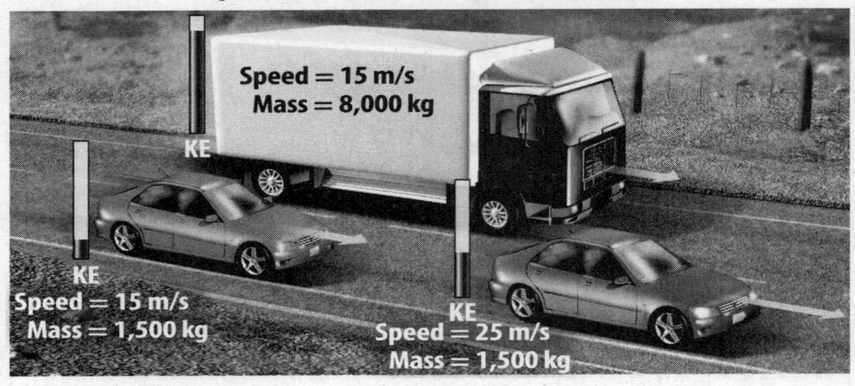

Kinetic Energy and Speed

Speed is one factor that affects kinetic energy. The faster an object moves, the more kinetic energy it has. The figure above shows two cars and a truck moving along a highway. All the vehicles have kinetic energy (KE) because they are moving. However, each vehicle's speed helps determine the amount of kinetic energy the vehicle has. The vertical bars show the kinetic energy of each vehicle.

Kinetic Energy and Mass

The kinetic energy of a moving object also depends on its mass. If two objects are moving at the same speed, the object with more mass has more kinetic energy.

Notice in the figure that the two cars have the same mass. The car in front has more kinetic energy because it is moving faster. The car in the back and the truck are moving at the same speed. The truck has more kinetic energy than the car in the back because the truck has more mass than that car.

Potential Energy—Stored Energy

An object can have energy even when it is not moving. If you hold a ball in your hand and then let it go, gravity will cause the ball to fall to Earth. The gravitational interaction between the ball and Earth causes a change to occur.

Before you dropped the ball, the ball had energy. This form of energy is called potential (puh TEN chul) energy. **Potential energy** *is stored energy due to the interactions between objects or particles.* Potential energy has different forms: gravitational potential energy, elastic potential energy, and chemical potential energy.

Gravitational Potential Energy

When you are holding a book, energy is stored between the book and Earth. This type of energy is called gravitational potential energy. If you lift the book higher, the gravitational potential energy between the book and Earth increases.

The gravitational potential energy stored between any object and Earth depends on the object's mass and its height above the ground. Dropping a bowling ball from a height of 1 m causes greater change than dropping a tennis ball from the same height. When two objects are at the same height, the one with more mass has more gravitational potential energy.

Gravitational Potential Energy

The two vases on the bookcase are identical; however, they have different potential energies because they are at different positions above the ground. The vase on the top shelf of the bookcase has more gravitational potential energy than the vase on the bottom shelf. An object that falls from a greater height can cause a greater change than an identical object that falls from a lower height.

Reading Check

4. Relate Gravitational potential energy depends on which two factors? (Circle the correct answer.)

a. speed and distance
b. mass and height above the ground
c. mass and speed

Visual Check

5. Identify The two vases on the bookcase have the same mass. Circle the vase that has the greater gravitational potential energy.

Elastic Potential Energy

Another form of potential energy is elastic (ih LAS tik) potential energy. Elastic potential energy is energy that is stored when an object is compressed or stretched. When you jump on a pogo stick, you compress the spring. This gives the spring elastic potential energy. When the spring decompresses, it pushes you into the air.

Stretching an object also stores elastic potential energy. When you stretch a rubber band, elastic potential energy is stored in the rubber band. When you release the rubber band, the stored elastic potential energy changes into kinetic energy. The kinetic energy causes the rubber band to snap back to its original shape.

Plucking the strings of a guitar, jumping on a trampoline, and pulling back on the string of a bow give these objects elastic potential energy by stretching. When the strings and the trampoline return to their original positions, they cause change.

Think it Over

6. Consider Which is ***not*** an example of how an object gains elastic potential energy by stretching? (Circle the correct answer.)

a. jumping on a pogo stick
b. pulling on a rubber band
c. jumping on a trampoline

Chemical Potential Energy

When you eat, you take in another form of potential energy. Food and other substances are made of atoms joined together by chemical bonds. Chemical potential energy is energy stored in the bonds between atoms.

Look at the figure below. The small balls in the figure represent atoms that make up a glucose molecule. The lines between the atoms represent chemical bonds. Chemical potential energy is stored in these bonds. When you eat food, chemical reactions within your body release chemical potential energy stored in the food. Your body uses chemical potential energy in foods for all its activities, such as moving, thinking, and growing. Bonds between atoms in other substances, such as gasoline, also store chemical potential energy. People use the chemical potential energy in gasoline to drive cars.

Key Concept Check
7. Compare In what way are all forms of potential energy the same?

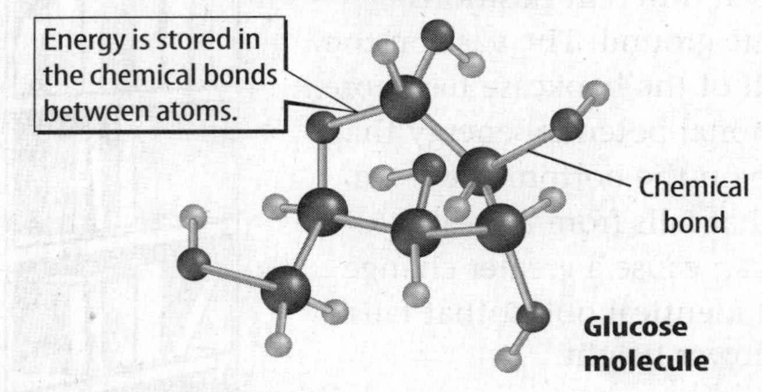

Visual Check
8. Identify Highlight some of the chemical bonds in the figure.

Energy and Work

A force is a push or a pull. When a force is applied to an object, the object's kinetic and potential energy can change. You can transfer energy by doing work. **Work** *is the transfer of energy that occurs when a force makes an object move in the direction of the force while the force is acting on the object.*

Look at the figure at the top of the next page. The girl does work on the box as she lifts it. As she lifts the box onto the shelf, the energy of the box changes. The work she does transfers energy to the box. The energy of the box increases because of the gravitational interaction between the box and Earth. The box's potential energy increases as she lifts the box higher. The vertical bars in the figure show the work that the girl does (W) and the box's potential energy (PE).

Work depends on force and distance. You do work on an object only if the object moves. Suppose the girl shown in the figure tries to lift the box but cannot lift it off the floor. She transfers no energy, so she does no work on the box.

Work

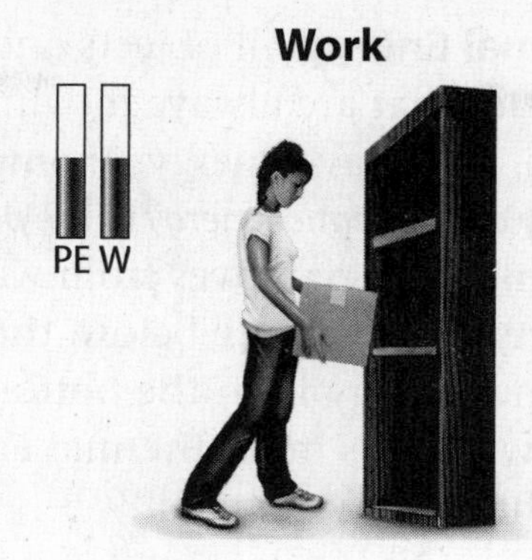

An object that has energy can also do work. What will happen if the girl drops the box as she is moving it onto the shelf? When the box hits the floor, it does work on the floor. Some of the box's kinetic energy is transferred to the floor. The girl will hear some of the energy as a loud crash and feel some of the energy near her feet as the energy travels through the floor. Because energy and work are connected, energy is sometimes described as the ability to do work.

Other Forms of Energy

You have just learned about two forms of energy—kinetic energy and potential energy. Kinetic energy is energy due to motion. Potential energy is stored energy. There are other forms of energy as well. All forms of energy are measured in units called joules (J). A softball dropped from a height of about 0.5 m has about 1 J of kinetic energy just before it hits the floor.

Mechanical Energy *The sum of potential energy and kinetic energy in a system of objects is* **mechanical energy.** When you do work on an object, you change the object's mechanical energy.

Think again about the girl moving the box shown above. At what point did the mechanical energy of the box change? The mechanical energy of the box increased when the girl lifted it off the ground. Now think about a basketball game. The mechanical energy of a basketball increases when a player shoots the ball.

Sound Energy Musical instruments are just a few of the many things that produce sound. When you pluck a guitar string, the string vibrates and creates sound. You hear a sound when sound waves produced by the vibrating guitar string reach your ears. *The energy that sound carries is* **sound energy.** Sound energy is produced by objects that vibrate. Sound energy cannot travel through a vacuum such as the space between Earth and the Sun.

Visual Check

9. Determine When did the transfer of energy take place between the girl and the box?

____

____

Key Concept Check

10. Analyze How is energy related to work?

____

____

____

____

Think it Over

11. Apply Imagine that you push on a large rock. At what point does your effort change the rock's mechanical energy?

____

____

____

Thermal Energy All objects and materials are made of particles that are always moving. Because these particles move, they have energy. **Thermal energy** *is the sum of kinetic energy and potential energy of the particles that make up an object.* Thermal energy moves from warmer objects to colder objects. In the figure below, the hot chocolate has more thermal energy than the bottle of cold water. The bottle of cold water has more thermal energy than the block of ice that has the same mass.

Visual Check

12. Apply Circle the object with the least thermal energy. What is causing the ice block to melt?

Thermal Energy

CHOCOLATE

WATER

Electric Energy An electrical fan uses another form of energy—electric energy. When you turn on a fan, an electric current flows through the fan's motor. **Electric energy** *is the energy an electric current carries.* Electrical appliances, such as fans and dishwashers, change electric energy into other forms of energy.

Reading Check

13. Define What are electromagnetic waves? (Circle the correct answer.)

a. waves that form on a beach

b. waves that can travel through a vacuum

c. waves that can be heard

Radiant Energy—Light Energy The Sun gives off energy that travels to Earth as electromagnetic waves. Unlike sound waves, electromagnetic waves can travel through a vacuum. Light waves, microwaves, and radio waves are electromagnetic waves. *The energy that electromagnetic waves carry is **radiant energy.*** Sometimes radiant energy is called light energy.

Key Concept Check

14. Describe three forms of energy.

Nuclear Energy A nucleus is at the center of every atom. **Nuclear energy** *is energy that is stored and released in the nucleus of an atom.* In the Sun, nuclear energy is released when nuclei join together. In a nuclear power plant, nuclear energy is released when the nuclei of uranium atoms are split apart.

Mini Glossary

electric energy: energy that an electric current carries

energy: the ability to cause change

kinetic (kuh NEH tik) energy: energy due to motion

mechanical energy: the sum of potential energy and kinetic energy in a system of objects

nuclear energy: energy stored and released in the nucleus of an atom

potential (puh TEN chul) energy: stored energy due to the interactions between objects or particles

radiant energy: the energy that electromagnetic waves carry

sound energy: the energy that sound carries

thermal energy: the sum of kinetic energy and potential energy of the particles that make up an object

work: the transfer of energy that occurs when a force makes an object move in the direction of the force while the force is acting on the object

1. Review the terms and their definitions in the Mini Glossary. Write a sentence that describes how energy is related to work.

2. Use what you have learned about energy to complete the table.

Form of Energy	Definition	Form of Energy	Definition
Kinetic		Potential	
Electric		Thermal	
Mechanical		Radiant	
Nuclear		Sound	

What do you think NOW?

Reread the statements at the beginning of the lesson. Fill in the After column with an A if you agree with the statement or a D if you disagree. Did you change your mind?

Log on to ConnectED.mcgraw-hill.com and access your textbook to find this lesson's resources.

Energy and Energy Transformations

Energy Transformations

Key Concepts 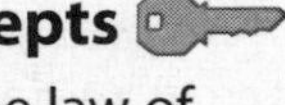
- What is the law of conservation of energy?
- How does friction affect energy transformations?
- How are different types of energy used?

Before You Read

What do you think? Read the three statements below and decide whether you agree or disagree with them. Place an A in the Before column if you agree with the statement or a D if you disagree. After you've read this lesson, reread the statements to see if you have changed your mind.

Before	Statement	After
	4. Energy can change from one form to another.	
	5. Energy is destroyed when you apply the brakes on a moving bicycle or a moving car.	
	6. The Sun releases radiant energy.	

 Mark the Text

Identify Main Ideas Highlight the sentences in this lesson that talk about how energy changes form. Use the highlighted sentences to review.

Visual Check

1. Identify Which energy transformation pops the corn kernels?

Read to Learn

Changes Between Forms of Energy

Have you ever made popcorn in a microwave oven to eat while watching television? Energy changes form when you make popcorn, as shown in the figure below. A microwave oven changes electric energy into radiant energy. Radiant energy changes into thermal energy in the popcorn kernels. These changes from one form of energy to another are called energy transformations. As you watch TV, energy transformations occur in the television. A television transforms electric energy into sound energy and radiant energy.

Energy Transformation

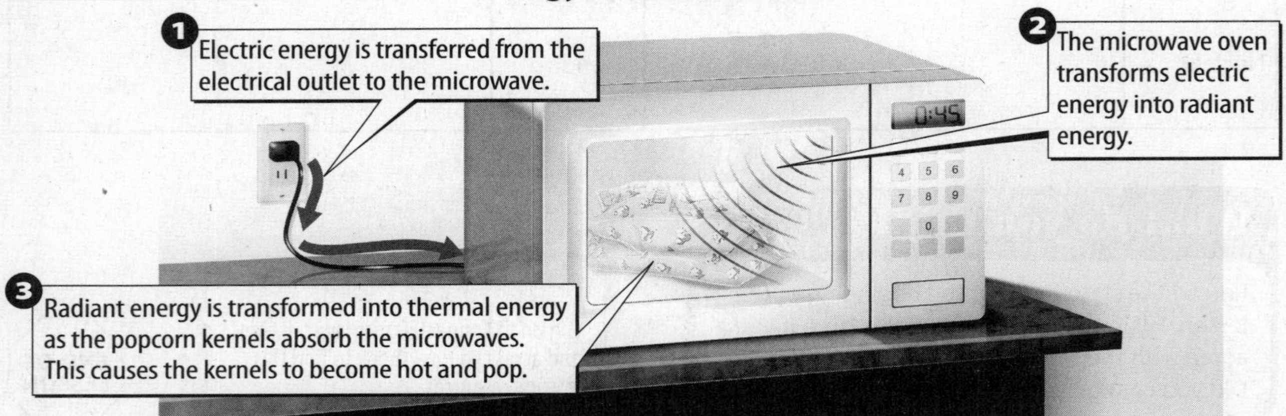

Changes Between Kinetic and Potential Energy

Energy transformations also occur when you toss a ball upward. The ball slows down as it rises and speeds up as it falls. The ball's speed and height change as the energy changes from one form to another.

Kinetic Energy to Potential Energy

In the figure to the right, notice that the ball is moving fastest and has the most kinetic energy (KE) as it leaves the girl's hands. As the ball moves upward, its speed and kinetic energy decrease. However, the ball's potential energy (PE) increases because the ball's height increases. The ball's kinetic energy is changing to potential energy. At the ball's highest point, its gravitational potential energy is greatest, and its kinetic energy is lowest.

Potential Energy to Kinetic Energy

As the ball moves downward, its potential energy decreases. At the same time, the ball's kinetic energy increases because its speed increases. As the ball drops, potential energy changes to kinetic energy. When the ball reaches the boy's hands, the ball's kinetic energy is once again at its highest value. Energy changes between kinetic energy and potential energy as the ball moves. The bars in the figure show that the ball's total energy does not change.

The Law of Conservation of Energy

The total energy in the universe is the sum of all the different forms of energy everywhere. *According to the* **law of conservation of energy,** *energy can be transformed from one form into another or transferred from one region to another, but energy cannot be created or destroyed.* The total amount of energy in the universe does not change.

Friction and the Law of Conservation of Energy

Sometimes it seems as if the law of conservation of energy is not accurate. Imagine riding a bicycle. The moving bicycle has mechanical energy. What happens to this mechanical energy when you apply the brakes?

When you apply the brakes, the bicycle's mechanical energy is transformed into thermal energy as the brake pads rub against the bicycle's wheels. The total amount of energy never changes. The additional thermal energy warms the brakes, the wheels, and the air around the bicycle.

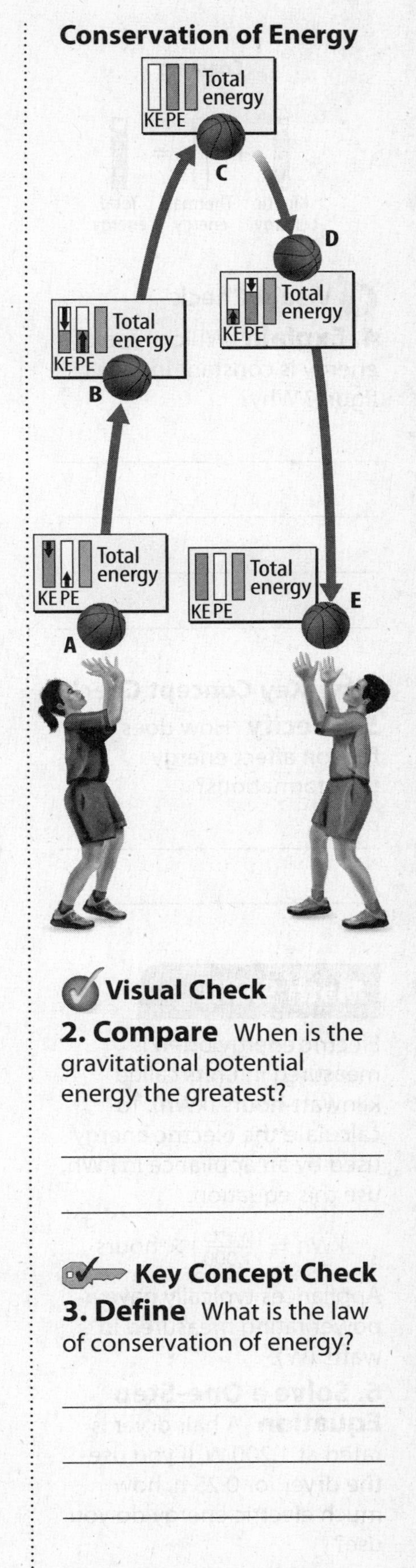

Visual Check

2. Compare When is the gravitational potential energy the greatest?

Key Concept Check

3. Define What is the law of conservation of energy?

Friction and Thermal Energy

Coasting

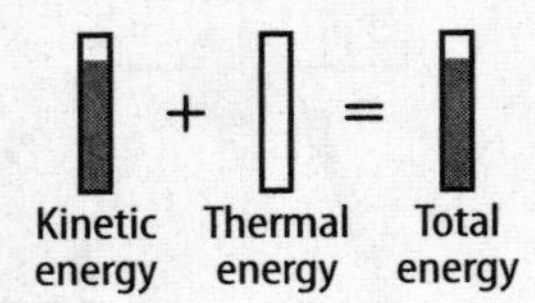

Applying brakes

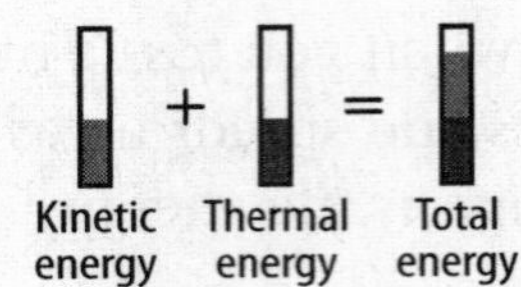

Stopped

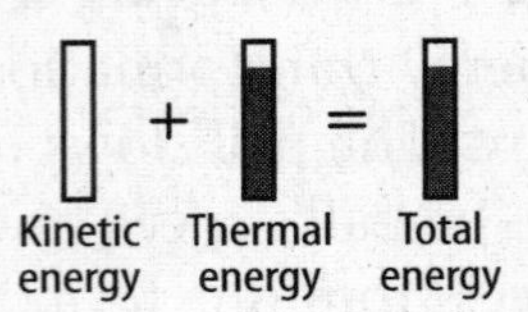

Visual Check

4. Explain Which type of energy is constant in the figure? Why?

____

____

____

____

Key Concept Check

5. Specify How does friction affect energy transformations?

____

____

Math Skills

Electric energy often is measured in units called kilowatt-hours (kWh). To calculate the electric energy used by an appliance in kWh, use this equation:

$$\text{kWh} = \left(\frac{\text{watts}}{1{,}000}\right) \times \text{hours}$$

Appliances typically have a power rating measured in watts (W).

6. Solve a One-Step Equation A hair dryer is rated at 1,200 W. If you use the dryer for 0.25 h, how much electric energy do you use?

____

Friction When the bicycle's brake pads rub against the moving wheels, friction occurs. **Friction** *is a force that resists the sliding of two surfaces that are touching*. Friction between the brake pads and the moving wheels transforms the bicycle's mechanical energy into thermal energy as shown in the figure above.

There is always some friction between any two surfaces that are rubbing against each other. As a result, some mechanical energy is always transformed into thermal energy when two surfaces rub against each other.

Reducing Friction It is easier to pedal a bicycle if there is less friction between the bicycle's parts. With less friction, less of the bicycle's mechanical energy is transformed into thermal energy. One way to reduce friction is to apply a lubricant such as oil to surfaces that rub against each other.

Using Energy

You use different forms of energy to do different things. You use radiant energy from a lamp to light a room. You use chemical energy stored in your body to run a race. Energy usually changes from one form to another when you use it. For example, a lamp changes electric energy to radiant energy that lights a room. Some of the electric energy also changes to thermal energy. Thermal energy causes the lamp's bulb to become warm to the touch.

Using Thermal Energy

All forms of energy can be transformed into thermal energy. People often use thermal energy to cook food or provide warmth. A gas stove transforms the chemical energy stored in natural gas into the thermal energy that cooks the food. An electric space heater transforms the electric energy from a power plant into the thermal energy that warms a room. In a jet engine, burning fuel releases thermal energy that the engine transforms into mechanical energy.

Using Chemical Energy

During photosynthesis, a plant transforms the Sun's radiant energy into chemical energy that it stores in chemical compounds. Some of these compounds become food for other organisms. Your body transforms the chemical energy from food into the kinetic energy that you use for movement. Your body also transforms chemical energy into the thermal energy that keeps you warm.

Using Radiant Energy

A cell phone sends and receives radiant energy using microwaves. When you speak into a cell phone, the phone transforms sound energy into electric energy and then into radiant energy. Sound waves from your voice carry energy into the phone. The phone converts the sound energy into electric energy and then into radiant energy. Microwaves carry the radiant energy away. When you listen to someone on a cell phone, the phone is transforming radiant energy into electric energy and then into sound energy.

Using Electric Energy

Many of the devices you use every day, such as a handheld video game, an MP3 player, and a hair dryer, use electric energy. Some devices, such as hair dryers, use electric energy from electric power plants. Other devices, such as handheld video games, transform the chemical energy stored in batteries into electric energy.

Waste Energy

When energy changes form, some thermal energy is always released. For example, a lightbulb converts some electric energy into radiant energy. However, the lightbulb also transforms some electric energy into thermal energy. This is what makes the lightbulb hot. Some of this thermal energy moves into the air and cannot be used.

Scientists often refer to thermal energy that cannot be used as waste energy. Whenever energy is used, some energy is transformed into useful energy and some is transformed into waste energy. For example, the chemical energy in gasoline makes cars move. However, most of that chemical energy ends up as waste energy—thermal energy that moves into the air.

FOLDABLES

Make a side-tab book to organize your notes on energy transformations.

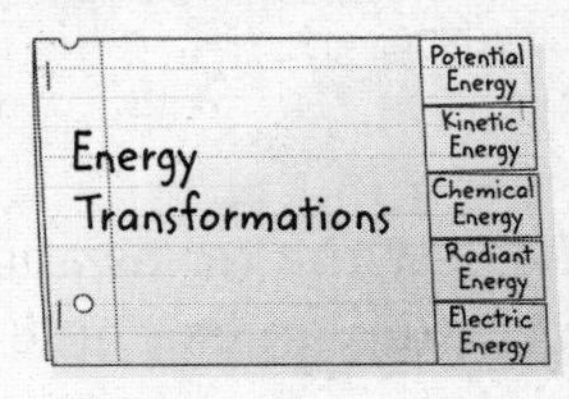

Key Concept Check

7. Specify How are different types of energy used?

Reading Check

8. Define What is waste energy?

After You Read

Mini Glossary

friction: a force that resists the sliding of two surfaces that are touching

law of conservation of energy: energy can be transformed from one form into another or transferred from one region to another, but energy cannot be created or destroyed

1. Review the terms and their definitions in the Mini Glossary. Write a sentence that provides an example of friction.

2. Use what you have learned about energy to complete the table.

Energy Change	Example	Energy Change	Example
Kinetic to potential	Throw a ball into the air.	Chemical to thermal	
Electric to thermal		Sound to radiant	A call is received on a cell phone.
Potential to kinetic	An object falls to the ground.	Electric to radiant	
Chemical to electric		Chemical to kinetic	Food is digested to help a person move.

3. Explain one concept you learned from the sentences that you highlighted.

What do you think NOW?

Reread the statements at the beginning of the lesson. Fill in the After column with an A if you agree with the statement or a D if you disagree. Did you change your mind?

Log on to ConnectED.mcgraw-hill.com and access your textbook to find this lesson's resources.

Waves, Light, and Sound

Waves

Before You Read

What do you think? Read the two statements below and decide whether you agree or disagree with them. Place an A in the Before column if you agree with the statement or a D if you disagree. After you've read this lesson, reread the statements to see if you have changed your mind.

Before	Statement	After
	1. Waves carry matter from place to place.	
	2. All waves move with an up-and-down motion.	

Key Concepts 

- What are waves, and how are waves produced?
- How can you describe waves by their properties?
- What are some ways in which waves interact with matter?

Read to Learn

What are waves?

A flag waves in the breeze. Ocean waves break onto a beach. You wave your hand at a friend. All of these actions have something in common. Waves always begin with a source of energy that causes a back-and-forth or up-and-down disturbance. For example, wind causes a disturbance in the flag. This disturbance moves along the length of the flag as a wave. A wave is a disturbance that transfers energy from one place to another without transferring matter.

Study Coach

Building Vocabulary Write each vocabulary term in this lesson on an index card. Shuffle the cards. After you have studied the lesson, take turns picking cards with a partner. Each of you should define the term using your own words.

Key Concept Check

1. Define What are waves?

Energy Transfer

Wind transfers energy to the fabric of the flag. The flag ripples back and forth as the energy travels along the fabric. Each point on the flag moves back and forth, but the fabric doesn't move along with the wave. Waves transfer energy, not matter, from place to place.

When you lift a pebble, you transfer energy to it. Suppose you drop the pebble into a pond. The pebble's energy transfers to the water.

Waves carry the energy away from the point where the pebble hit the water. The water itself moves up and down as the wave passes, but the water does not move along with the wave.

Two Main Types of Waves

The ways in which waves transport energy differ. Some waves carry energy only through matter. Others can carry energy through matter or through empty space.

Mechanical Waves *A wave that travels only through matter is a* **mechanical wave.** A medium is the matter through which a mechanical wave travels. A mechanical wave forms when a source of energy causes particles that make up a medium to vibrate. A pebble falling into water transfers its kinetic energy to particles of the water. The water particles vibrate and push against nearby particles, transferring the energy outward. After each particle pushes the next particle, it returns to its original rest position. Energy is transferred, but the water particles are not.

Electromagnetic Waves *A wave that can travel through empty space or through matter is an* **electromagnetic wave.** This type of wave forms when a charged particle, such as an electron, vibrates. For example, electromagnetic waves transfer the Sun's energy to Earth through empty space. Once the waves reach Earth, they travel through matter, such as the atmosphere or a glass window in a house.

Key Concept Check
2. Explain How are waves produced?

______

______

______

______

______

Describing Wave Motion

Some waves move particles of a medium up and down or side to side, perpendicular to the direction the wave travels. For example, the waves in a flag move side to side, perpendicular to the direction of the wind. Other wave disturbances move particles of the medium forward, then backward in the same direction, or parallel, to the motion of the wave. And some waves are a combination of both types of motion. The table below summarizes these three types of wave motion—transverse, longitudinal, and a combination of both.

REVIEW VOCABULARY
perpendicular
at right angles

Interpreting Tables
3. Contrast How is the motion of electromagnetic waves different from the motion of mechanical waves?

______

______

______

______

______

Types of Wave Motion

Type of Wave Motion	Mechanical Waves	Electromagnetic Waves
Transverse— perpendicular to the direction the wave travels	example: flag waving in a breeze	example: light waves
Longitudinal— parallel to the direction the wave travels	example: sound waves	
Combination— both transverse and longitudinal	example: water waves	

Transverse Waves *A wave in which the disturbance is perpendicular to the direction the wave travels is a* **transverse wave.** You can make a transverse mechanical wave by attaching one end of a rope to a hook and holding the other end, as shown in the figure below. When you move your hand up and down, transverse waves travel along the rope. High points on a wave are called crests, and low points are called troughs. 

Transverse Wave

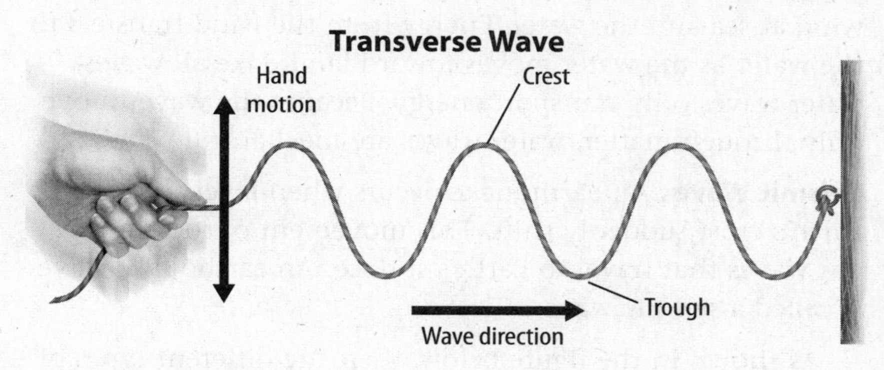

A vibrating charge, such as an electron, produces an electromagnetic wave. Electromagnetic waves are transverse waves. The electric and magnetic wave disturbances are perpendicular to the motion of the vibrating charge. Light is a form of energy transferred by transverse electromagnetic waves. X-rays and radio waves are electromagnetic waves.

Longitudinal Waves *A wave that makes the particles of a medium move back and forth parallel to the direction the wave travels is a* **longitudinal wave.** Longitudinal waves are mechanical waves. Like a transverse wave, a longitudinal wave disturbance passes energy from particle to particle of a medium. For example, when you knock on a door, energy of your hand transfers to the particles that make up the door. The energy of the vibrating particles of the door is transferred to the air in the next room.

You can also make a longitudinal wave by pushing or pulling on a coiled spring toy. Pushing moves the coils closer together. Pulling spreads the coils apart. The back-and-forth motion of your hand causes a back-and-forth motion in the spring. The longitudinal waves move parallel to your hand's motion.

Waves in Nature

Waves are common in nature because so many different energy sources produce waves. Two common waves in nature are water waves and seismic, or earthquake, waves.

Reading Check

4. Identify What is the direction of the disturbance in a transverse wave?

____

 **Visual Check**

5. Apply If the hand moved side to side instead of up and down, what direction would the wave travel?

____

____

FOLDABLES

Make a vertical three-tab Venn book to compare and contrast transverse and longitudinal waves.

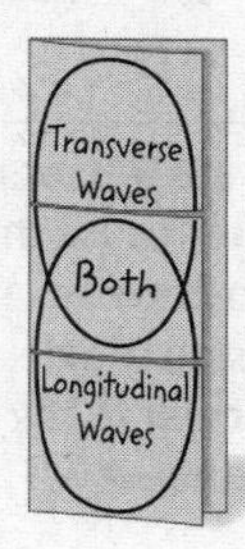

Water Waves Although water waves look like transverse waves, water particles move in circles. Water waves are a combination of transverse and longitudinal waves. Water particles move forward and backward. They also move up and down. The result is a circular path that gets smaller as the waves approach land.

Water waves form because there is friction between the wind at sea and the water. Energy from the wind transfers to the water as the water moves toward land. Like all waves, water waves only transport energy. Because the waves move only through matter, water waves are mechanical waves. ✓

Reading Check

6. Recognize How does a water wave get its energy?

Seismic Waves An earthquake occurs when layers of rock of Earth's crust suddenly shift. This movement of rock sends out waves that travel to Earth's surface. An earthquake wave is called a seismic wave.

As shown in the figure below, there are different types of seismic waves. Seismic waves can be longitudinal, transverse, or a combination of the two. Seismic waves are mechanical waves because they move through matter.

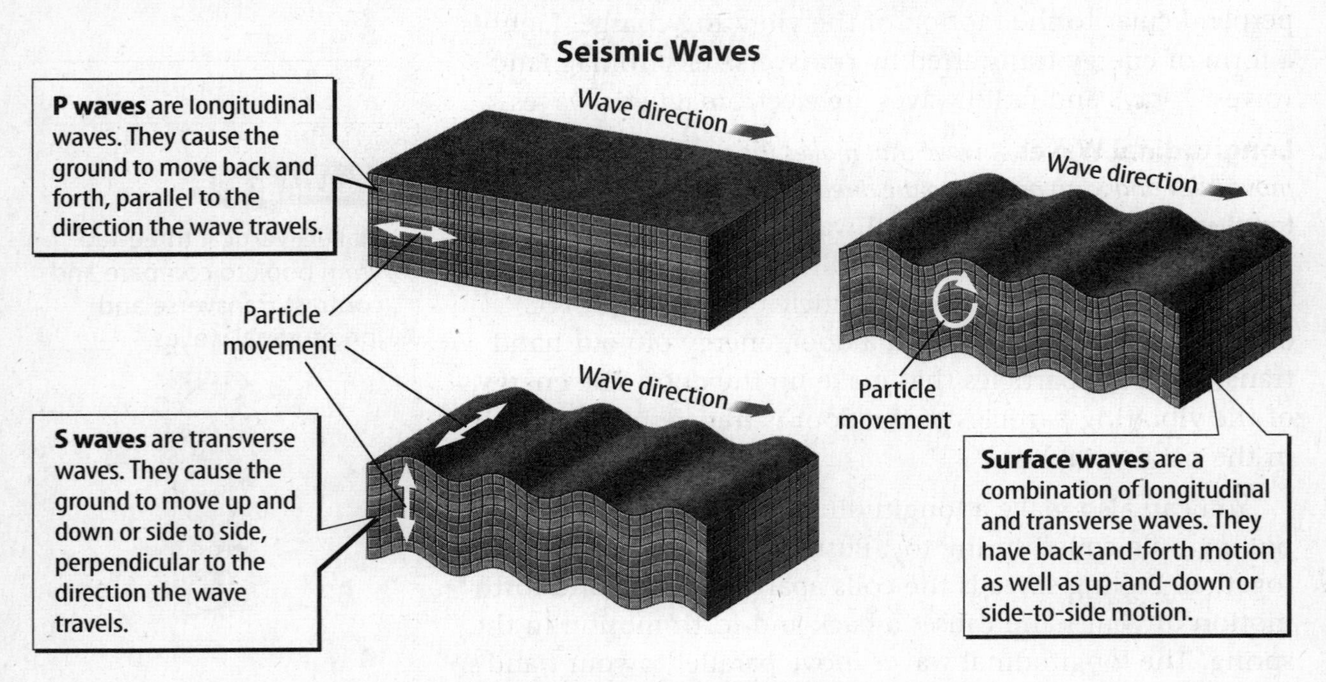

Visual Check

7. Compare Which seismic wave is similar to a water wave?

Properties of Waves

How could you describe water waves at a beach? You might describe how tall the waves are or how fast the waves move toward shore. When scientists describe waves, they describe the properties of wavelength and frequency.

Wavelength

Waves have high points called crests and low points called troughs. As shown in the figure below, the distance between a point on one wave and the same point on the next wave is called the wavelength. Different types of waves can have wavelengths that range from thousands of kilometers to less than the size of an atom!

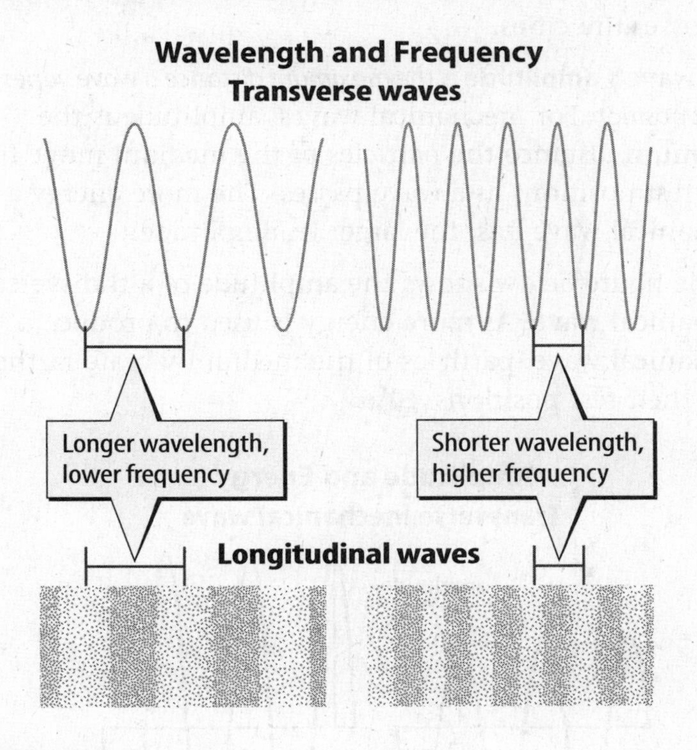

Frequency

The number of wavelengths that pass a point each second is a wave's **frequency.** Frequency is measured in hertz (Hz). One hertz equals one wave per second. As shown in the figure above, the longer the wavelength, the lower the frequency. As the distance between the crests decreases, the number of waves passing a point each second increases.

Wave Speed

The speed of a wave depends on the medium, or type of material, through which it travels. Electromagnetic waves always travel through empty space at a speed of 3×10^8 m/s. That's 300 million meters each second! They travel slower through a medium, or matter, because they must interact with particles. Mechanical waves travel only through matter. They travel slower than electromagnetic waves travel through space. Sound waves travel about one-millionth the speed of light waves. The speed of water waves depends on the strength of the wind that produces them. The table to the right compares the speeds of different types of waves.

Visual Check

8. Examine Based on their appearance, how can you tell that the waves on the right have a shorter wavelength than those on the left?

Reading Check

9. Recognize What is frequency?

Interpreting Tables

10. Identify Circle the wave speed for a sound wave in air.

Wave Speeds

Type of Wave	Typical wave speed (m/s)
Ocean wave	25
Sound wave in air	340
Transverse seismic wave (S wave)	1,000 to 8,000
Longitudinal seismic wave (P wave)	1,000 to 14,000
Electromagnetic wave through empty space	300,000,000

Amplitude and Energy

Different waves carry different amounts of energy. Some earthquakes, for example, can be catastrophic because they carry so much energy. A shift in Earth's crust can cause particles in the crust to vibrate back and forth very far from their rest position, producing seismic waves. In January 2010, seismic waves in Haiti transferred enough energy to destroy entire cities.

A wave's **amplitude** *is the maximum distance a wave varies from its rest position.* For mechanical waves, amplitude is the maximum distance the particles of the medium move from their rest positions as a wave passes. The more energy a mechanical wave has, the larger its amplitude.

The figure below shows the amplitude of a transverse mechanical wave. As more energy is used to produce a mechanical wave, particles of the medium vibrate farther from their rest positions.

Key Concept Check
11. Identify How can you describe waves?

Visual Check
12. Recognize How are amplitude and energy of a mechanical wave related?

Amplitude and Energy
Transverse mechanical wave

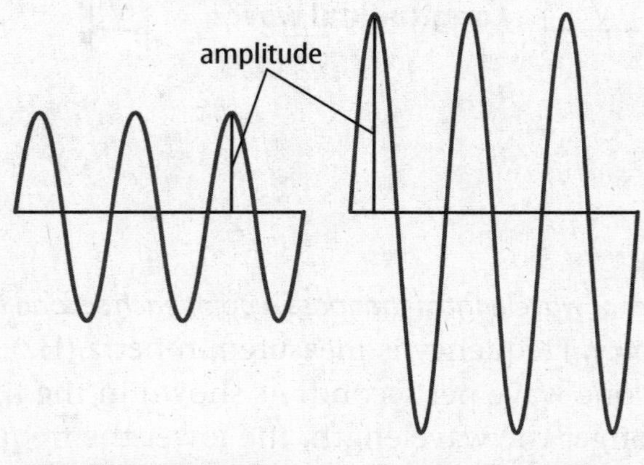

Smaller amplitude, lower energy

Larger amplitude, higher energy

Wave Interaction with Matter

You have read that when you knock on a door, longitudinal sound waves transfer the energy of the knock through the door. However, when a person in the next room hears the knock, it is not as loud as the sound on your side of the door. The sound is weaker after it passes through the door because the waves interact with the matter that makes up the door.

Transmission

Some of the sound from your knock passes through the door. The waves transmit, or carry, the energy all the way through the door. The energy then passes into air particles, and the person on the other side hears the knock.

Absorption

Some of the sound is absorbed by the particles that make up the door. Instead of passing through the door, the energy increases the motion of the particles of the wood. The sound energy changes to thermal energy within the door. So, less sound energy passes into the air in the next room. ✓

Reflection

Some of the energy you used to knock on the door reflects, or bounces back, into the room you are in. Sound waves in the air transfer sound back to your ears. The energy of electromagnetic waves also can be transmitted, absorbed, or reflected. ✓

Law of Reflection

You can predict how waves will reflect from a smooth surface. A light wave approaching a surface at an angle is called the incident wave. The wave that leaves the surface is the reflected wave. Picture a dotted line drawn perpendicular to the surface at the point where the wave hits the surface. This dotted line is called the normal.

The law of reflection states that the angle between the incident wave and the normal always equals the angle between the reflected wave and the normal. If the incident angle increases, the reflected angle also increases.

Refraction

The change in direction of a wave as it changes speed, moving from one medium into another, is called **refraction.** Refraction causes an object to appear to be in a place different from its real location. The image of the fish in the figure at right is an example. Light reflects off the fish in all directions. The reflected light waves move faster after they leave the water and move into the air. Recall that the normal is an imaginary line perpendicular to the surface at the point where a wave hits a surface. As light rays pass from the water into the air, they refract away from the normal, changing direction. In the figure, the boy's brain assumes the light traveled in a straight line. The light rays seem to come from the position of the image he sees.

✓ Reading Check

13. Describe How does the absorbed energy affect the wood?

____

____

____

✓ Reading Check

14. Define What are transmission, absorption, and reflection?

____

____

____

____

____

Visual Check

15. Identify Circle the place where the light refracts away from the normal.

Refraction

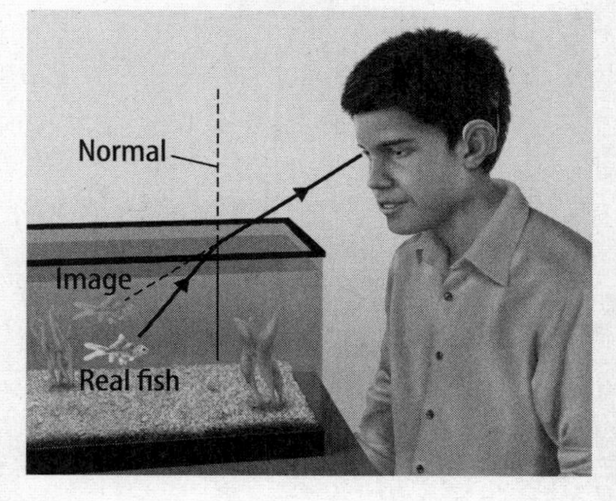

When Waves Refract Note that waves refract only if they move at an angle into another medium. They do not refract if they move straight into a medium. Waves refract toward the normal if they move slower after entering a medium and away from the normal if they move faster.

Diffraction

Diffraction is the change in direction of a wave when it travels past the edge of an object or through an opening. If you are walking in a school hallway and hear sound coming from an open classroom door, the sound waves have diffracted around the corner to your ears. Diffraction causes waves to spread around barriers and through openings.

Key Concept Check
16. Summarize What are some ways in which waves interact with matter?

After You Read

Mini Glossary

amplitude: the maximum distance a wave varies from its rest position

electromagnetic wave: a wave that can travel through empty space or through matter

frequency: the number of wavelengths that pass a point each second

longitudinal wave: a wave that makes the particles of a medium move back and forth parallel to the direction the wave travels

mechanical wave: a wave that travels only through matter

refraction: the change in direction of a wave as it changes speed, moving from one medium into another

transverse wave: a wave in which the disturbance is perpendicular to the direction the wave travels

1. Review the terms and their definitions in the Mini Glossary. Write a sentence that compares electromagnetic waves and mechanical waves.

____

____

2. The straight arrows in the diagram below illustrate a light wave hitting and bouncing off a surface. Name the waves that each arrow represents. Then predict the degree of the angle at which the wave will reflect from the surface.

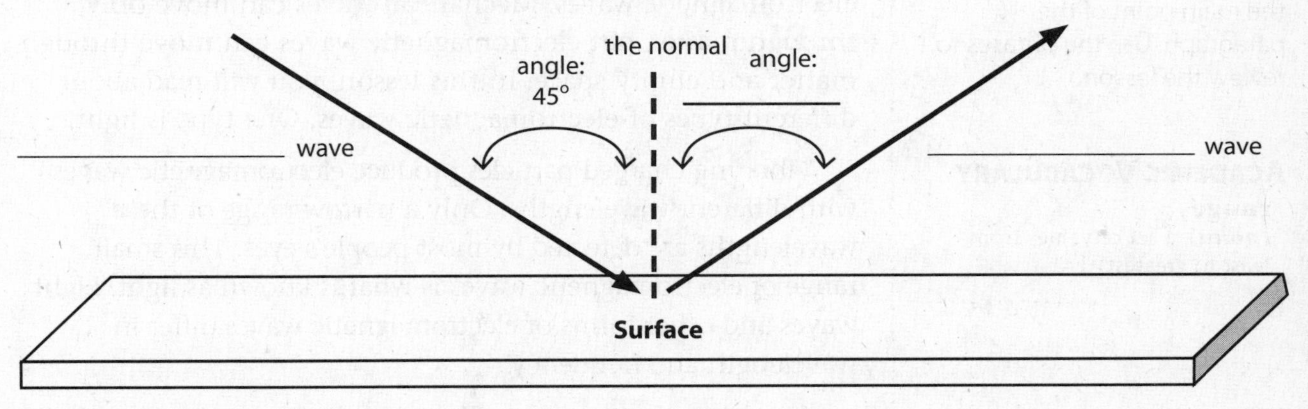

3. What is the difference between P waves, S waves, and surface waves in an earthquake?

____

____

What do you think NOW?

Reread the statements at the beginning of the lesson. Fill in the After column with an A if you agree with the statement or a D if you disagree. Did you change your mind?

Log on to ConnectED.mcgraw-hill.com and access your textbook to find this lesson's resources.

CHAPTER 14
LESSON 2

Waves, Light, and Sound

Light

Key Concepts

- How does light differ from other forms of electromagnetic waves?
- What are some ways in which light interacts with matter?
- How do eyes change light waves into the images you see?

Mark the Text

Identify the Main Ideas Write a phrase beside each paragraph that summarizes the main point of the paragraph. Use the phrases to review the lesson.

ACADEMIC VOCABULARY
range
(noun) a set of values from least to greatest

Key Concept Check
1. Distinguish How does light differ from other forms of electromagnetic waves?

Before You Read

What do you think? Read the two statements below and decide whether you agree or disagree with them. Place an A in the Before column if you agree with the statement or a D if you disagree. After you've read this lesson, reread the statements to see if you have changed your mind.

Before	Statement	After
	3. Light is the only type of wave that can travel through empty space.	
	4. Only shiny surfaces reflect light.	

Read to Learn

What are light waves?

There are two main types of waves—mechanical waves and electromagnetic waves. Mechanical waves can move only through matter, but electromagnetic waves can move through matter and empty space. In this lesson, you will read about different types of electromagnetic waves. One type is light.

Vibrating charged particles produce electromagnetic waves with different wavelengths. Only a narrow range of these wavelengths are detected by most people's eyes. This small range of electromagnetic waves is what is known as light. Light waves and other forms of electromagnetic waves differ in wavelength and frequency.

An object that produces light is called a luminous object. The Sun is Earth's major source of visible light. Almost half the Sun's energy that reaches Earth is visible light. Other luminous objects include lightbulbs, stars, and the flames of burning objects.

The Electromagnetic Spectrum

Light is just one type of electromagnetic wave. There is a wide range of electromagnetic waves that make up the electromagnetic spectrum, as shown in the figure on the next page. Besides light, you encounter several other types of electromagnetic waves every day.

The Electromagnetic Spectrum

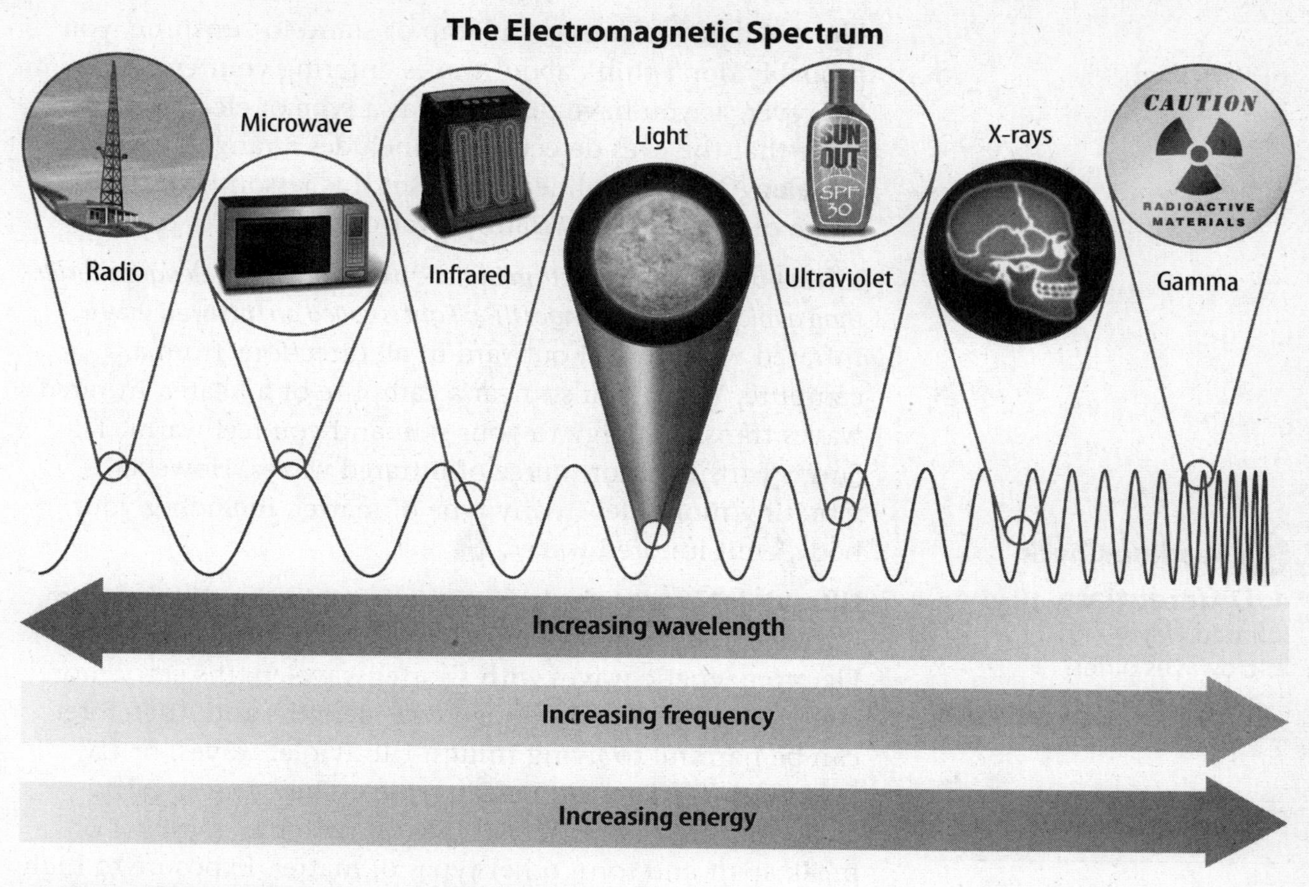

Types of Electromagnetic Waves

The electromagnetic spectrum consists of seven main types of waves. These waves range from low-energy, long-wavelength radio waves to very high-energy, short-wavelength gamma rays. Notice the relationship between wavelength, frequency, and energy shown by the arrows in the figure above. As the wavelength of electromagnetic waves decreases, the wave frequency increases. Low-frequency electromagnetic waves have low energy, and high-frequency waves have high energy.

Radio Waves *A low-frequency, low-energy electromagnetic wave that has a wavelength longer than about 30 cm is called a* **radio wave.** Radio waves have the least amount of energy of any electromagnetic wave. On Earth, radio and television transmitters produce radio waves that carry radio and television signals.

Microwaves Not only do microwaves cook your food, but they also carry cell phone signals. Wavelengths of microwaves range from about 1 mm to 30 cm. Because microwaves easily pass through smoke, light rain, and clouds, they are used to transmit information by satellites. Weather radar systems reflect microwaves off clouds to detect and calculate their distance and motion. This information is used in weather maps. ✓

✓ Visual Check

2. Name a type of electromagnetic wave that has a short wavelength, high frequency, and high energy.

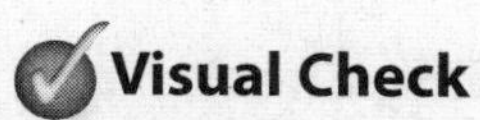

✓ Reading Check

3. Recognize What characteristic of microwaves makes them useful for satellite transmissions?

Light When you turn on a lamp or stand in sunshine, you probably don't think about waves entering your eyes. However, as you have read, light is a type of electromagnetic wave that the eyes detect. Light includes a range of wavelengths. You will learn later in this lesson how this range of wavelengths relates to various properties of light.

Infrared Waves *An electromagnetic wave with a wavelength shorter than a microwave but longer than light is called an* ***infrared wave.*** Infrared waves travel outward in all directions from a campfire. When you sit near a campfire or a heater, infrared waves transfer energy to your skin and you feel warm. The Sun is Earth's major source of infrared waves. However, vibrating molecules in any type of matter, including your body, emit infrared waves.

Reading Check
4. Differentiate How do infrared waves and microwaves differ?

Ultraviolet Waves *An electromagnetic wave with a slightly shorter wavelength and higher frequency than light is an* ***ultraviolet wave.*** Electromagnetic waves with shorter wavelengths carry more energy than those with longer wavelengths and, therefore, can be harmful to living things. Ultraviolet waves, or UV rays, from the Sun can be dangerous. These waves carry enough energy to cause particles of matter to combine or break apart and form other types of matter. Exposure to high levels of ultraviolet waves can damage your skin.

Reading Check
5. Recognize Why can ultraviolet waves be dangerous?

Ultraviolet waves from the Sun are sometimes labeled UV-A, UV-B, or UV-C based on their wavelengths. UV-A have the longest wavelengths and least energy. UV-C are the most dangerous because they have the shortest wavelengths and carry the most energy.

The ozone layer in Earth's atmosphere blocks the Sun's most harmful UV rays, keeping them from reaching Earth's surface. The ozone layer absorbs all of the UV-C and 95 percent of the UV-B from the Sun. It absorbs just 5 percent of the UV-A.

Reading Check
6. State How does the ozone layer protect Earth?

X-rays High-energy electromagnetic waves that have slightly shorter wavelength and higher frequencies than ultraviolet waves are X-rays. These waves can be very powerful. They have enough energy to pass through skin and muscle, but denser bone can stop them. This makes them useful for taking pictures of the inside of the body. Airport scanners sometimes use X-rays to take pictures of the contents of luggage.

Gamma Rays Vibrations within the nucleus of an atom produce electromagnetic waves called gamma rays. They have shorter wavelengths and higher frequencies than any other form of electromagnetic wave. Gamma rays carry so much energy that they can penetrate up to 10 cm of lead, one of the densest elements. On Earth, gamma rays are produced by radioactive elements and nuclear reactions.

Electromagnetic Waves from the Sun

The Sun produces an enormous amount of energy that travels outward in all directions as electromagnetic waves. Because Earth is so far from the Sun, Earth receives less than one billionth of the Sun's energy. However, if all the Sun's energy that reaches Earth in a 20-min period could be transformed to useful energy, that energy could power everything on Earth for a year!

About 44 percent of the Sun's energy that reaches Earth is carried by light waves. About 49 percent is carried by infrared waves, and about 7 percent is carried by ultraviolet waves. Less than 1 percent of the energy from the Sun is carried by radio waves, microwaves, X-rays, and gamma rays.

Speed, Wavelength, and Frequency

How could you describe the light from stars or the lights in a city at night? You might use words like *bright* or *dim*, or you might describe the color of the lights. You also could say how easily the light moves through a material. People use properties to describe light and to distinguish one color of light from another.

Like all types of electromagnetic waves, light travels at a speed of 3×10^8 m/s in empty space. When light enters a medium or matter, it slows. This is because of the interactions between the waves and the particles that make up the matter. The wavelength and frequency of a light wave determine the color of the light. The average human eye can distinguish among millions of wavelengths, or colors. Reds have the longest wavelengths and the lowest frequencies of light. Colors at the violet end of the visible light spectrum have the shortest wavelengths and the highest frequencies.

Light and Matter Interact

In Lesson 1, you read that matter can transmit, absorb, or reflect waves. How do these interactions affect light that travels from a source to your eyes?

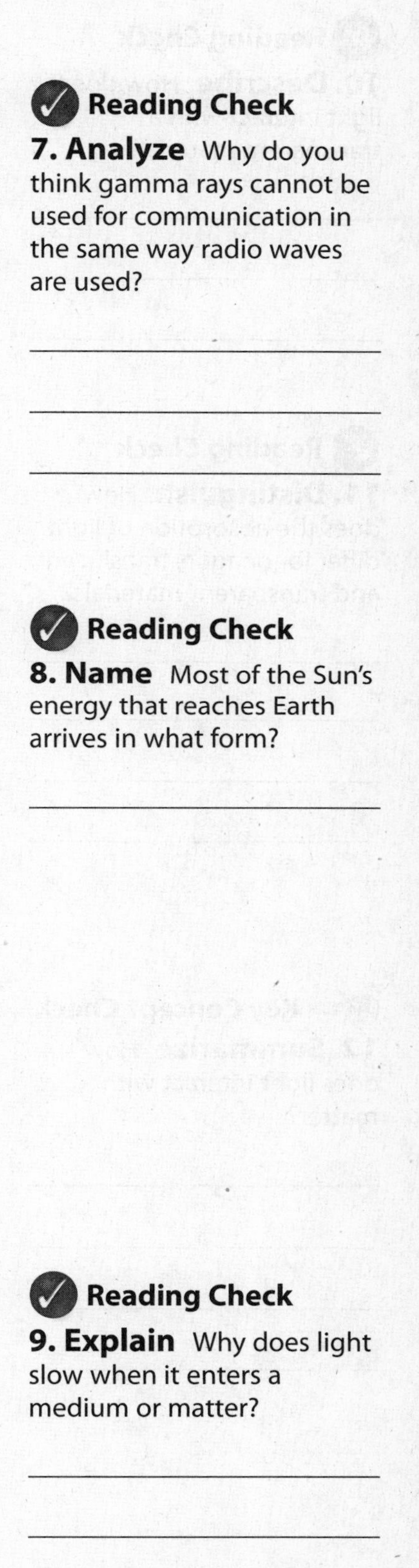

Reading Check

10. Describe How does light interact with a transparent material?

Reading Check

11. Distinguish How does the absorption of light differ for opaque, translucent, and transparent materials?

Key Concept Check

12. Summarize How does light interact with matter?

Transmission

Air and clear glass transmit light with little or no distortion. *A material that allows almost all of the light striking it to pass through, and through which objects can be seen clearly is* ***transparent.*** Light moves through the material without being scattered.

Materials such as waxed paper or frosted glass also transmit light, but you cannot see objects through the materials clearly. Light that moves through the material is scattered. *A material that allows most of the light that strikes it to pass through, but through which objects appear blurry is* ***translucent.***

Absorption

Some materials absorb most of the light that strikes them. These materials transmit no light. Therefore, you cannot see objects through them. *A material through which light does not pass is* ***opaque.***

Reflection

Why can you see your reflection clearly in a mirror but not in the wall of your room? Recall that waves reflect off surfaces according to the law of reflection. Parallel rays that reflect from a smooth surface remain parallel and form a clear image. Light that reflects from a bumpy surface scatters in many directions. A wall seems smooth, but up close it is too bumpy to form a clear image.

Different types of matter interact with light in different ways. For example, a window transmits and reflects light. Although you can see through a transparent window, you also can see your reflection. Some of the light that strikes an opaque object, such as a book, is absorbed and reflected at the same time. Reflected light allows an object to be seen.

Color

The colors of an object depend on the wavelengths of light that enter the eye. A luminous object, such as a campfire, is the color of light that it emits. If an object is not luminous, its perceived color depends on other factors.

Opaque Objects An opaque object is the color it reflects. The reflected wavelengths enter your eyes. The object absorbs all other colors of light. White objects reflect all colors of light. Black objects absorb all colors of light.

When white light strikes the American flag, the blue background absorbs all wavelengths of light except blue. The blue wavelengths reflect back to your eye from that part of the flag. You see blue. The red stripes absorb all colors except red. The red wavelengths reflect to your eye. The white stars and stripes reflect all colors, so you see white. The color of an opaque object is the color it reflects. ✓

Transparent and Translucent Objects If you look at a white lightbulb through a filter of red plastic wrap, only red wavelengths are transmitted through the plastic. The red plastic absorbs other wavelengths. The lightbulb appears to be red. ✓

Intensity of Light

Another property used to describe light is intensity. **Intensity** *is the amount of energy that passes through a square meter of space in one second.* Intensity depends on the amount of energy a source emits. Light from a flashlight, for example, has a much lower intensity than light from the Sun.

Intensity also depends on the light's distance from the source. When you are near a lamp, you probably notice that the intensity of the light is greater closer to the lamp than it is farther away. Many of the stars emit as much energy as the Sun. However, the light from those stars is less intense than light from the Sun because the stars are so much farther away than the Sun.

The brightness of a light is a person's perception of intensity. One person's eyes might be more sensitive to light than someone else's eyes. As a result, different people might describe the intensity of a light differently. In addition, eyes are more sensitive to some colors than others. ✓

The environment also can affect the brightness of a light. If you were to stand in a dark field and look up at the night sky, you could see many more stars than if you were viewing the same night sky from a brightly lit city street. Light from nearby buildings and other sources can prevent you from seeing stars in the sky.

✓ **Reading Check**

13. Explain Why do the stars on the American flag appear white when lit with white light?

____

____

____

✓ **Reading Check**

14. Identify What determines the color of a transparent or translucent object?

____

____

____

✓ **Reading Check**

15. Recognize When someone describes the brightness of a light, what property of light are they describing?

____

FOLDABLES

Make a vertical two-tab book to organize your notes on scattering and refraction.

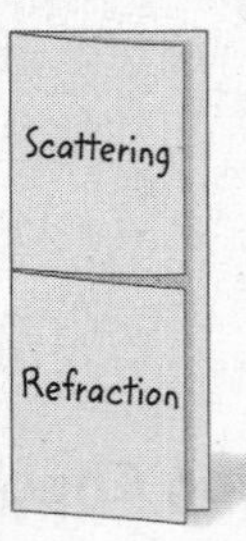

Interaction of Sunlight and Matter

Have you ever wondered why the sky is blue or the Sun is yellow? The interaction of light and matter causes these interesting effects when sunlight travels through air.

Scattering of Sunlight

As sunlight moves through Earth's atmosphere, most of the light reaches the ground. However, blue wavelengths are shorter than red wavelengths. The particles that make up air scatter the shorter blue wavelengths more than they scatter longer wavelengths. The sky appears blue because the blue wavelengths spread out in every direction. They eventually reach the eye from all parts of the sky.

A light source, such as the Sun, that emits all colors of light should appear white. Why does the Sun often appear yellow instead of white? After the blue wavelengths of light scatter, the remaining colors appear yellow.

Reading Check

16. Explain Why is the sky blue? Why is the Sun yellow?

Refraction of Sunlight

Another interesting effect of sunlight occurs because of refraction. Recall that light changes speed as it travels from one medium into another. If light enters a different medium at an angle, the light wave refracts, or changes direction.

As shown in the figure below, the refraction of light can affect the appearance of the setting Sun. The Sun's rays travel more slowly when they enter Earth's atmosphere. The light rays refract toward Earth's surface. The brain assumes the rays that reach your eyes have traveled in a straight line, and the Sun seems to be higher in the sky than it actually is. This refraction is why you can see the Sun even after it has set below the horizon.

Visual Check

17. Point Out Why does the Sun appear higher in the sky than it really is?

Refraction of Sunlight

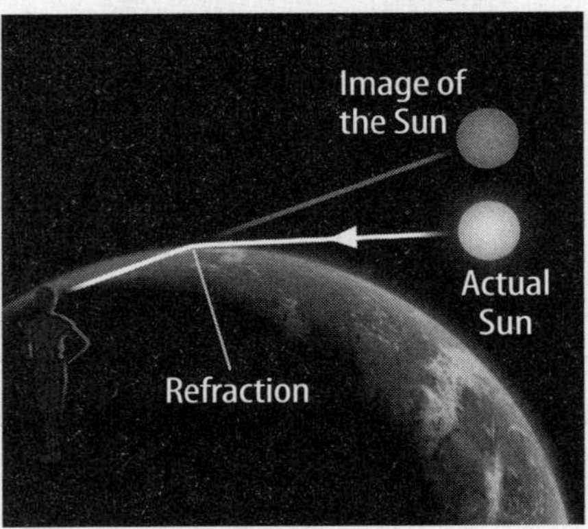

Vision and the Eye

Light makes it possible for you to see objects. Light from luminous objects travels directly from the object to the viewer. Objects also are seen when they reflect light to the eyes. What happens to light after it enters the eyes? How do eyes and the brain transform light waves into information about people, places, and things?

As shown in the figure below, light enters the eye through the cornea. The cornea and the lens focus light onto the retina. Cells in the retina absorb the light and send signals about the light to the brain. Read the steps in the figure to learn more about how the eye works.

Key Concept Check

18. Explain How do eyes change light waves into the images you see?

____

____

____

____

____

The Eye

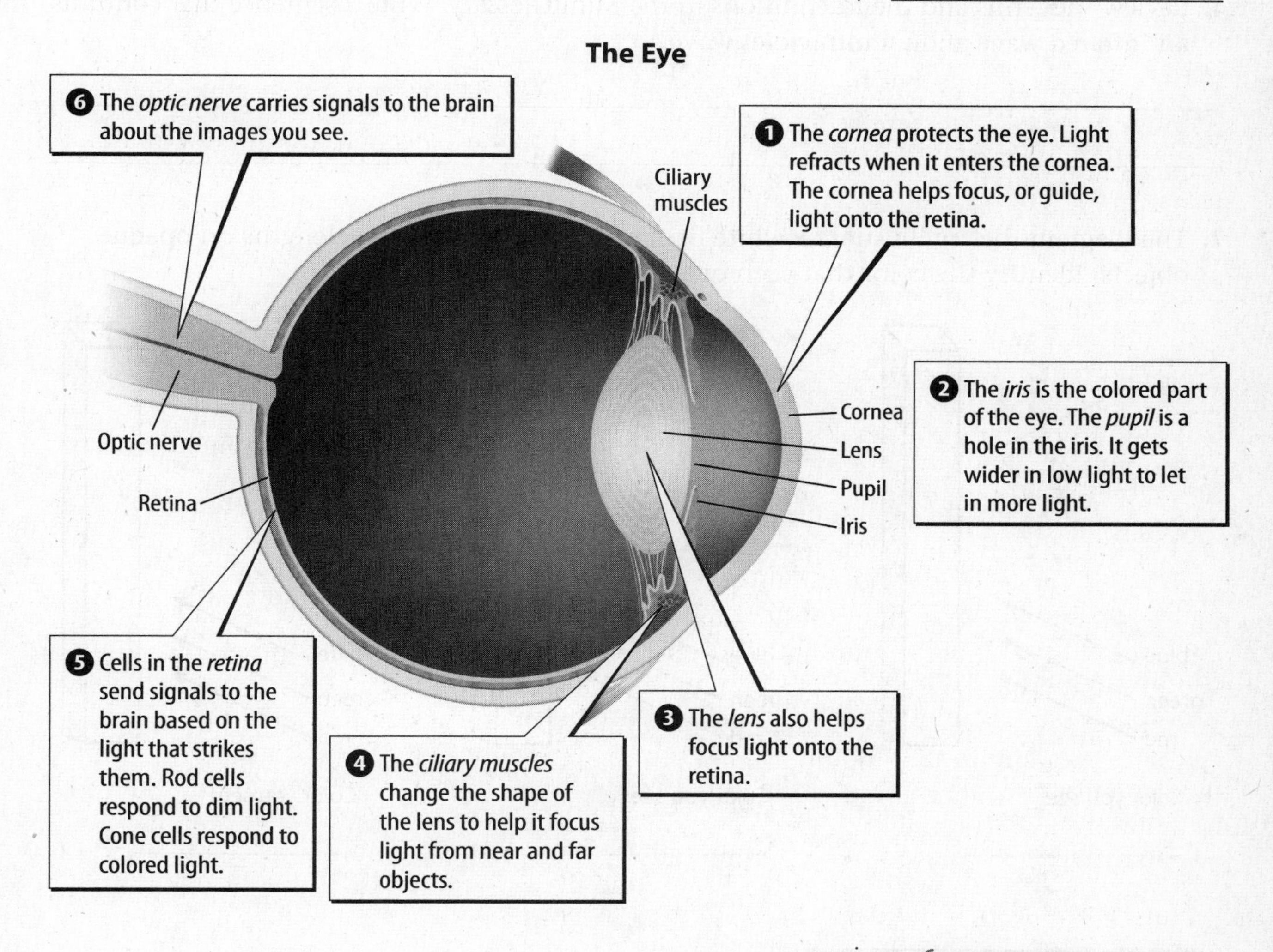

Visual Check

19. Identify What part of the eye responds to color?

____

After You Read

Mini Glossary

infrared wave: an electromagnetic wave with a wavelength shorter than a microwave but longer than light

intensity: the amount of energy that passes through a square meter of space in one second

opaque: a material through which light does not pass

radio wave: a low-frequency, low-energy electromagnetic wave that has a wavelength longer than about 30 cm

translucent: a material that allows most of the light that strikes it to pass through, but through which objects appear blurry

transparent: a material that allows almost all of the light striking it to pass through, and through which objects can be seen clearly

ultraviolet wave: an electromagnetic wave with a slightly shorter wavelength and higher frequency than light

1. Review the terms and their definitions in the Mini Glossary. Write a sentence that contrasts an infrared wave and an ultraviolet wave.

____

____

2. The diagrams below illustrate a white light shining a range of wavelengths on opaque objects. Identify the color that each opaque object will appear to you.

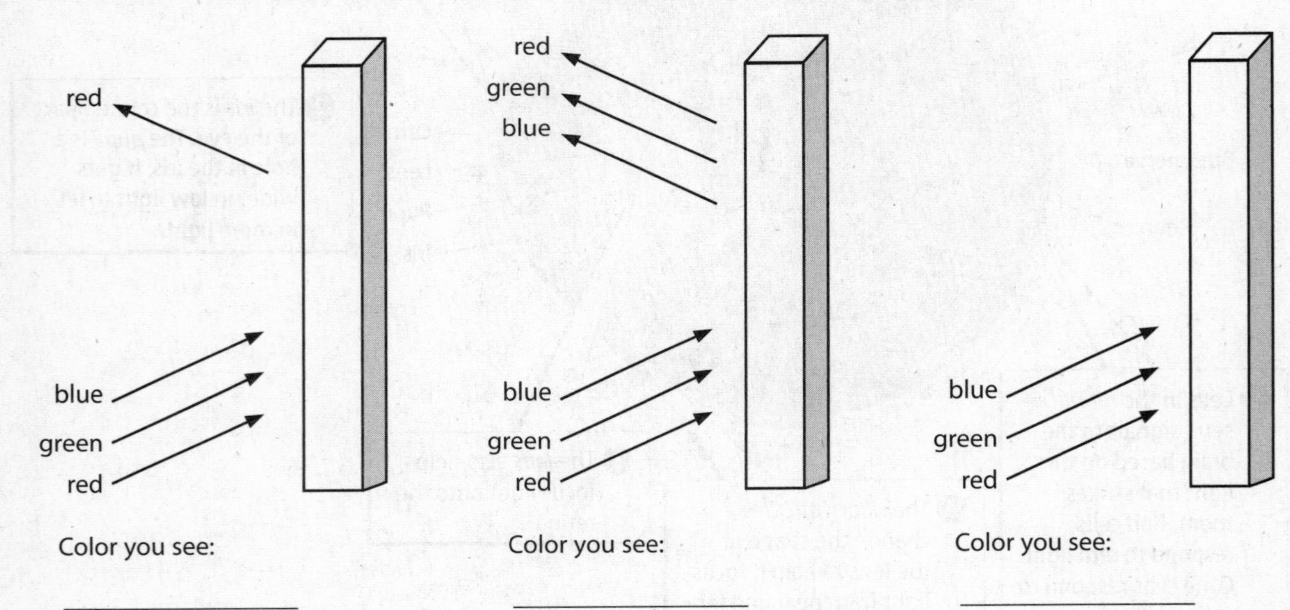

What do you think NOW?

Reread the statements at the beginning of the lesson. Fill in the After column with an A if you agree with the statement or a D if you disagree. Did you change your mind?

Log on to ConnectED.mcgraw-hill.com and access your textbook to find this lesson's resources.

Waves, Light, and Sound

Sound

Before You Read

What do you think? Read the two statements below and decide whether you agree or disagree with them. Place an A in the Before column if you agree with the statement or a D if you disagree. After you've read this lesson, reread the statements to see if you have changed your mind.

Before	Statement	After
	5. Sound travels faster through solid materials than through air.	
	6. The more energy used to produce a sound, the louder the sound.	

Key Concepts 
- What are some properties of sound waves?
- How do ears enable people to hear sounds?

Read to Learn

What are sound waves?

Just as light is a type of wave that can be seen, sounds are a type of wave that can be heard. Sound waves are longitudinal, mechanical waves. Unlike light waves, sound waves must travel through a medium.

Study Coach

Create a Quiz about sound. Exchange quizzes with a partner. After taking the quizzes, discuss your answers. Read more about the topics you don't understand.

Audible Vibrations

What would you hear if you struck two metal pans together? Now suppose you strike two pillows together. How would the two sounds differ? Sound waves are audible vibrations—vibrations the ear can detect. You hear a loud sound when you hit the pans together because they vibrate so much. You barely hear a sound when you hit the pillows together because they vibrate so little. Healthy young humans can hear sound waves produced by vibrations with frequencies between about 20 Hz and 20,000 Hz. As people age, their ability to hear the higher and the lower frequencies of sound decreases. The human ear is most sensitive to frequencies between 1,000 Hz and 4,000 Hz. ✓

Animals have ranges of hearing that help them live in their environment. For example, elephants hear sounds as low as 15 Hz. Chickens hear sounds between 125 Hz and 2,000 Hz. Porpoises can hear sounds between 75 Hz and 150,000 Hz.

Reading Check

1. Explain Why would you hear a louder sound if you dropped a book onto a wooden floor than if you dropped it onto a pillow?

____

____

____

Compressions and Rarefactions

Sound waves usually travel to your ears through air. Air particles are in constant motion. As the particles bounce off objects, they exert a force, or pressure. The figure below shows how sound waves moving through air change air pressure by causing air particles to move toward and then away from each other.

Reading Check
2. Distinguish How do compressions and rarefactions differ?

____

____

____

Suppose you pluck a guitar string. As the string springs back, it pushes air particles forward, forcing them closer together. This increases the air pressure near the string. *The region of a longitudinal wave where the particles of the medium are closest together is a* ***compression.*** As the string vibrates, it moves in the other direction. This leaves behind a region with lower pressure. *A* ***rarefaction*** *is the region of a longitudinal wave where the particles are farthest apart.*

Sound Waves

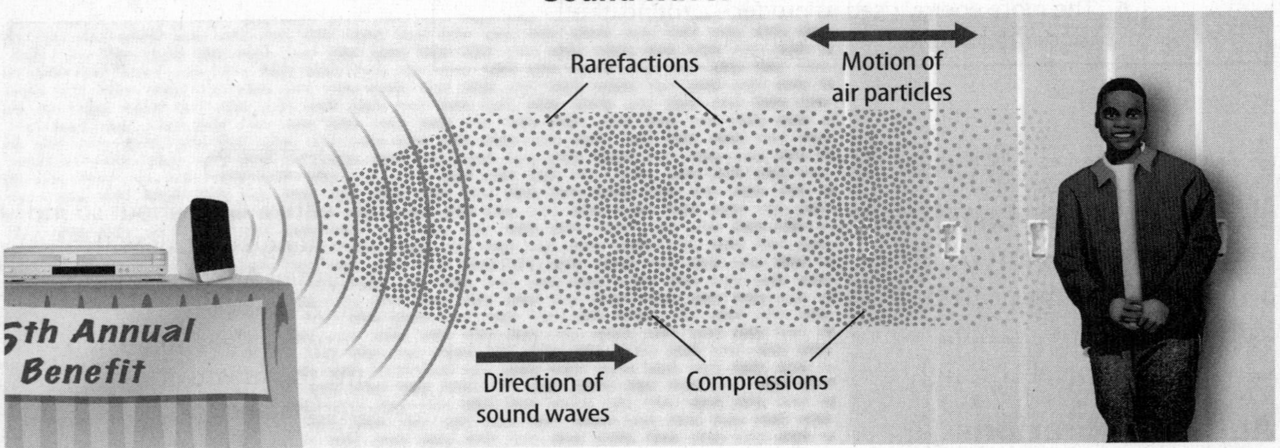

Visual Check
3. Recognize Why is pressure lower in rarefactions than in compressions?

____

____

____

Properties of Sound Waves

A sound wave is described by its wavelength, frequency, amplitude, and speed. These properties of sound waves depend on the compressions and rarefactions of the sound waves.

Wavelength, Frequency, and Pitch

Recall that the wavelength of a wave gets shorter as the wave's frequency increases. How does the frequency of a sound wave affect what is heard?

The perception of how high or low a sound seems is called **pitch.** The higher the frequency, the higher the pitch of the sound. For example, a female voice generally produces higher-pitched sounds than a male voice. This is because the female voice has a higher range of frequencies.

Amplitude and Energy

You use more energy to shout than to whisper. The more energy you put into your voice, the farther the particles of air move as they vibrate. The distance a vibrating particle moves from its rest position is the amplitude. The more energy used to produce the sound wave, the greater the amplitude.

Speed

Sound waves travel much slower than electromagnetic waves. With sound, the transmitted energy must pass from particle to particle. Two factors that affect the speed of sound are the type of medium and the temperature.

Type of Medium In a gas, the particles are far apart. They collide less often than particles in a liquid or a solid. Therefore, a gas takes longer to transfer sound energy from one particle to another, as shown in the table to the right.

The Speed of Sound

Material	Speed (m/s)
Air (0°C)	331
Air (20°C)	343
Water (20°C)	1,481
Water (0°C)	1,500
Seawater (25°C)	1,533
Ice (0°C)	3,500
Iron	5,130
Glass	5,640

Temperature The temperature of the medium also affects the speed of sound. As the temperature of a gas increases, the particles move faster and collide more often. This increase in the number of collisions transfers more energy in less time.

Temperature has the opposite effect on liquids and solids. As liquids and solids cool, the molecules move closer together. They collide more often and transfer energy faster.

Intensity and Loudness

You might think that the greater the amplitude of a sound wave is, the louder it will sound. That is true if you stay at the same distance from the source. However, as you move away, the wave's amplitude decreases and the sound seems quieter. This is because as a sound wave moves farther from its source, more and more particles collide, and the energy from the wave spreads out among more particles. Therefore, as you move farther from the source of the sound waves, less of the waves' energy is present in the same area of space. Recall that the amount of energy that passes through a square meter of space in one second is the intensity of a wave. Loudness is your ear's perception of intensity.

FOLDABLES®

Make a horizontal four-tab book to review properties of sound waves.

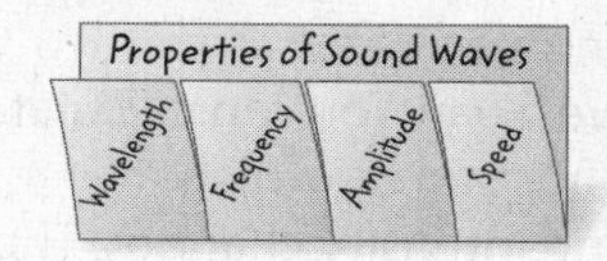

Interpreting Tables

4. Interpret At 20°C, which transmits sound faster: air or water? Why?

Key Concept Check

5. Identify What are some properties of sound waves?

Reading Check

6. Recognize Why does a sound seem quieter as you move farther from the source of the sound?

Math Skills

Because sound energy travels out in all directions from the source, the intensity of the sound decreases as you move away. You can calculate the fraction by which the sound intensity changes. The fraction is $\left(\frac{r_1}{r_2}\right)^2$, where r_1 is the starting distance and r_2 is the ending distance from the source. For example, by what fraction does sound intensity decrease when you move 3 m to 6 m from a source?

a. Replace the variables with given values.

$$\text{fraction} = \left(\frac{3}{6}\right)^2$$

b. Solve the problem.

$\left(\frac{3}{6}\right)^2 = \left(\frac{1}{2}\right)^2 = \frac{1}{4}$, so the intensity decreases to $\frac{1}{4}$ of its original value.

7. Use a Fraction You stand 2 m from a sound source. How does the sound intensity change if you move to a distance of 6 m?

____

Visual Check

8. Identify What is the highest decibel level to which you can listen without risking permanent hearing loss?

____

Key Concept Check

9. Identify How do your ears enable you to hear sounds?

____

____

____

The Decibel Scale

The unit used to measure sound intensity, or loudness, is the **decibel (dB).** The decibel levels of common sounds are shown in the figure below. Each increase of 10 dB results in a sound about twice as loud. As the decibel level goes up, the amount of time you can listen to the sound without risking hearing loss gets shorter and shorter. People who work around loud sounds wear protective hearing devices to prevent hearing loss.

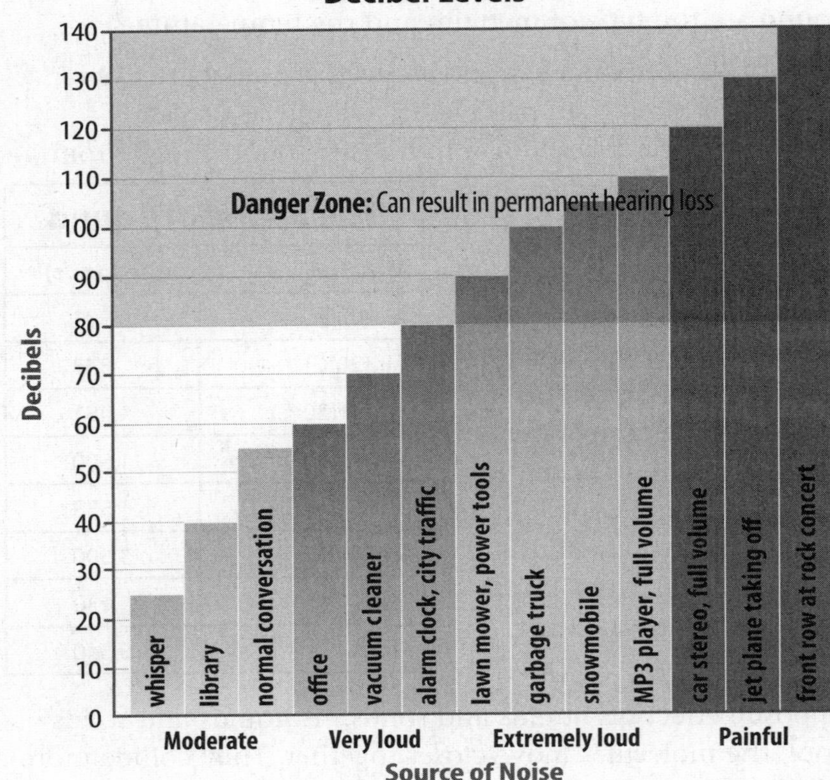

Hearing and the Ear

Typically, objects are seen when light waves enter the eyes. Similarly, sound waves enter the ears with information about the environment. The human ear has three main parts, as shown in the figure on the next page. The external outer ear, which includes the part that you can see, collects sound waves. The middle ear amplifies, or intensifies, the sound waves. The middle ear includes the eardrum and three tiny bones—the hammer, the anvil, and the stirrup. The inner ear contains the cochlea (KOH klee uh). The cochlea changes the sound waves to nerve signals. The brain then can process these signals to create the perception of sound.

Parts of the Human Ear

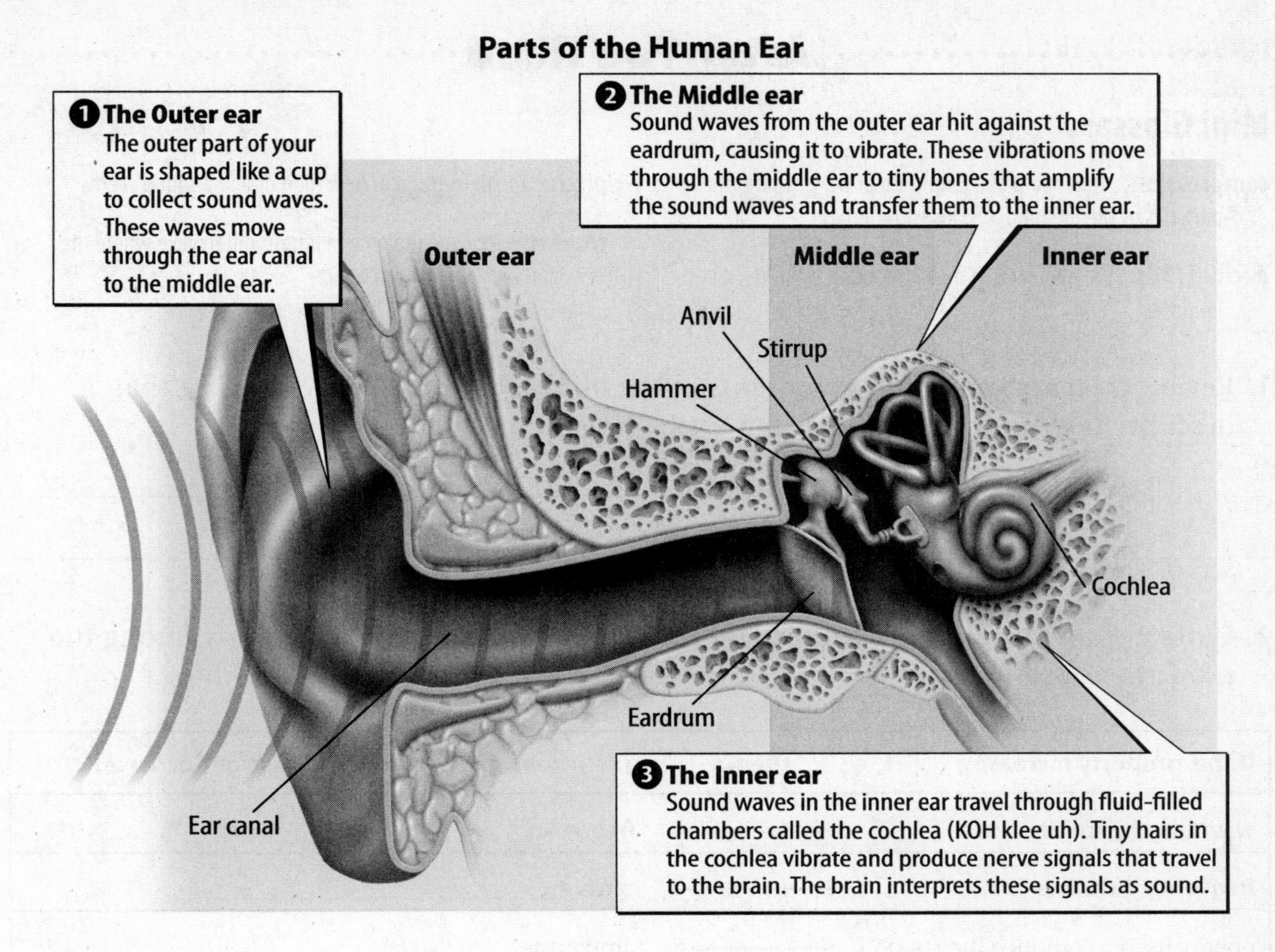

Visual Check

10. Name Which part of the ear has a spiral shape?

____

After You Read

Mini Glossary

compression: the region of a longitudinal wave where the particles of the medium are closest together

decibel (dB): the unit used to measure sound intensity, or loudness

pitch: the perception of how high or low a sound seems

rarefaction: the region of a longitudinal wave where the particles are farthest apart

1. Review the terms and their definitions in the Mini Glossary. Write a sentence that describes how an increase in decibel level affects sound intensity.

____

____

2. Write *increases* or *decreases* in the blanks below to describe the relationships among the properties of sound.

If this property increases,	then...	does this property increase or decrease?
wavelength increases	→	frequency ________
frequency increases	→	pitch ________
energy used to produce the sound increases	→	amplitude ________
temperature of a gas increases	→	the speed of sound through the gas ________
temperature of a liquid or solid increases	→	the speed of sound through the liquid or solid ________
distance from the sound source increases	→	intensity ________

3. Record a question from your partner's quiz that was difficult for you. Then answer it.

____

____

What do you think NOW?

Reread the statements at the beginning of the lesson. Fill in the After column with an A if you agree with the statement or a D if you disagree. Did you change your mind?

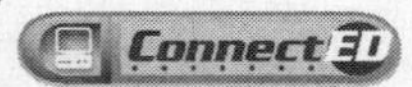

Log on to ConnectED.mcgraw-hill.com and access your textbook to find this lesson's resources.

Electricity and Magnetism

Electric Charges and Electric Forces

Before You Read

What do you think? Read the two statements below and decide whether you agree or disagree with them. Place an A in the Before column if you agree with the statement or a D if you disagree. After you've read this lesson, reread the statements to see if you have changed your mind.

Before	Statement	After
	1. Electrically charged objects always attract each other.	
	2. Electric fields apply magnetic forces on other electric fields.	

Key Concepts 

- How do electrically charged objects differ?
- How do objects become electrically charged?
- How do electrically charged objects interact?

Read to Learn

Electric Charges

Have you ever walked across a carpeted floor, reached for a metal doorknob, and received a small shock? The shock comes from electric charges jumping between your fingers and the doorknob.

What are electric charges? Where do they come from? Why do they jump from one object to another? In this lesson, you will learn the answers to these questions.

Atoms are the tiny particles that make up all the matter around you. An atom has a nucleus made up of two kinds of smaller particles. These particles are protons and neutrons.

An atom also is made up of electrons. Electrons move around the atom's nucleus, as shown in the figure to the right. Protons and electrons have a property called electric charge. Neutrons do not have electric charge.

Atom

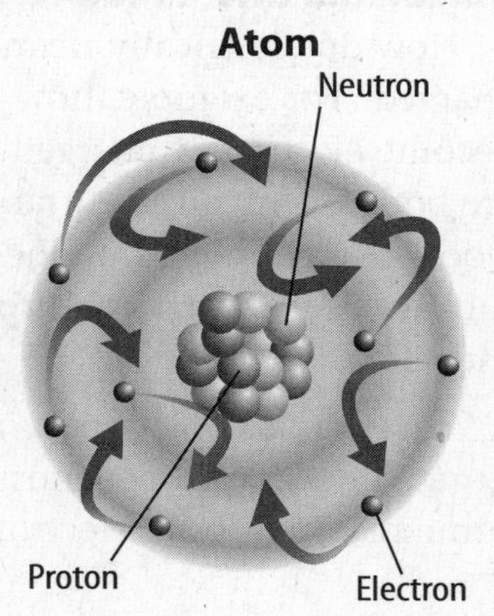

Mark the Text

Building Vocabulary
As you read, underline the words and phrases that you do not understand. When you finish reading, discuss these words and phrases with another student or your teacher.

ACADEMIC VOCABULARY
nucleus
(noun) basic or essential part; core

Visual Check
1. Identify Circle the electrons.

Positive and Negative Charge

There are two types of electric charge—positive and negative. Here, *positive* and *negative* do not mean more and less. The terms are simply names scientists use to talk about the two types of electric charge.

Protons have positive charge. Electrons have negative charge. The amount of positive charge of a proton equals the amount of negative charge of an electron.

Atoms have equal numbers of positive protons and negative electrons. *A particle with equal amounts of positive charge and negative charge is* **electrically neutral.** Electrically neutral atoms make up all objects. Therefore, objects are normally electrically neutral, too. However, electrons sometimes transfer between objects. How does transferring electrons affect objects?

Reading Check
2. Explain Why are atoms electrically neutral?

When electrons move from one electrically neutral object to another, both objects become electrically charged. *An object is* **electrically charged** *when it has an unbalanced amount of positive charge or negative charge.* Objects can be either positively charged or negatively charged.

Reading Check
3. State What happens when electrons move from one electrically neutral object to another?

Positively Charged An object that has lost one or more electrons as it comes in contact with another object has an unbalanced electric charge. An object that has lost electrons has more protons than electrons. Thus, the object has more positive charge than negative charge. The object is positively charged.

Negatively Charged An object that gains one or more electrons also has an unbalanced electric charge. An object that has gained electrons has more electrons than protons. Thus, the object has more negative charge than positive charge. The object is negatively charged.

Key Concept Check
4. Differentiate How do electrically charged objects differ?

Materials and Electric Charge

How do electrically neutral objects become electrically charged? For example, how can a balloon and a stuffed toy become electrically charged? At first, the balloon and the toy are both electrically neutral. You rub the balloon against a wool-covered toy. Wool does not hold electrons as tightly as rubber. Electrons transfer from the toy to the balloon when they are in contact.

Whether an object becomes positively charged or negatively charged depends on the material it contacts. Some materials hold electrons more tightly than others.

Electron Movement in Common Items
Becomes Positive
Glass
Nylon
Wool
Silk
Aluminum
Paper
Cotton
Wood
Rubber
Copper
Polyester
Polystyrene
Polyvinyl chloride
Becomes Negative

The table above lists common materials in order of how tightly they hold electrons. Notice glass is above wool in the table. This means that if you rub a glass cup against a wool toy, electrons transfer from the glass onto the wool. The glass becomes positively charged, and the wool becomes negatively charged.

Electric Discharge

You read that objects can become electrically charged. However, an electrically charged object tends to lose its unbalanced charge after a period of time. *The loss of an unbalanced electric charge is an* **electric discharge.**

Some electric discharges happen slowly. For example, electrons on negatively charged objects discharge, or move, from the object onto water molecules in the air. You may have noticed that the static cling of electrically charged clothing lasts longer on dry days than on humid days when there is more water vapor in the air.

Some electric discharges happen quickly. For example, lightning is the sudden loss of unbalanced electric charges that build up in thunderstorm clouds. Another example is the continual electric discharge through a fluorescent light that causes a powder inside the tube to glow brightly.

Interpreting Tables

5. Interpret Is it easier for electrons to move from aluminum or paper?

____

Key Concept Check

6. Explain How do the balloon and the stuffed toy become charged?

____

____

____

____

Reading Check

7. Define What is an electric discharge?

____

____

____

FOLDABLES

Make a horizontal three-tab book with an extended tab. Use it to organize your notes on the relationships between electric forces.

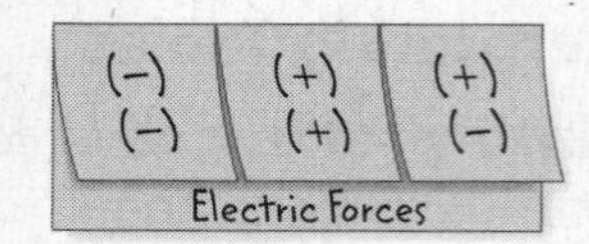

Think it Over

8. Apply Name one other material that might be a good electric insulator.

____

Reading Check

9. Explain What is an electric field?

____

____

____

____

Electric Insulators and Conductors

Objects of different materials become electrically charged as they come in contact. The charge can remain where objects touch, or the charge can spread over the entire object.

Think of a balloon rubbing against a sweater. Charges from the sweater stay in the area of the balloon that touched the sweater. However, as you walk across a carpet, charges from the carpet spread over your entire body. Your hand receives an electric shock as you reach for a metal doorknob.

Electric charges do not spread over the balloon because electrons cannot easily move in rubber. *A material in which electric charges cannot easily move is an* **electric insulator.** Plastic, wood, and glass also are electric insulators.

Materials in which electric charges easily move are **electric conductors.** Some of the best electric conductors are metals such as iron, copper, gold, and aluminum.

Electric Fields and Electric Forces

Suppose you rub two balloons on a wool sweater. Electrons transfer from the sweater to the balloons. Both balloons become negatively charged. The sweater becomes positively charged. You notice that the sweater attracts, or applies a pulling force on, the balloons. However, the balloons repel, or apply a pushing force to, each other. *The force that two electrically charged objects apply to each other is an* **electric force.**

Electric fields surround charged objects.

To open a door, your hand must touch the door to apply a force to it. However, an electrically charged object does not have to touch another charged object to apply an electric force to it. For example, the two charged balloons in the example above repel each other even though they do not touch.

How do charged objects apply electric forces to each other without touching? The answer is a bit of a mystery. However, scientists know there is a region around a charged object that applies an electric force to other charged objects. *This invisible region around any charged object where an electric force is applied is an* **electric field.**

Charged Objects

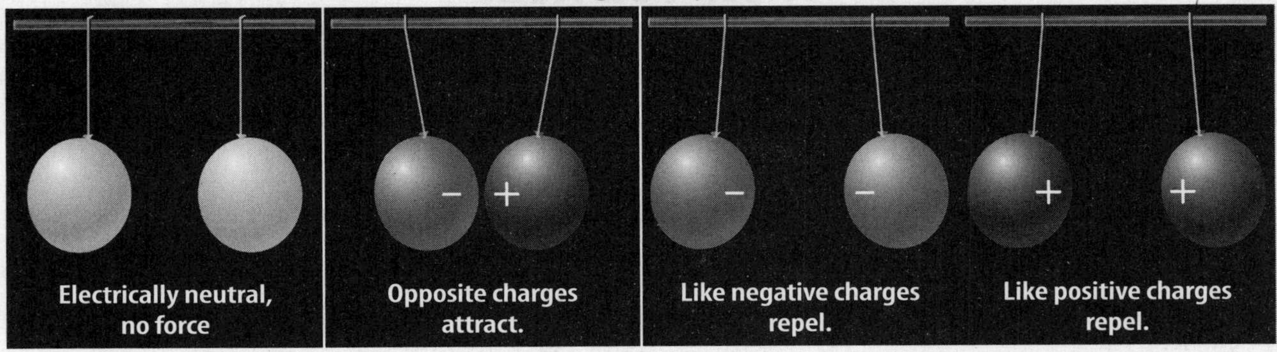

Electric force depends on the types of charge.

You read that an electric force can be a push or a pull. Whether the force is a push or a pull depends on the types of charge on the objects, as shown in the figure above. If both objects are positively charged or if both objects are negatively charged, the two objects push each other away. In other words, objects with the same type of charge repel each other.

If one object is positively charged and the other object is negatively charged, the two objects pull each other together. In other words, objects with opposite types of electric charge attract each other.

Visual Check

10. State why the negative and positive balloons attract each other.

____

____

____

____

Key Concept Check

11. Summarize How do electrically charged objects interact?

____

____

____

____

After You Read

Mini Glossary

electrically charged: when an object has an unbalanced amount of positive charge or negative charge

electrically neutral: when a particle has equal amounts of positive charge and negative charge

electric conductor: a material in which electric charges easily move

electric discharge: the loss of an unbalanced electric charge

electric field: the invisible region around any charged object where an electric force is applied

electric force: the force that two electrically charged objects apply to each other

electric insulator: a material in which electric charges cannot easily move

1. Review the terms and their definitions in the Mini Glossary. Write a sentence explaining the difference between an electric conductor and an electric insulator.

____

____

2. On the first line in the ovals, identify the three particles that make up an atom. On the second line, indicate whether the particle has a positive, negative, or neutral electric charge.

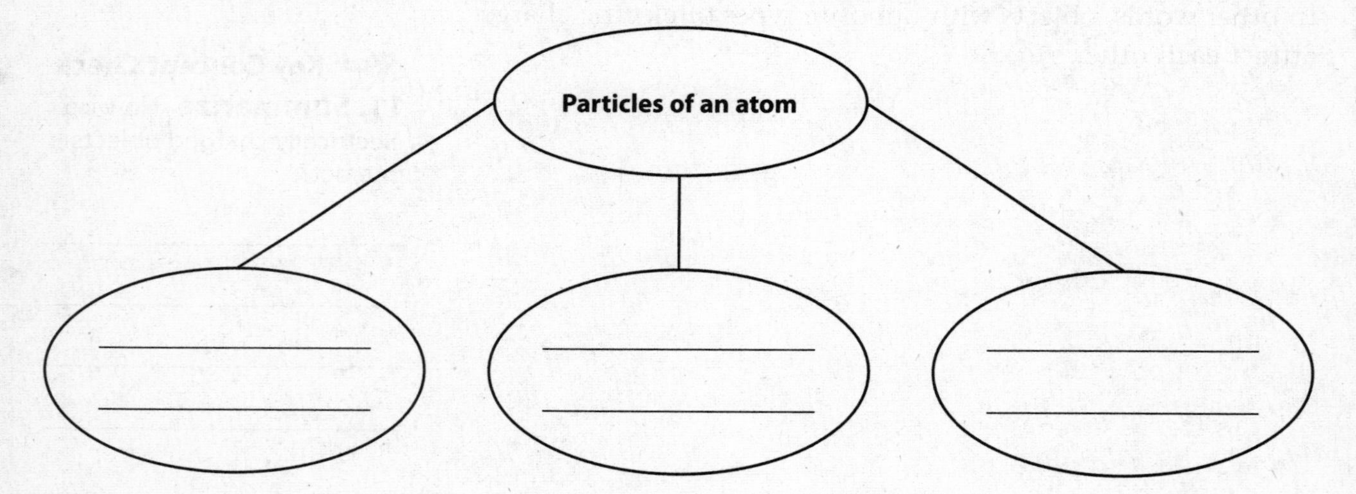

3. Suppose you rub a nylon scarf against a polyester scarf. Which will become positively charged? Which will become negatively charged?

____

What do you think NOW?

Reread the statements at the beginning of the lesson. Fill in the After column with an A if you agree with the statement or a D if you disagree. Did you change your mind?

Log on to ConnectED.mcgraw-hill.com and access your textbook to find this lesson's resources.

Electricity and Magnetism

Electric Current and Electric Circuits

Before You Read

What do you think? Read the statement below and decide whether you agree or disagree with it. Place an A in the Before column if you agree with the statement or a D if you disagree. After you've read this lesson, reread the statement to see if you have changed your mind.

Before	Statement	After
	3. A battery in an electric circuit produces an electric current.	

Key Concepts 

- How are electric current and electric charge related?
- What are the parts of a simple electric circuit?
- How do the two types of electric circuits differ?

Read to Learn

Electric Current—Moving Electrons

Negatively charged electrons are the tiny particles that move around the nuclei of atoms. Recall that many of the electrons of an electric conductor, such as copper, are free to move from atom to atom. When free electrons move in the same direction, an electric current is produced. *An* **electric current** *is the movement of electrically charged particles.*

Like all moving objects, moving electrons have kinetic energy. As electrons move from atom to atom, their kinetic energy transforms to other useful energy forms, such as light and thermal energy. Moving electrons, or an electric current, is one of the most common forms of energy.

Study Coach

Make Flash Cards For each head in this lesson, write a question on one side of a flash card and the answer on the other side. Quiz yourself until you know all of the answers.

Key Concept Check

1. Compare How are electric current and electric charge related?

Two Types of Electric Current

Recall that an electric current is the movement of electrons. An electric current carries energy at about the speed of light. However, the negatively charged electrons themselves move more slowly.

Imagine a tube filled with marbles. When a marble is pushed into one end of the tube, it causes another marble to pop out the other end of the tube. Each marble does not instantly move the length of the tube. Similarly, as electrons move into one end of a wire, other electrons leave the other end of the wire almost instantly. Each electron does not suddenly move the length of the wire.

FOLDABLES

Create a horizontal three-tab book and use it to explain the components of a circuit.

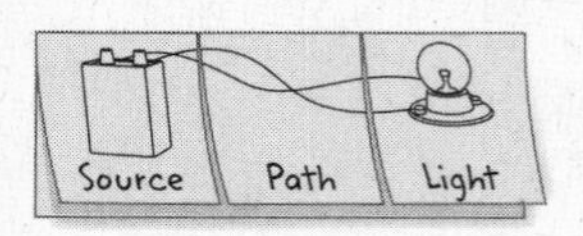

Direct Current In the previous example, marbles added continually to one end of the tube produce a steady stream of marbles flowing out the other end of the tube. In a similar way, electrons continually added to one end of a wire create a constant one-way flow of electrons. This is known as direct current. Some energy sources, including batteries, produce only direct current. Many portable devices, such as flashlights and radios, operate using direct current.

Alternating Current If marbles are repeatedly added to one end of the tube and then to the other end, the marbles in the tube would move back and forth, never moving far from their original positions. An electric current that continually reverses direction is known as an alternating current. Large generators in power plants supply homes and businesses with alternating current.

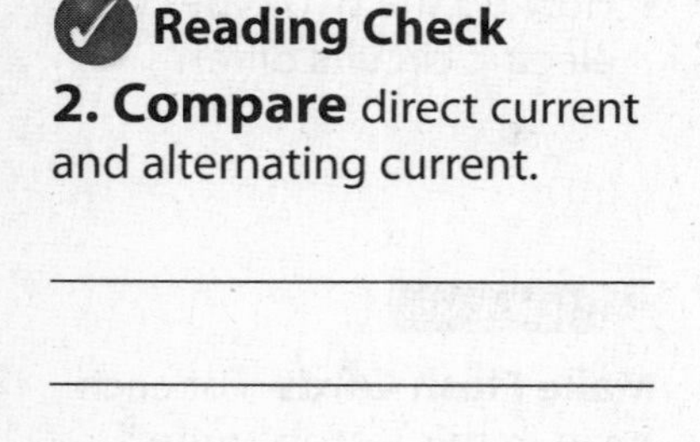

Reading Check

2. Compare direct current and alternating current.

____

____

____

____

The Circuit—A Path for Electric Current

Electric circuits transform the energy of an electric current to useful forms of energy. *An* **electric circuit** *is a closed, or complete, path in which an electric current flows.* Electric circuits are all around you.

A Useful Circuit

Electric circuits are designed to transform electric energy to specific forms. Electric circuits in a microwave oven transform electric energy to radiant energy that cooks food.

The figure below illustrates an electric circuit designed to transform the electric energy of a battery into the light energy emitted by a lightbulb. As shown, the circuit is complete, or closed, and the lightbulb is lit. When the circuit is broken, or open, at any point, the electric current stops and the lightbulb does not light.

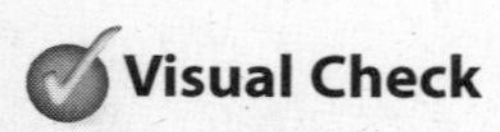

Visual Check

3. Locate Circle the part in the circuit that provides electric energy.

Simple Circuit

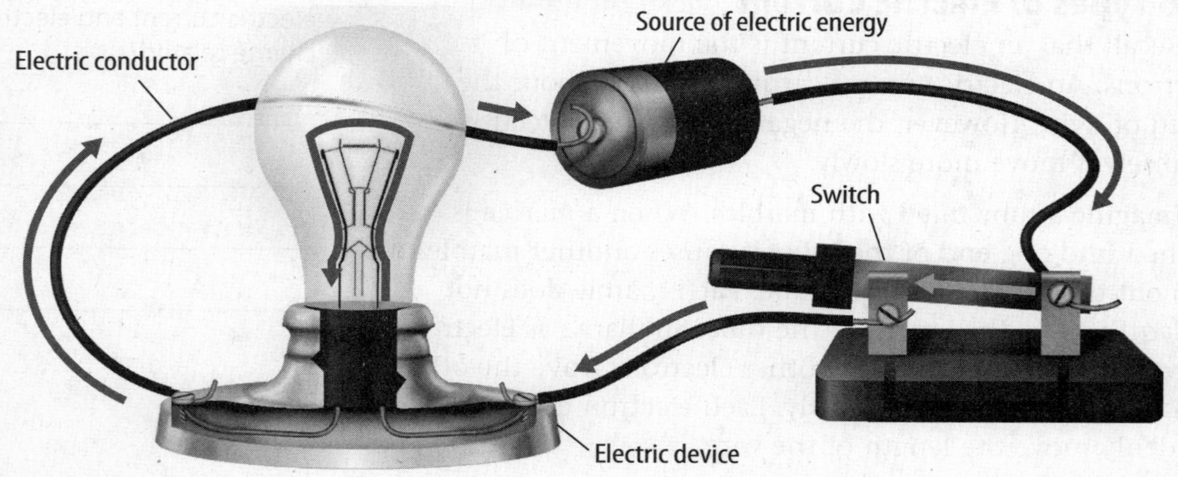

A Simple Circuit

Some electric circuits, such as those in computers, are complicated and have hundreds of parts. However, many common and useful circuits have only a few components.

Simple circuits are used in flashlights, doorbells, and many kitchen appliances. All simple circuits contain: 1) a source of electric energy, such as a battery; 2) an electric device, such as a lightbulb; and 3) an electric conductor, such as a wire. In addition to these basic components, many circuits often include a switch. How do these basic components interact to make a useful electric current?

Sources of Electric Energy There are many uses of electric energy. Most uses require specific types of sources of electric energy. For example, a flashlight requires a small, portable source. Cities need sources that produce large amounts of electric energy that are nonpolluting. Some of the technologies now being developed and improved to help meet the world's growing demand for electric energy are discussed below and on the next page.

Batteries When an electric energy source needs to be small and portable, batteries often are the energy source used. A battery is simply a can of chemicals. Chemical reactions within a battery move electrons from one end of the battery (the positive terminal) to the other end (the negative terminal). Outside the battery, the electrons flow through a closed circuit from the negative terminal back to the positive terminal. As the chemical reactions continue, electrons keep moving through the battery and circuit.

Generators *Machines that transform mechanical energy to electric energy are* **generators.** Many power plants use fossil fuels or nuclear energy to power large generators. These fuels provide thermal energy to boil water into steam. The steam flows through and rotates a turbine that, in turn, rotates a generator. These types of turbine-powered generators provide most of the electric energy used in the United States. Wind or moving water is the power source for other generators. You will read more about generators in the next lesson.

Key Concept Check

4. Identify What are the parts of a simple electric circuit?

____

____

____

____

Reading Check

5. State What is a battery?

____

____

____

____

____

Reading Check

6. Explain How do most U.S. generators work?

____

____

____

____

Solar Cells These cells change sunlight into electric energy. Solar panels are made of a large number of connected solar cells. Simple solar cells power calculators and many other small devices. More complicated systems have enabled humans to learn about the solar system and beyond.

Fuel Cells Like batteries, fuel cells produce electric energy by a chemical reaction. But, unlike batteries, fuel cells need a constant flow of fuel, such as hydrogen gas. An advantage of using fuel cells as a source of electric energy is that they produce no pollution. Fuel cells have generated electric energy on space flights. Now, scientists and engineers are developing fuel cells that people could use everyday. ✓

Reading Check

7. Consider What is an advantage of using fuel cells?

____

____

____

Electric devices transform energy. An electric device is a part of a circuit. An electric device is designed to transform electric energy to another useful form of energy. For example, a lightbulb is designed to transform electric energy to light. Transformation of electric energy occurs wherever there is electric resistance in a circuit. **Electric resistance** *is a measure of how difficult it is for an electric current to flow in a material.* Electric devices with greater electric resistance transform greater amounts of electric energy. What causes a transformation of electric energy?

Think about an electric lightbulb. As electrons move in the high-resistance wire filament of the lightbulb, they collide with atoms of the filament. The atoms absorb some of the electrons' kinetic energy, then release the energy as light. ✓

Reading Check

8. Point Out What causes energy transformations in electric devices?

____

____

Electric Conductors and Electric Circuits An electric conductor, such as a metal wire, is used to complete the circuit by connecting the energy source to the electric device. Copper and aluminum are good materials for wires in electric circuits because they are excellent conductors. A good conductor has little electric resistance.

Recall that an electric current easily flows through an electric conductor. However, even the best conductors, such as copper wire, have some resistance to an electric current. All conductors, including a device's power cord, have some electric resistance. Small amounts of electric energy in a circuit's conductors always transform to wasted thermal energy. ✓

Reading Check

9. Explain Why are wires in an electric circuit often made of copper?

____

____

____

____

Series and Parallel Circuits

An electric circuit can have more than one device. For example, a string of holiday lights is a circuit that has many lightbulbs, or devices. Some holiday lights are circuits in which all of the lightbulbs go out when one of the bulbs is removed from its socket.

Now, think of the electric lights in the rooms of your home. These lights are devices connected in an electric circuit, too. What happens to the light in the kitchen when you remove the lightbulb from the lamp in your room? Nothing. The kitchen light remains lit.

How can you explain this difference in the two circuits? The answer is that there are two types of electric circuits.

Series Circuit In the previous examples, the string of holiday lights is a series circuit. A series circuit is an electric circuit that has only one path through which an electric current can flow. In other words, all of the devices in a series circuit are connected end-to-end.

As shown in the figure below, the same electric current flows through all the lightbulbs in the string. Breaking, or opening, a series circuit causes the electric current to stop flowing through the entire circuit.

Series Circuit

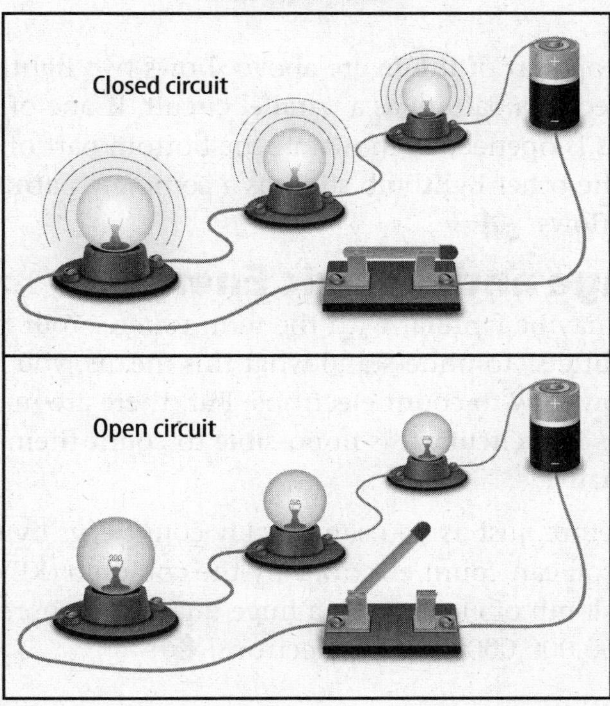

Reading Check

10. Select An electric circuit _____. (Circle the correct answer.)

a. has no devices

b. has only one device

c. can have more than one device

Reading Check

11. Describe a series circuit.

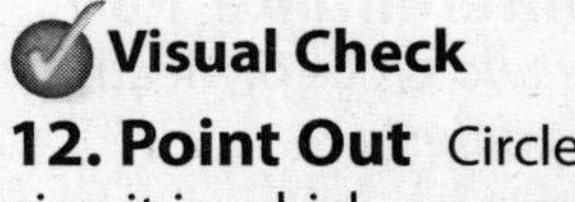

Visual Check

12. Point Out Circle the circuit in which no current can flow.

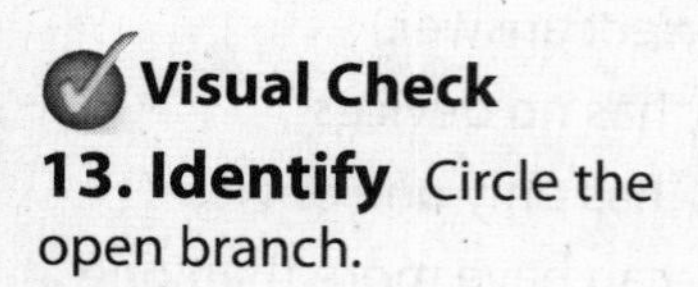

Visual Check
13. Identify Circle the open branch.

Parallel Circuit

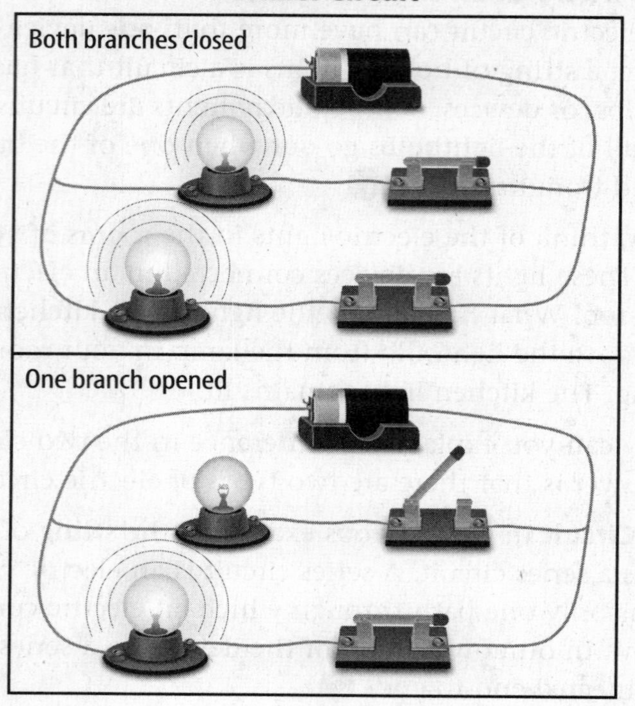

Parallel Circuit A different type of circuit connects the devices in your home. Houses are not wired with series circuits. Instead, they are wired with parallel circuits. A parallel circuit is an electric circuit in which each device connects to the electric source with a separate path, or branch.

The top part of the figure above shows two lightbulbs connected to a battery as a parallel circuit. If one of the branches is opened, as shown in the bottom part of the figure, the other lightbulb still has a complete path in which current flows.

Key Concept Check
14. Differentiate How do the two types of electric circuits differ?

____

____

____

____

Voltage and Electric Energy

You may be familiar with the term *voltage*. Your home has 120-V outlets. To understand what this means, you must first know how to count electrons. But, there are many electrons in a circuit. It is impossible to count them individually.

Therefore, just as you can quickly count eggs by the dozen, you can count electrons by the coulomb (KEW lahm). One coulomb of electrons is a huge quantity—approximately 6,000,000,000,000,000,000 electrons!

Reading Check
15. Name How are electrons counted?

____

Voltage of an Entire Circuit

Recall that all parts of an electric circuit have electric resistance. Because a circuit has electric resistance, energy is required to move electrons through a circuit. *The* ***voltage*** *of an electric circuit is the amount of energy used to move one coulomb of electrons through the circuit.* ✓

Think of two identical lightbulbs. One lightbulb is powered by a 3-V battery. The other is powered by a 6-V battery. As you might expect, the lightbulb in the 6-V circuit is lit brighter than the lightbulb in the 3-V circuit. But why?

The definition of voltage tells you that the 6-V battery uses twice as much energy as the 3-V battery as it produces a current. Thus, the 6-V circuit transforms twice the electric energy to light.

✓ Reading Check

16. Define What is voltage?

______

______

______

______

Voltage of Part of a Circuit

You also can measure the voltage of part of a circuit. The voltage measured across a part of a circuit tells you how much energy is used by moving electrons through that part of the circuit. The figure below shows the voltages across a wire and a lightbulb in the same circuit. The higher voltage across the lightbulb tells you that the lightbulb transforms more electric energy than the wire.

17. Examine Which part of the circuit is transforming most of the battery's energy into some other form?

______

______

The sum of the voltages across all parts of an electric circuit equals the voltage of the energy source. This means that an electric circuit transforms all of the energy of an electric current.

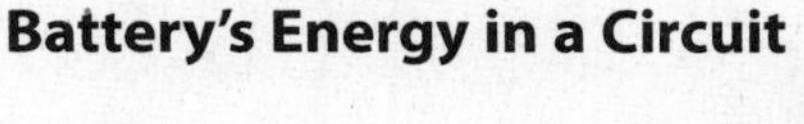

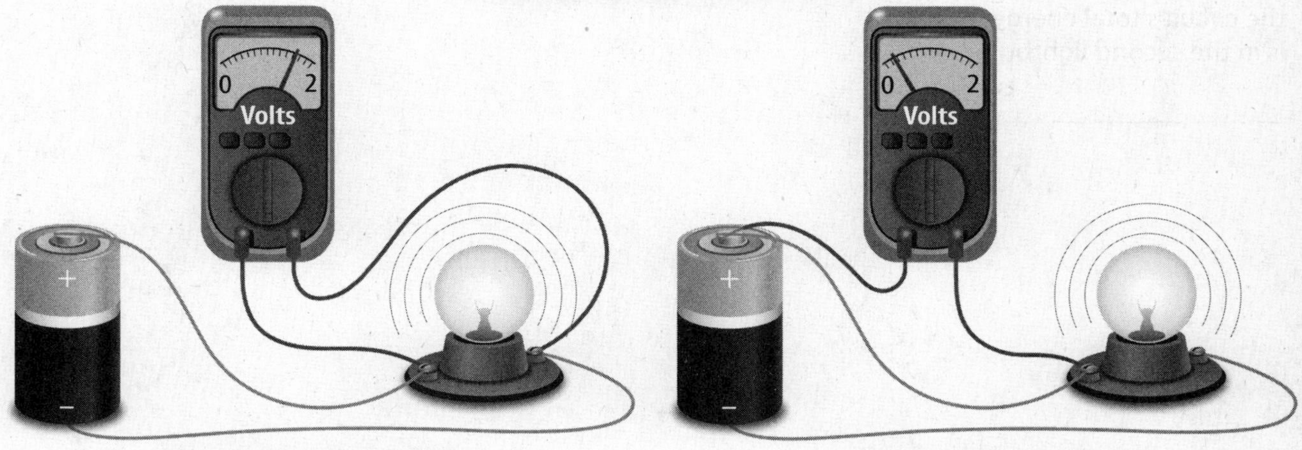

Higher voltage across lightbulb

Lower voltage across wire

Math Skills

Imagine a 9-V battery and two lightbulbs in a series circuit. The voltage across one lightbulb is 6 V. The second lightbulb reads 3 V. What part of the circuit's total energy is used by each lightbulb?

Divide the voltage reading across one of the lightbulbs by the voltage across the entire circuit (across the battery).

First bulb: $\frac{6\text{ V}}{9\text{ V}} = \frac{2}{3}$

Second bulb: $\frac{3\text{ V}}{9\text{ V}} = \frac{1}{3}$

If you add the fractions together, they equal one.

For example: $\frac{2}{3} + \frac{1}{3} = 1$

This is because the sum of the energies used by each device in a circuit equals the total energy in the circuit.

18. Using Fractions

A 12-V battery powers a series circuit that contains two lightbulbs. The voltage across one of the lightbulbs is 8 V. What fractional part of the circuit's total energy used is in the second lightbulb?

____

A Practical Electric Circuit

Recall that a simple circuit can function with only a few basic parts—a lightbulb can be lit with just a battery and a couple of wires. However, most useful circuits include additional components to make them more useful and safer. A hair dryer, for example, uses these components:

- a temperature-sensitive safety cutoff switch that automatically turns off the hair dryer if it becomes too hot
- an electric motor that transforms electric energy to the mechanical energy of the fan that blows air over your hair
- a heating element that transforms electric energy to the thermal energy that dries your hair
- a switch that allows you to conveniently start and stop the hair dryer
- a wall outlet that provides a source of energy for the hair dryer as well as many other electric devices in your home

After You Read

Mini Glossary

electric circuit: a closed, or complete, path in which an electric current flows

electric current: the movement of electrically charged particles

electric resistance: a measure of how difficult it is for an electric current to flow in a material

generator: a machine that transforms mechanical energy to electric energy

voltage: the amount of energy used to move one coulomb of electrons through a circuit

1. Review the terms and their definitions in the Mini Glossary. Write a sentence in your own words to explain what a generator is.

2. Use the graphic organizer to identify and describe the two types of electric circuits.

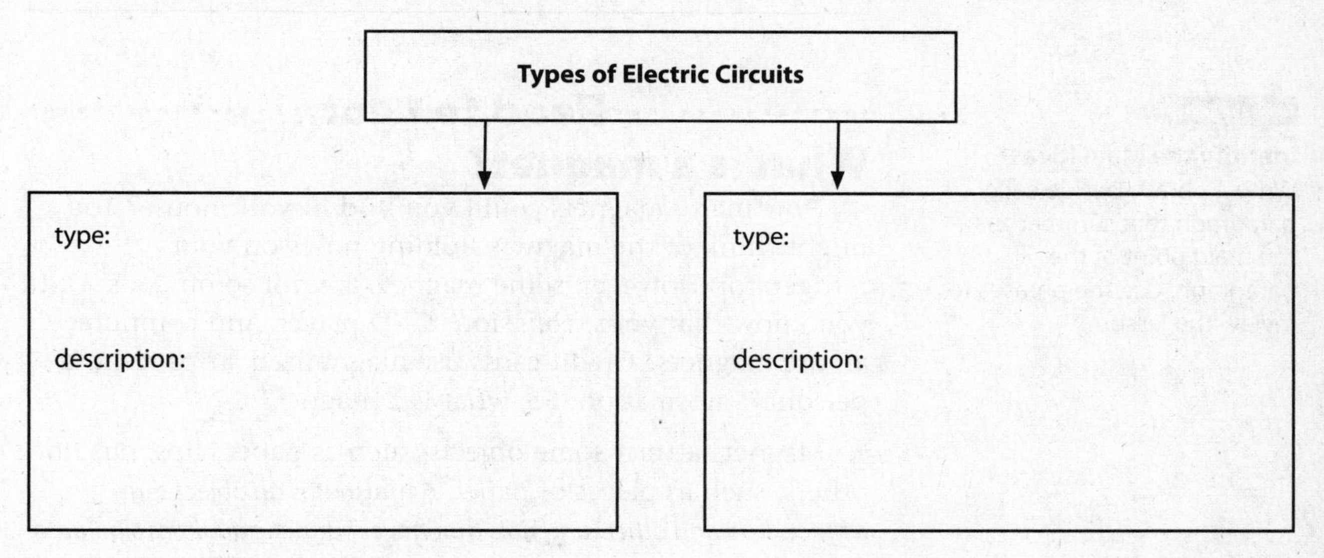

3. Select and define a word from one of the flash cards you created as you read the lesson.

What do you think NOW?

Reread the statements at the beginning of the lesson. Fill in the After column with an A if you agree with the statement or a D if you disagree. Did you change your mind?

Log on to ConnectED.mcgraw-hill.com and access your textbook to find this lesson's resources.

Electricity and Magnetism

Magnetism

Key Concepts
- What causes a magnetic force?
- How are magnets and magnetic domains related?
- How are electric currents and magnetic fields related?

Before You Read

What do you think? Read the three statements below and decide whether you agree or disagree with them. Place an A in the Before column if you agree with the statement or a D if you disagree. After you've read this lesson, reread the statements to see if you have changed your mind.

Before	Statement	After
	4. Every magnet has one magnetic pole.	
	5. Earth is magnetic but is not a magnet.	
	6. A magnet moving within a wire loop produces an electric current.	

Mark the Text

Identify the Main Ideas Write a phrase beside each paragraph that summarizes the main point of the paragraph. Use the phrases to review the lesson.

Read to Learn

What is a magnet?

How many magnets could you find in your house? You might think of the magnets holding notes on your refrigerator. However, some magnets are not so obvious. Did you know that your television, DVD player, and computer all use magnets? Credit cards use magnetized strips to hold personal information. So, what is a magnet?

Magnets attract some objects, such as paper clips, but not others, such as pieces of paper. *A* **magnet** *is an object that attracts iron and other materials that have magnetic qualities similar to iron.* A magnet attracts paper clips and some nails because they contain iron. Magnets also attract other metals, such as nickel, cobalt, and alnico, an aluminum-nickel-cobalt alloy. *Any material that a magnet attracts is a* **magnetic material.**

Reading Check

1. Recognize Why is the metal cobalt a magnetic material?

____

____

Magnetic Fields and Magnetic Forces

An invisible electric field surrounds an electrically charged object. In the same way, an invisible magnetic field surrounds a magnet and an electric current. Even though magnetic fields are invisible, they can be detected by the forces they apply. *A* **magnetic force** *is a push or a pull a magnetic field applies to either a magnetic material or an electric current.*

Seeing a Magnetic Field

A magnet's magnetic field applies a magnetic force to a magnetic material even when the magnet and the magnetic material do not touch. A magnetic field and its force are stronger closer to the magnet and weaker farther away from the magnet.

Magnetic Field

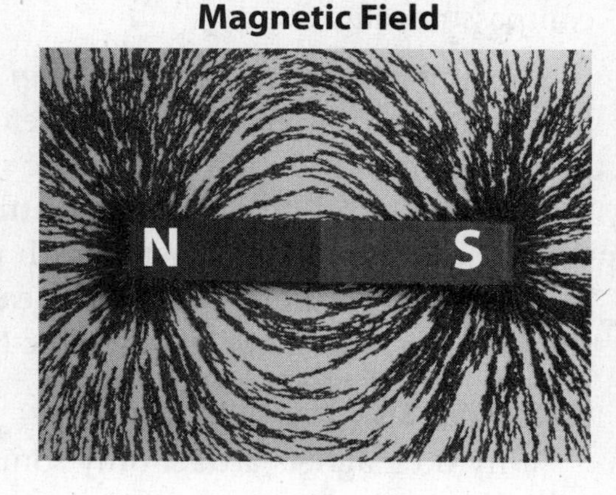

The figure at right helps you to visualize a magnetic field. Iron is a magnetic material. When iron filings are sprinkled around a magnet, the filings will line up with the magnet's magnetic field. The iron filings form a pattern of curved lines. These curved lines are the magnet's magnetic field lines.

Visual Check
2. Summarize How can iron filings illustrate a magnetic attraction?

Magnetic Poles

Magnets are made in many sizes and shapes. However, all magnets have something in common. Every magnet has two magnetic poles. One pole of a magnet is called the magnetic north pole. The other pole is called the magnetic south pole. The magnetic poles are the two places on a magnet where the magnetic field lines are closest together. This is also where the magnetic field applies the strongest force. ✓

Reading Check
3. Name What do all magnets have in common?

Magnetic field lines point away from the magnet's magnetic north pole and toward the magnet's magnetic south pole. For a bar magnet, as shown in the figure above, the ends of the magnet are the magnetic poles.

Key Concept Check
4. Explain What causes the forces applied by magnets?

Magnetic Poles and Magnetic Forces

The forces that magnets apply to each other depend on which magnetic poles are near each other. If two magnetic south poles or two magnetic north poles are close to each other, the magnets repel, or push away from each other. In disc magnets, this repulsion causes one magnet to "float" on the invisible magnetic field of another disc magnet. If a magnet's magnetic north pole is near another magnet's magnetic south pole, the magnets attract each other. This attraction causes the magnets to come together. In other words, similar poles repel, opposite poles attract.

Earth as a Magnet

How does a magnetic compass help you find Earth's geographic North Pole? A compass needle is a small bar magnet. Like all magnets, a magnetic field surrounds a compass needle.

Flowing molten iron and nickel in Earth's outer core create a magnetic field around Earth. Therefore, Earth has a magnetic north pole and a magnetic south pole. Recall that the opposite poles of two magnets attract each other. Thus, the compass needle's magnetic north pole points toward Earth's magnetic south pole. This means that Earth's magnetic south pole is near Earth's geographic North Pole.

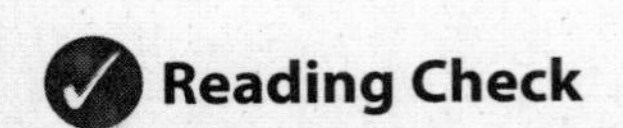

5. State Which of Earth's geographic poles is about in the same location as Earth's magnetic south pole?

Magnets

Why do magnets attract only some materials? Remember, all matter is made of atoms. A magnetic field surrounds each atom. In some materials, atoms are grouped in magnetic domains. *A* ***magnetic domain*** *is a region in a magnetic material in which the magnetic fields of the atoms all point in the same direction.* The magnetic fields of the atoms within a domain combine into a single field around the domain. Think of a magnetic domain as a tiny magnet within a material.

Nonmagnetic Materials

Most materials, including aluminum and plastic, do not have atoms grouped in magnetic domains. The atoms point in many different directions. The random magnetic fields cancel out the magnetic effects of each other. These nonmagnetic materials cannot be made into magnets.

FOLDABLES®

Create a horizontal two-tab book and use it to describe and to collect examples of magnetic and nonmagnetic materials.

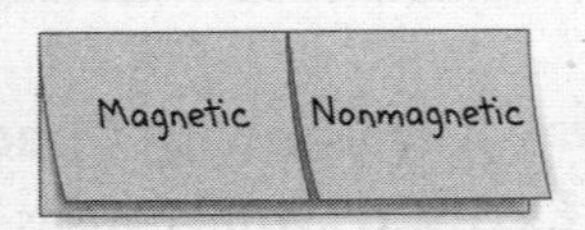

Magnetic Materials

In some materials, such as iron and steel, atoms are grouped in magnetic domains. These materials are called magnetic materials. However, not all magnetic materials are magnets. For example, the magnetic fields of the domains of a steel nail point in different directions. The magnetic fields of these domains cancel out the magnetic effects of each other. Here, the magnetic material is not a magnet.

A magnetic material becomes a magnet as the magnetic fields of the material's magnetic domains line up to point in the same direction. The aligned magnetic fields of the domains combine to form a single magnetic field around the entire object. In this case, the magnetic material is a magnet.

Key Concept Check

6. Compare How are magnets and magnetic domains related?

Temporary and Permanent Magnets

Some magnetic materials lose their magnetic fields quickly. Other materials keep their magnetic fields for a long time. How long a magnet remains a magnet depends partly on the material from which it is made.

Soft magnetic materials are not soft to the touch. They are called soft because they quickly lose their magnetic fields. Materials that keep their magnetic fields for long periods of time are hard magnetic materials.

Reading Check

7. Name the two types of magnetic materials.

Temporary Magnets Placing a soft magnetic material, such as iron, in a strong magnetic field causes the material's magnetic domains to line up. This makes the material a magnet. When the material is moved away from the magnetic field, its domains return to their random positions, and the material is no longer a magnet. In part *a* of the figure to the right, the nail is not a magnet.

Temporary Magnet

a)

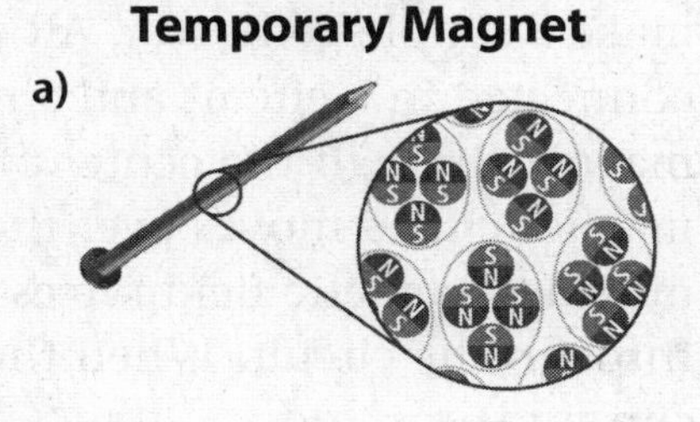

b)

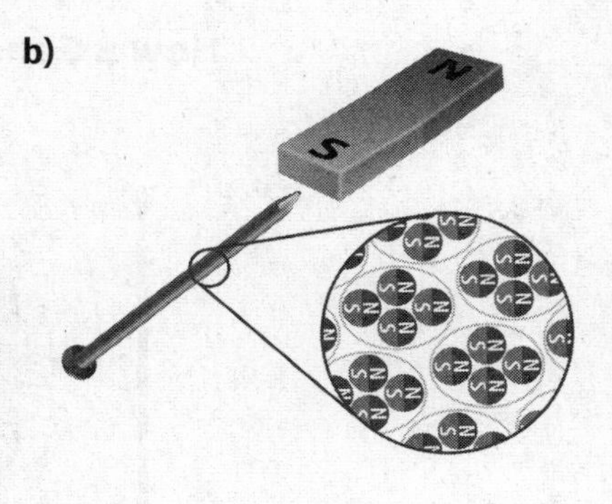

Visual Check

8. Identify Circle the nail that shows magnetic domains in alignment.

However, in part *b*, the nail is a magnet. This happens because the magnet's magnetic field causes the magnetic field of the nail's magnetic domains to line up. Thus, the nail becomes a magnet. The nail is a temporary magnet because it attracts other magnetic materials only if it is in the magnetic field of another magnet.

Reading Check

9. Explain Why do soft magnetic materials make temporary magnets?

Permanent Magnets Hard magnetic materials are mixtures of iron, nickel, and cobalt combined with other elements. When a hard magnetic material is placed in an extremely strong magnetic field, the material's magnetic domains align and lock into place.

When a magnet made in this way is removed from the strong magnetic field, the object remains a magnet permanently, unlike a temporary magnet. Lodestone, a naturally occurring permanent magnet, is found in Earth's crust. Other permanent magnets can be made with electric devices called magnetizers.

Combining Electricity and Magnetism

In 1820, Danish scientist Hans Christian Ørsted noticed that a compass needle moved when a nearby electric current was switched on. He was convinced there was a relationship between electricity and magnetism. Today, we call this relationship electromagnetism. Almost all the electrical devices in your home, if they use an electric motor, depend on electromagnetism.

Reading Check
10. Define What is electromagnetism?

Magnetic fields produce electric currents.

Recall that a generator is a machine that produces an electric current. As shown in the figure below, you can make a simple generator. All you need is a small wire coil connected in a circuit and a magnet. When you move the magnet through the center of the coil, the magnet's magnetic field moves over the loops of the coil. The moving magnetic field forces an electric current to flow through the circuit. When the magnet stops moving, the current stops, too.

Visual Check
11. Identify Circle the magnetic field.

How a Generator Works

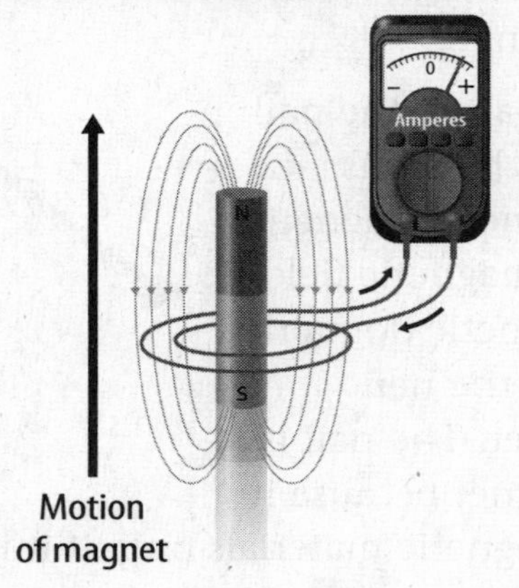

More complex generators use wire coils with more loops and stronger magnets that rotate in place. You can produce an electric current by using a hand-cranked generator to rotate a magnet within a small wire coil. The hand-cranked generator produces only a small amount of electric current. Huge generators use coils with several kilometers of wire and giant magnets to produce the electric current that is supplied to homes, buildings, and cities.

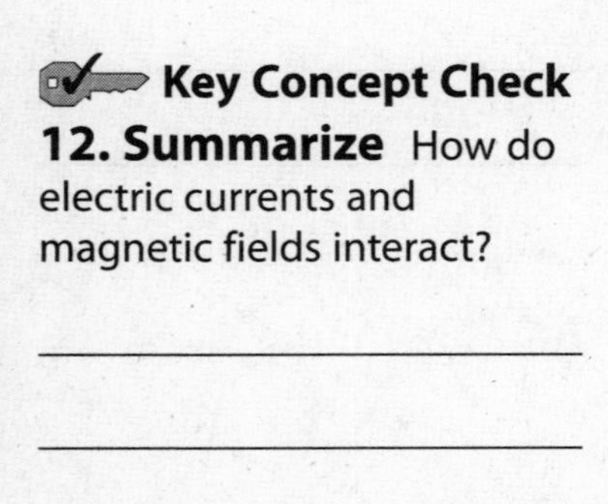

Key Concept Check
12. Summarize How do electric currents and magnetic fields interact?

Electric currents produce magnetic fields.

Recall that some magnetic materials become temporary magnets when placed in the magnetic field of another magnet. There is another type of temporary magnet that is very common and useful.

Hans Ørsted discovered that a magnetic field surrounds a current-carrying wire. The magnetic field is shown as a series of circles in the figure to the right. If a current-carrying wire is wound into a coil, the magnetic field becomes stronger. When a soft magnetic material is placed within the coil, the magnetic field becomes even stronger. *A temporary magnet made with a current-carrying wire coil wrapped around a magnetic core is an* **electromagnet.**

Magnetic Field

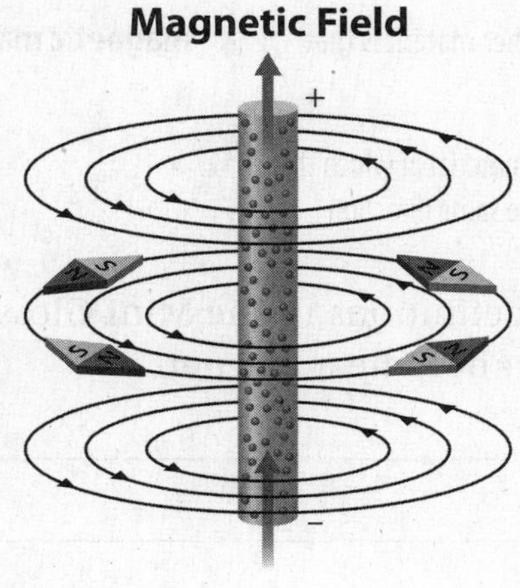

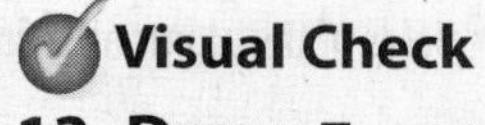

Visual Check

13. Draw Trace the flow of electric current in the wire.

Electromagnets are useful because they can be controlled in ways other magnets cannot. First, an electromagnet's magnetic field can be turned off and on. Turning off the electric current in the coil turns off the magnetic field. Second, the north and south poles of the electromagnet reverse when the current reverses. And finally, the strength of an electromagnet can be controlled with the number of loops in the coil and the amount of electric current in the coil. ✓

✓ Reading Check

14. Summarize How can the strength of an electromagnet be controlled?

____

____

____

____

____

____

____

After You Read

Mini Glossary

electromagnet: a temporary magnet made with a current-carrying wire coil wrapped around a magnetic core

magnet: an object that attracts iron and other materials that have magnetic qualities similar to iron

magnetic domain: a region in a magnetic material in which the magnetic fields of the atoms all point in the same direction

magnetic force: a push or a pull a magnetic field applies to either a magnetic material or an electric current

magnetic material: any material that a magnet attracts

1. Review the terms and their definitions in the Mini Glossary. Write a sentence explaining how a magnetic material can become a magnet.

2. Use the graphic organizer to identify the two types of magnetic materials. Then state one similarity and one difference between the two types.

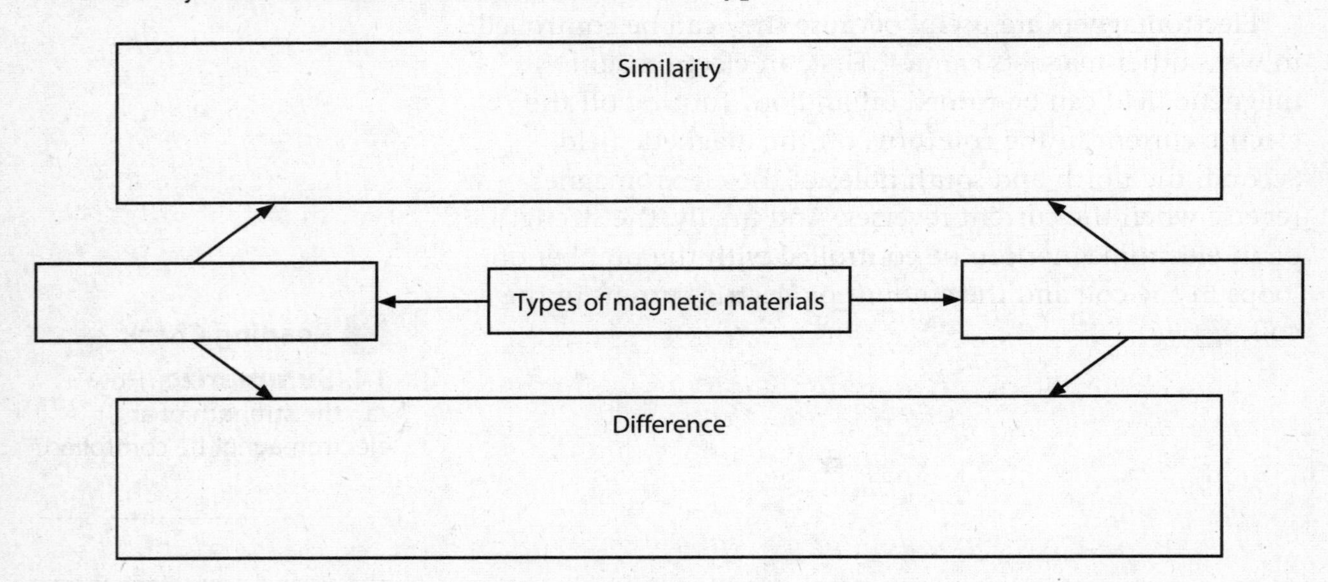

3. You hold two bar magnets close together; however, instead of sticking to teach other, they push apart. Explain why.

What do you think NOW?

Reread the statements at the beginning of the lesson. Fill in the After column with an A if you agree with the statement or a D if you disagree. Did you change your mind?

Log on to ConnectED.mcgraw-hill.com and access your textbook to find this lesson's resources.

PERIODIC TABLE OF THE ELEMENTS

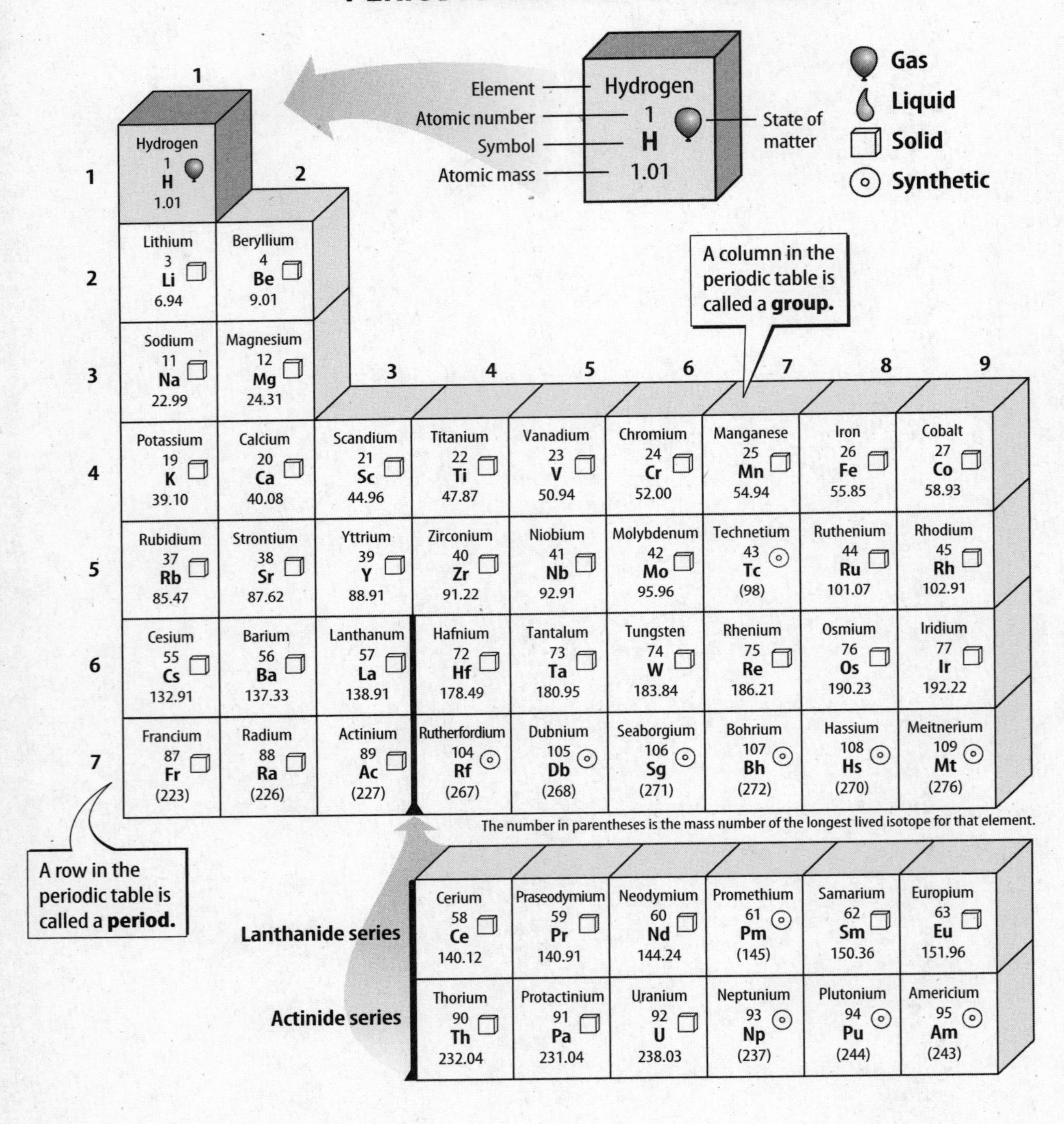